ATLAN
SALMON
IN MAINE

Committee on Atlantic Salmon in Maine
Board on Environmental Studies and Toxicology
Ocean Studies Board
Division on Earth and Life Studies

NATIONAL RESEARCH COUNCIL
OF THE NATIONAL ACADEMIES

THE NATIONAL ACADEMIES PRESS
Washington, D.C.
www.nap.edu

THE NATIONAL ACADEMIES PRESS 500 Fifth Street, NW Washington, D.C. 20001

NOTICE: The project that is the subject of this report was approved by the Governing Board of the National Research Council, whose members are drawn from the councils of the National Academy of Sciences, the National Academy of Engineering, and the Institute of Medicine. The members of the committee responsible for the report were chosen for their special competences and with regard for appropriate balance.

This project was supported by Grant No. 01-0008 between the National Academy of Sciences and the National Fish and Wildlife Foundation. Any opinions, findings, conclusions, or recommendations expressed in this publication are those of the author(s) and do not necessarily reflect the view of the organizations or agencies that provided support for this project.

International Standard Book Number 0-309-09135-7 (book)
International Standard Book Number 0-309-52977-8 (PDF)

Library of Congress Control Number 2004110692

Cover design: The National Academies Press. The image is of a hen salmon from the Penobscot River painted by Walter H. Rich and is reprinted with permission of the Museum of Science, Boston, from Kendall, W.C., The Fishes of New England: The Salmons, 1935.

Additional copies of this report are available from:

The National Academies Press
500 Fifth Street, NW
Box 285
Washington, DC 20055

800-624-6242
202-334-3313 (in the Washington metropolitan area)
http://www.nap.edu

Printed in the United States of America

THE NATIONAL ACADEMIES

Advisers to the Nation on Science, Engineering, and Medicine

The **National Academy of Sciences** is a private, nonprofit, self-perpetuating society of distinguished scholars engaged in scientific and engineering research, dedicated to the furtherance of science and technology and to their use for the general welfare. Upon the authority of the charter granted to it by the Congress in 1863, the Academy has a mandate that requires it to advise the federal government on scientific and technical matters. Dr. Bruce M. Alberts is president of the National Academy of Sciences.

The **National Academy of Engineering** was established in 1964, under the charter of the National Academy of Sciences, as a parallel organization of outstanding engineers. It is autonomous in its administration and in the selection of its members, sharing with the National Academy of Sciences the responsibility for advising the federal government. The National Academy of Engineering also sponsors engineering programs aimed at meeting national needs, encourages education and research, and recognizes the superior achievements of engineers. Dr. Wm. A. Wulf is president of the National Academy of Engineering.

The **Institute of Medicine** was established in 1970 by the National Academy of Sciences to secure the services of eminent members of appropriate professions in the examination of policy matters pertaining to the health of the public. The Institute acts under the responsibility given to the National Academy of Sciences by its congressional charter to be an adviser to the federal government and, upon its own initiative, to identify issues of medical care, research, and education. Dr. Harvey V. Fineberg is president of the Institute of Medicine.

The **National Research Council** was organized by the National Academy of Sciences in 1916 to associate the broad community of science and technology with the Academy's purposes of furthering knowledge and advising the federal government. Functioning in accordance with general policies determined by the Academy, the Council has become the principal operating agency of both the National Academy of Sciences and the National Academy of Engineering in providing services to the government, the public, and the scientific and engineering communities. The Council is administered jointly by both Academies and the Institute of Medicine. Dr. Bruce M. Alberts and Dr. Wm. A. Wulf are chair and vice chair, respectively, of the National Research Council

www.national-academies.org

COMMITTEE ON ATLANTIC SALMON IN MAINE

Members

MICHAEL T. CLEGG *(Chair)*, University of California, Riverside
PAUL K. BARTEN, University of Massachusetts, Amherst
IAN A. FLEMING, Oregon State University, Corvallis
MART R. GROSS, University of Toronto, Toronto, Ontario
LEWIS S. INCZE, University of Southern Maine, Portland
ANNE R. KAPUSCINSKI, University of Minnesota, St. Paul
PATRICK Y. O'BRIEN, ChevronTexaco Energy Technology Company, Richmond, CA
BARBARA NEIS, Memorial University of Newfoundland, St. John's, Newfoundland
NILS RYMAN, Stockholm University, Stockholm, Sweden
PETER E. SMOUSE, Rutgers University, New Brunswick, NJ
JENNIFER L. SPECKER, University of Rhode Island, Narragansett
ROBERT R. STICKNEY, Texas A & M University, College Station
JON G. SUTINEN, University of Rhode Island, Kingston

Staff

DAVID POLICANSKY, Project Director
SUSAN ROBERTS, Director, Ocean Studies Board
RUTH E. CROSSGROVE, Senior Editor
MIRSADA KARALIC-LONCAREVIC, Research Associate
LEAH L. PROBST, Research Associate
JENNIFER SAUNDERS, Research Associate
BRYAN SHIPLEY, Research Associate
JOHN BROWN, Senior Program Assistant
SAMMY BARDLEY, Library Assistant

Sponsor

NATIONAL FISH AND WILDLIFE FOUNDATION

BOARD ON ENVIRONMENTAL STUDIES AND TOXICOLOGY

RAYMOND A. WASSEL, Senior Program Officer for Environmental Sciences and Engineering
KULBIR BAKSHI, Senior Program Officer for Toxicology
ROBERTA M. WEDGE, Senior Program Officer for Risk Analysis
K. JOHN HOLMES, Senior Program Officer
SUSAN N.J. MARTEL, Senior Program Officer
SUZANNE VAN DRUNICK, Senior Program Officer
EILEEN N. ABT, Senior Program Officer
ELLEN K. MANTUS, Senior Program Officer
RUTH E. CROSSGROVE, Senior Editor

JODI BACHIM, Senior Program Assistant
NANCY CAPUTO, Senior Program Assistant
SARAH CAPOTE, Program Assistant

OTHER REPORTS OF THE BOARD ON ENVIRONMENTAL STUDIES AND TOXICOLOGY

Air Quality Management in the United States (2004)
Endangered and Threatened Fishes in the Klamath River Basin: Causes of Decline and Strategies for Recovery (2004)
Cumulative Environmental Effects of Alaska North Slope Oil and Gas Development (2003)
Estimating the Public Health Benefits of Proposed Air Pollution Regulations (2002)
Biosolids Applied to Land: Advancing Standards and Practices (2002)
The Airliner Cabin Environment and Health of Passengers and Crew (2002)
Arsenic in Drinking Water: 2001 Update (2001)
Evaluating Vehicle Emissions Inspection and Maintenance Programs (2001)
Compensating for Wetland Losses Under the Clean Water Act (2001)
A Risk-Management Strategy for PCB-Contaminated Sediments (2001)
Acute Exposure Guideline Levels for Selected Airborne Chemicals (4 volumes, 2000–2004)
Toxicological Effects of Methylmercury (2000)
Strengthening Science at the U.S. Environmental Protection Agency (2000)
Scientific Frontiers in Developmental Toxicology and Risk Assessment (2000)
Ecological Indicators for the Nation (2000)
Waste Incineration and Public Health (1999)
Hormonally Active Agents in the Environment (1999)
Research Priorities for Airborne Particulate Matter (4 volumes, 1998–2004)
Arsenic in Drinking Water (1999)
The National Research Council's Committee on Toxicology: The First 50 Years (1997)
Carcinogens and Anticarcinogens in the Human Diet (1996)
Upstream: Salmon and Society in the Pacific Northwest (1996)
Science and the Endangered Species Act (1995)
Wetlands: Characteristics and Boundaries (1995)
Biologic Markers (5 volumes, 1989–1995)
Review of EPA's Environmental Monitoring and Assessment Program (3 volumes, 1994–1995)
Science and Judgment in Risk Assessment (1994)
Pesticides in the Diets of Infants and Children (1993)
Dolphins and the Tuna Industry (1992)

Science and the National Parks (1992)
Human Exposure Assessment for Airborne Pollutants (1991)
Rethinking the Ozone Problem in Urban and Regional Air Pollution (1991)
Decline of the Sea Turtles (1990)

Copies of these reports may be ordered from the National Academies Press
(800) 624-6242 or (202) 334-3313
www.nap.edu

OTHER REPORTS OF THE OCEAN STUDIES BOARD

Elements of a Science Plan for the North Pacific Research Board (2004)
A Geospatial Framework for the Coastal Zone: National Needs for Coastal Mapping and Charting (2004)
Future Needs in Deep Submergence Science: Occupied and Unoccupied Vehicles in Basic Ocean Research (2004)
Cooperative Research in the National Marine Fisheries Service (2004)
Non-Native Oysters in the Chesapeake Bay (2004)
Enabling Ocean Research in the 21st Century: Implementation of a Network of Ocean Observatories (2003)
The Decline of the Steller Sea Lion in Alaskan Waters: Untangling Food Webs and Fishing Nets (2003)
Oil in the Sea III: Inputs, Fates, and Effects (2003)
Ocean Noise and Marine Mammals (2003)
Environmental Information for Naval Warfare (2003)
Exploration of the Seas: Voyage into the Unknown (2003)
Spills of Emulsified Fuels: Risks and Responses (2002)
Science and Its Role in the National Marine Fisheries Service (2002)
Marine Biotechnology in the Twenty-First Century: Problems, Promise, and Products (2002)
Effects of Trawling and Dredging on Seafloor Habitat (2002)
A Review of the Florida Keys Carrying Capacity Study (2002)
Marine Protected Areas: Tools for Sustaining Ocean Ecosystems (2001)
Recruiting Fishery Scientists: Workshop on Stock Assessment and Social Science Careers (2001)
Challenges in Ocean Policy (2001)
Abrupt Climate Change: Inevitable Surprises (2001)
Clean Coastal Waters (2000)
Marine Mammals and Low-Frequency Sound: Progress Since 1994 (1996, 2000)
Oceanography and Mine Warfare (2000)
Opportunities for Environmental Applications of Marine Biotechnology: Proceedings of the October 5–6, 1999, Workshop (2000)
Improving the Collection, Management and Use of Fisheries Data (2000)
Illuminating the Hidden Planet: The Future of Seafloor Observatory Science (2000)
Bridging Boundaries Through Regional Marine Research (2000)
50 Years of Ocean Discovery (2000)
The Community Development Quota Program in Alaska and Lessons for the Western Pacific (1999)
Sustaining Marine Fisheries (1999)
Sharing the Fish (1999)

Science for Decisionmaking (1999)
Global Ocean Science (1999)
From Monsoons to Microbes: Understanding the Ocean's Role in Human Health (1999)
Building Ocean Science Partnerships: The United States and Mexico Working Together (1999)
Review of Northeast Fishery Stock Assessments (1998)
Opportunities in Ocean Sciences: Challenges on the Horizon (1998)
Improving Fish Stock Assessments (1998)
The Global Ocean Observing System (1997)
Oceanography and Naval Special Warfare (1997)
Proceedings of the Symposium on Tactical Meteorology and Oceanography (1996)
Expanding the Uses of Naval Ocean Science and Technology (1996)
Science, Policy, and the Coast (1995)
Improving Interactions Between Coastal Science and Policy: Proceedings of the Gulf of Maine Symposium (1995, 1996)

Copies of these reports may be ordered from the National Academies Press
(800) 624-6242 or (202) 334-3313
www.nap.edu

Preface

Atlantic salmon are symbolic of a time when the biological endowment of the eastern United States was richer and more diverse than today. These impressive fish were once abundant in the rivers of Maine, and their range extended south and west as far as the Connecticut River and perhaps even to the Hudson River. Today the distribution of wild Atlantic salmon in the eastern United States is restricted to a few rivers of Maine where total annual runs are numbered in hundreds of fish rather than the tens or perhaps hundreds of thousands of the past.

In the year 2000, the U.S. Fish and Wildlife Service and the National Marine Fisheries Service listed Maine Atlantic salmon as an endangered species under the distinct population segment language of the Endangered Species Act (ESA). This decision has the potential to trigger regulatory actions that might have economic impacts on agriculture, aquaculture, forestry, and other activities in Maine.

This National Research Council committee was asked to describe what is known about the genetic makeup of Atlantic salmon in Maine, and we did so in a report issued in January 2002. We were also asked to assess the causes of decline and to suggest strategies for the rehabilitation of Atlantic salmon in Maine. This document responds to that latter charge.

To set the stage, we must consider the remarkably complex life cycle of Atlantic salmon. This is a species that is exquisitely adapted to two very different environments. Adult fish mature in an ocean environment and then return to their natal freshwater streams to breed. The newly emerged fry develop, and after a period in their natal streams, the fish go down river to the ocean where they must transition to a very different physiological state as they enter the ocean environment. The fish migrate as far as the western coast of Greenland, and after approximately 2 years the surviving adult fish return to their natal streams to breed, thus completing the cycle. Unlike their Pacific relatives, a small number of Atlantic salmon may once again return to the ocean to repeat the cycle. The basic problem is that too few adult fish now return to maintain a stable population, and the present demographic trajectory appears to predict extinction of Atlantic salmon in Maine.

Obviously, the current plight of Atlantic salmon in Maine cannot be discussed in isolation. They are embedded in a larger biological and physical system that is highly dynamic and is in part responding to imperfectly understood global processes of change. Global climate systems have fluctuated over the past 100,000 years, and the range of Atlantic salmon must have expanded and contracted repeatedly in response to those larger forces. The history of European colonization of North America coincides first with a cooling of climate associated with the little ice age and then with a subsequent warming period that is still in progress. These global factors are superposed on many regional and local influences. For example, most Maine rivers are dammed, often multiple times, denying suitable spawning habitat. Industrial logging and agricultural activities have influenced local watersheds, stream quality, and stream flow regimes. Acid rain from regional industrial activity has also affected stream quality. Yet many of these factors have improved over the past quarter century in response to environmental legislation and to changing industrial and land-use patterns. The biological community has also changed, owing to introductions of nonnative species and concomitant ecological shifts driven by the changing physical setting. Fishing was once a serious cause of mortality, but that is now much reduced by prohibitions on fishing, both in Maine's rivers and in the ocean, although some mortality might still occur as the result of bycatch with other fishing activities. Thus, the picture appears to be one of gradual environmental improvement. Despite that, populations of Atlantic salmon in Maine are still declining.

Clearly, a large number of potentially interacting factors impinge on the fate of Maine Atlantic salmon. There are very few quantitative data on the impacts of most of these factors on salmon reproduction and survival. Our committee has worked diligently to assess the threats to salmon in Maine and to make reasonable inferences about how these might be mitigated in the face of limited information. We are faced with a problem of ecosystem management where management choices are shrouded in uncertainty. Many variables affect the system, some yet to be identified, but only a small subset of these variables is subject to manipulation. In short, the challenge faced by our committee was to address a real-world problem in applied biology. This challenge has humbled us. It presented us with a case study in the manifold complexities of biological and human systems, and it taught us that there are few strong levers available to the real-world system manager.

The practical question of what is happening to the fish and to the ecosystems in which they are embedded remains. What factors are responsible for the observed decline and what levers are available to the manager to move the system toward a more sustainable end point? Our committee has debated this central question at length. From the management perspective, we review the governance system and consider the success or lack thereof of hatchery supplementation. We also consider the likely effects of dam removal as options worthy of special attention. We do not claim to have a magic solution to the long-standing problem facing Atlantic salmon in Maine. What we offer instead is an analytical framework for evaluating management options and for establishing priorities. This framework is drawn from risk and decision analyses; at its best, those techniques can assist in setting priorities and in identifying those actions most likely to be effective. Like any algorithm, risk and decision analyses are no better than the quality of the available data, but the application of the framework does assist in identifying crucial missing information, and it can reveal those factors of minimal importance, because the larger system is insensitive to their manipulation. We appreciate the difficult and challenging tasks faced by those charged with the implementation of a recovery plan for salmon, and we hope that the analytical framework described in this report will serve to assist both managers and recovery planners.

The task of our committee was made much easier by the many individuals who testified before us during the course of our open meetings in the State of Maine. We are grateful for their willingness to share their collective wisdom. Space does not permit the individual acknowledgment here of everyone who helped the committee through testimony, advice, and sharing of written information, although we are extremely grateful to all of them, and the wiser for their efforts. They and others who helped the committee in various ways range from elected officials to volunteers and from government biologists to members of private organizations. They and the experts who reviewed this report, to whom we also are extremely grateful, are listed in the Acknowledgments. However, Ed Baum deserves special mention here for his long service in generating knowledge of Atlantic salmon in Maine and for the wealth of information that he shared with this committee through his book and presentations to us.

Chairing a committee faced with a difficult and controversial charge can be very hard work. In this case the burden was made light by an exceptionally knowledgeable and cooperative committee. I learned a great deal from my colleagues, and I approach the end of this report with a sense of regret that the fellowship associated with our work has reached an end. The superb staff of the National Research Council also eased our task. Our study director, Dr. David Policansky, is himself an expert in fish biology, genetics, and conservation policy. Dr. Policansky contributed to the final report in numerous ways, and the report is much better because of his professional touch. Dr. Susan Roberts and Leah Probst contributed their expertise and skill to the report's quality as well, and we are grateful to Jennifer Saunders, Bryan Shipley, Dominic Brose, and John Brown for their attention to the needs of the committee throughout its meetings and for their attention to detail in production of the report. We also thank Ruth Crossgrove and Mirsada Karalic-Loncarevic for their editorial and research efforts.

Michael T. Clegg
Chair, Committee on Atlantic Salmon in Maine

Acknowledgments

This report has been enhanced by the contributions of many people. They are listed below and we are grateful to them all.

This report has been reviewed in draft form by individuals chosen for their diverse perspectives and technical expertise, in accordance with procedures approved by the NRC's Report Review Committee. The purpose of this independent review is to provide candid and critical comments that will assist the institution in making its published report as sound as possible and to ensure that the report meets institutional standards for objectivity, evidence, and responsiveness to the study charge. The review comments and draft manuscript remain confidential to protect the integrity of the deliberative process. We wish to thank the following individuals for their review of this report:

Edward T. Baum, Atlantic Salmon Unlimited
Barry Costa-Pierce, University of Rhode Island
Perry R. Hagenstein, Institute for Forest Analysis, Planning, and Policy
William K. Hershberger, National Center for Cool and Cold Water Aquaculture
Irving Kornfield, University of Maine
Kai N. Lee, Williams College
Rosamond Lee Naylor, Stanford University
Ralph Pisapia, Meridith, New Hampshire
Jack Schmidt, Utah State University

Nancy Targett, University of Delaware
James Wilson, University of Maine

Although the reviewers listed above have provided many constructive comments and suggestions, they were not asked to endorse the conclusions or recommendations, nor did they see the final draft of the report before its release. The review of this report was overseen by John Dowling, Harvard University (review monitor), and John Burris, Beloit College (review coordinator). Appointed by the NRC, they were responsible for making certain that an independent examination of this report was carried out in accordance with institutional procedures and that all review comments were carefully considered. Responsibility for the final content of this report rests entirely with the authoring committee and the institution.

We are grateful to Randall Peterman for his helpful advice on methods of risk assessment and risk analysis.

Presenters at Public Sessions

John Banks, Penobscot Indian Nation
Ed Baum, Atlantic Salmon Unlimited
Ken Beland, Maine Atlantic Salmon Commission
David Bell, Wild Blueberries Commission
Sebastian Belle, Maine Aquaculture Association
Russell W. Brown, NMFS Northeast Fisheries Science Center
Elizabeth Butler, Pierce Atwood
Mary Colligan, National Marine Fisheries Service
David Courtemanch, Maine Department of Environmental Protection
Laura Rose Day, Natural Resources Council of Maine
Scott Dickerson, Coastal Mountains Land Trust
Norman R. Dube, Maine Atlantic Salmon Commission
Kevin Friedland, University of Massachusetts-National Oceanic and Atmospheric Administration
John Gold, Texas A&M University
Andy Goode, Atlantic Salmon Federation
Terry Haines, University of Maine-U.S. Geological Survey
Melissa Halsted, Kennebec Soil & Water Conservation District
James Hawkes, NOAA-Fisheries, Maine Field Station
The Honorable Angus King, Jr., Governor of Maine
Thomas King, Craig Brook National Fish Hatchery, U.S. Fish and Wildlife Service
Fred Kircheis, Maine Atlantic Salmon Commission
John F. Kocik, NOAA

Stephen Koenig, Project SHARE
Irv Kornfield, University of Maine
Donna Loring, Penobscot Nation
Jerry Marancik, Craig Brook National Fish Hatchery, U.S. Fish and Wildlife Service
Morten Moesswilde, Department of Conservation, Maine Forest Service
Paul Nickerson, U.S. Fish and Wildlife Service
Stephen Page, ContiSea Ducktrap River Fish Farm
Lee Perry, Maine Department of Inland Fisheries and Wildlife
Jeff Reardon, Trout Unlimited
Timothy Sheehan, NOAA
Alan Spear, Bangor Hydro-Electric Company
Joan Trial, Maine Atlantic Salmon Commission
Douglas Watts, Friends of Kennebec Salmon
Mark Whiting, Maine Department of Environmental Protection
Fred Whoriskey, Jr., Atlantic Salmon Federation

Site Visits

The committee visited the following sites during its meetings and is grateful to the many people who provided information, shared their facilities, provided hospitality, and provided valuable information there.

Coopers Mills Sheepscot Sites
Craig Brook Hatchery
Ducktrap Coalition program
Ducktrap Timber Preserve
Edwards Dam
Head Tide/Sheepscot watershed
Maine Atlantic Salmon, LLC facilities
Pleasant River Weir
Wild Blueberry Farms

Contents

ATLANTIC SALMON IN MAINE

Summary

Atlantic salmon in Maine, once abundant, are now seriously depleted. Hundreds of thousands of adults returned to Maine's rivers and streams each year in historical times. In 2002, it is estimated that only 871 salmon returned to spawn in all Maine rivers. Atlantic salmon were listed as endangered under the federal Endangered Species Act (ESA) in November 2000. The listing covers the wild fish in eight Maine rivers (Figure S-1) as a single "distinct population segment" (DPS). Only 33 fish returned to those eight rivers, often called the DPS rivers, in 2002. (These estimates of returning salmon are minimal estimates, and the actual numbers are probably greater; nonetheless, the decline in salmon numbers is real and very serious.)

The controversy in Maine that accompanied the ESA listing led Congress to request the National Research Council's (NRC's) advice on the science relevant to understanding and reversing the declines in Maine salmon populations. The charge to the NRC's Committee on Atlantic Salmon in Maine (Box S-1) included an interim report that focused on the genetic makeup of Maine Atlantic salmon populations; that report was published in January 2002. The charge for the final report included a broader look at factors that have caused Maine's salmon populations to decline and the options for helping them to recover. This is the final report.

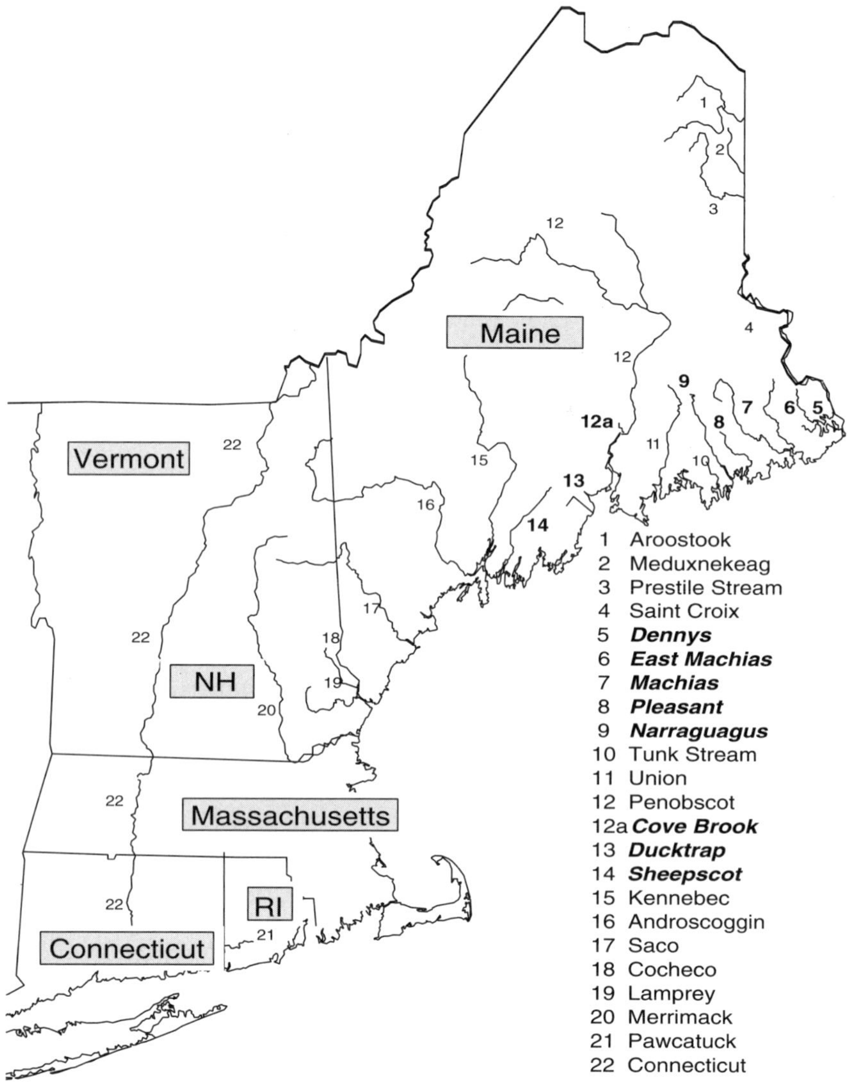

FIGURE S-1 USA Atlantic salmon rivers with active restoration and recovery programs in New England. The eight DPS rivers in Maine with Atlantic salmon listed as endangered under the ESA are (5) Dennys, (6) East Machias, (7) Machias, (8) Pleasant, (9) Narraguagus, (12a) Cove Brook, (13) Ducktrap, and (14) Sheepscot. SOURCE: Baum et al. 2002. Reprinted with permission of the author.

BOX S-1
Committee Statement of Task

A multidisciplinary committee will review the available scientific information on the status of Atlantic salmon populations in Maine and, where relevant, in adjacent areas. The committee will assess causes of the declines of their populations and the current threats to the continued survival of salmon, will evaluate the evidence on the population structure of those salmon, and will evaluate options for improving the survival of salmon. In assessing information, the committee will identify significant knowledge gaps and suggest additional research that would be important to the conservation and recovery of salmon populations.

Factors to be evaluated include the nature and distinctness of salmon populations in Maine rivers and surrounding areas; the interactions between aquaculture, hatchery, and wild populations; terrestrial and marine environmental factors affecting salmon populations; the effects on salmon of changes in the hydrology of Maine streams; and the effects on salmon of subsistence, recreational, and commercial fishing in freshwater and ocean areas in and around Maine.

A brief interim report will be produced within 9 months after formation of the committee. The interim report will address the genetic makeup of wild salmon populations in Maine and its possible relationship to recovery activities. A final report at the end of the study will describe and synthesize the information available on the biology of Atlantic salmon, the causes of their population declines, and the threats to their continued survival. It will evaluate and describe options for enhancing their continued survival and recovery and will provide some approximate estimates of the relative costs of the various options.

SALMON BIOLOGY

Naturally reproducing populations of Atlantic salmon occur in rivers and streams from southwestern Maine to northwestern Europe. Historically, they were found in the Hudson River in New York and north and east to the Canadian border, but today they are found only in Maine, from the Sheepscot River to the Canadian border. The populations have declined drastically, from perhaps half a million adults returning to all U.S. rivers each year in the early 1800s to a minimum estimate of 1,050 in 2001. Most U.S. Atlantic salmon are in Maine rivers, and 780 (90%) of those returned to only one river, the Penobscot, in 2002.

Salmon spawn in freshwater, where the young hatch and grow for 1–3 years before migrating to sea. At sea, they grow faster in the rich marine environment and then return as adults to the rivers where they hatched (called natal streams) to spawn—a life history called anadromy. Most adult salmon die after spawning, but some return to the ocean, and some of those fish return to spawn again. Some males mature early and survive spawning more often than adults do.

The homing of salmon provides an opportunity for the salmon to adapt to environmental conditions in their natal streams. The occasional straying of returning adults to streams other than their natal streams is probably important evolutionarily, because it allows recolonization of a stream if the local population dies out and provides for small infusions of new genetic material for continued evolutionary adaptation to changing conditions. The complex life-history pattern of anadromy exposes salmon both in the ocean and in streams to predation, fishing, habitat degradation, and other environmental perturbations. Understanding the causes of population decline is thus also complicated.

In addition to anadromous Atlantic salmon, Maine has populations of Atlantic salmon that complete their entire life history in freshwater. They are called landlocked salmon or ouananiche. They are the same species as the anadromous form, although there is some genetic difference between them. They are not endangered, but because they strongly resemble anadromous salmon and sometimes compete with them, they can complicate efforts to rehabilitate wild anadromous populations.

HATCHERIES AND AQUACULTURE

Augmentation of wild populations of Maine salmon with hatchery releases began in the early 1870s. At first, young fish were obtained from Lake Ontario. Later, the Craig Brook Hatchery in East Orland, Maine, using eggs from Penobscot River fish, was the stocking source. By the 1920s, Canadian eggs were being used, followed in the 1940s by eggs from the Machias, Penobscot, and Dennys rivers of Maine. In the 1950s and 1960s, some eggs of Canadian origin again were used, but by the late 1960s, eggs from Maine's Machias, Narraguagus, and Penobscot rivers were used. Fish reared in hatcheries derived from Penobscot River fish were used until late 1991, when the practice of river-specific stocking was adopted. The protocol used since involves catching young, actively feeding fish (parr) in the river, rearing them to maturity in the hatchery, mating them, and releasing the resulting fry into their native rivers before they start to feed.

Stocking, at least until 1992, added to rivers many fish (and eggs) whose genotypes did not reflect adaptation to the local environment. In addition, aquaculture (farming) of Atlantic salmon began in Maine in the 1980s, the first fish for market being produced in 1987. Derived in part from European Atlantic salmon, the genetic strains used for fish farming are even more different from native strains than are hatchery strains. Farm fish escape at all life stages, despite the efforts of producers to prevent escapes. In some years and in some rivers, more escaped farm fish return to spawn than wild fish. The impact of escapees on the genetics of wild populations is not well documented in Maine. Both hatchery- and pen-reared fish compete poorly with wild fish in other rivers that have been studied, but because there are so many escaped farm fish compared with wild fish in some rivers, some impact is likely to have occurred.

The addition of so many nonwild genotypes from hatcheries and from aquaculture escapees has led some to conclude that the fish returning to spawn in Maine's rivers could not possibly represent anything more than a mix of genotypes from Europe, Canada, and Maine. If that were true, then options for conservation might be considerably different from those that might be undertaken if the wild fish in Maine were genetically distinct, and that is why it is important to understand the genetic makeup of the wild salmon populations in Maine and the effects that hatcheries might have on it.

THE GENETICS OF MAINE SALMON

In its January 2002 interim report, the committee assessed how Maine salmon populations differ from other Atlantic salmon populations and among themselves. The committee addressed the question at three levels. First, are North American Atlantic salmon genetically different from European salmon? Second, are Maine salmon distinct from Canadian salmon? Third, to what degree are salmon populations in the eight Maine rivers in the ESA listing distinct from each other?

The committee concluded that North American Atlantic salmon are clearly distinct genetically from European salmon. In addition, despite the extensive additions of nonnative hatchery and aquaculture genotypes to Maine's rivers, the evidence is surprisingly strong that the wild salmon in Maine are genetically distinct from Canadian salmon. Furthermore, there is considerable genetic divergence among populations in the eight Maine rivers where wild salmon are found. The committee concluded that wild salmon in Maine do not reflect only (or even mainly) the result of decades of hatchery stocking. It is not possible to say whether or to what degree the genetic differences reflect adaptation to local conditions as opposed to random processes associated with small population sizes or some influence of stocking. However, the pattern of genetic variation seen among Maine streams is similar to patterns seen elsewhere in salmon and their relatives where no stocking has occurred.

HUMAN ALTERATION OF THE ENVIRONMENT

Maine's environment has been substantially altered by human use. Before humans arrived, the advance and retreat of continental ice sheets during the Pleistocene epoch (10,000 to about 1.5 million years ago) had a dominant influence on landforms and resulting stream networks and soils of Maine. Glaciers shaped mountains and valleys; left sand and gravel deposits; and carved out hundreds of lakes, ponds, and depressions that are now wetlands. The dominant soil types are a direct result of glaciation, a cold, wet climate, and forest succession over the past 10,000 years. In general, the soils are well drained, acidic, and relatively unfertile. The properties of the soils and watersheds generally yield high-quality streams and rivers.

Anthropogenic disturbance has occurred for centuries in New England's forests. Before European settlement, Native Americans used fire to alter wildlife habitat and enhance or maintain the productivity of wild foods and medicinal plants. Since the mid-1700s, Maine's environment has been altered by timber harvesting, clearing for agriculture, gradual abandonment of farmlands, industrial development, and more recently, residential land use. Maine was more than 92% forest in 1600. The forested area decreased dramatically as the combined effects of forest clearing for agriculture, industrial logging and milling, and subsequent forest fires reduced coverage to 53.2% by 1872. Forests have since regenerated on abandoned agricultural land and "cutover" areas, reversing the trend of deforestation of earlier centuries. In 1995, the Forest Service of the U.S. Department of Agriculture estimated that Maine's forest cover was 89.6%, but the composition of the vegetation was much different than it had been a few centuries ago.

By 1920, most of the forest left in the Penobscot, Kennebec, and Androscoggin watersheds had been altered by one or more logging cycles. By contrast, the Down East region (the part of Maine near and along the coast from roughly Penobscot Bay east to the border with Canada) still had areas of virgin forest exceeding 25,000 acres (10,125 hectares). A suite of socioeconomic and ecological factors might have contributed to the continued survival of wild Atlantic salmon in such rivers as the Narraguagus, Pleasant, Machias, East Machias, and Dennys. They include lower human population densities, less industrial use of the rivers, and a cooler climate.

One trend that has not been significantly reversed is the presence of dams placed on Maine's rivers for mills and other purposes. Most rivers there have one or more dams that reduce or eliminate fish passage and that alter riverine habitats. Some of the dams seem to have outlived their economic usefulness.

To a significant degree, salmon recovery will depend on changing human activities that are threatening the survival of salmon. Understanding the factors that affect human activities is a prerequisite for designing effective policies that will alleviate the threats that the activities pose to the survival of salmon. In addition, many governance organizations are involved with salmon management. They include agencies of the federal and state government as well as local and nongovernmental organizations. The large number of such organizations complicates understanding of how their actions affect salmon. It also means that their ability to work together depends on thoughtful and careful communication and agreements.

THREATS TO ATLANTIC SALMON IN MAINE

Human activities that directly or indirectly threaten salmon include dams and hydropower projects, Atlantic salmon aquaculture, water extraction for agriculture, fishing, hatcheries, logging, road construction, development of land sites, acidification of their streams, and research. Predation—always part of the environment of salmon—has been influenced by declines in the number of salmon and by changes in the numbers and kinds of their predators. Those factors interact with many other factors on land, in freshwater, and at sea. The difficulty is not only to identify factors that threaten salmon but also to decide which ones are most critical and which ones can be mitigated or reversed.

To address the difficulty of ranking the threats, the committee used a form of risk analysis. After threats have been identified and their severity and urgency ranked, decisions need to be made to address them. In some cases, legal or biological considerations might make the decisions obvious, but in most cases, decisions must be weighed against their likely effectiveness, cost, societal and political implications, and other consequences. The decision-making process should include people with local knowledge and people who must live with the consequences.

In this report, the committee has provided two decision analyses it conducted as examples: placement of dams and managing risks of salmon farms. These examples of decision analyses are not intended as conclusions, because people with local knowledge and people who must live with the consequences of the decisions did not take part in the analyses. The committee's conclusions focus on biological issues and on methods of gaining knowledge and understanding.

The committee's approach has been statewide, without a specific or exclusive focus on the eight DPS rivers or on the specific requirements of the ESA. That statewide approach was the committee's charge, and it has a sound scientific basis: much additional salmon habitat in other watersheds should be used in rebuilding salmon populations. By far the greatest natural environmental asset for salmon in Maine is the Penobscot River. It is the largest river wholly in Maine, and it has more than 90% of all the adult Atlantic salmon returns in Maine. For years, the Penobscot was the major source of brood stock for salmon hatcheries. The Kennebec, Androscoggin, Saco, St. Croix, St. John, and other non-DPS rivers also are environmental assets for salmon. Biologically, a restoration program for Maine salmon would not make sense if it did not take advantage of those rivers as well as the DPS rivers.

Dams

Dams obstruct adult and juvenile salmon passage and alter riverine habitats, including water quality. As a result, they degrade or eliminate spawning and rearing habitat for Atlantic salmon in Maine. Although dams are not as important a problem on the DPS as on other Maine rivers, they have made an enormous amount of habitat unavailable to Maine salmon and have affected much of the habitat that is still available. Fish-passage facilities help migrations to some degree, but they have no effect on the riverine habitat affected by dams, and they are inadequate or completely absent on many dams.

Hatcheries

Hatcheries have been used in Maine to attempt to increase the populations of salmon since the 1870s. At first, no attention was paid to genetics. Fish used for brood stock came from various Canadian and Maine rivers. Canadian fish or eggs were not used in Maine after 1967 except in 1985 and 1986, but many nonnative fish were introduced in the earlier decades. In 1992, river-specific stocking was instituted for the eight DPS rivers.

Even with river-specific stocking and the best available breeding protocols, hatcheries change the genetic makeup of salmon populations. Despite the efforts and money spent on rearing fish in hatcheries and stocking Maine's rivers, salmon populations are now at the lowest levels ever recorded. The available information is not sufficient to conclude whether hatcheries in Maine can actually help to rehabilitate salmon populations, whether they might even be harming them, or whether other factors are affecting salmon so strongly that they overwhelm any good that hatcheries might do.

Aquaculture

Salmon farms rear salmon from eggs in hatcheries and then grow them to market size in net-pens near the coast. The salmon farms were established in Maine in the 1980s. Risks to wild populations from salmon farms include the transmission of disease, the concentration of parasites (sea lice) and predators around the net-pens to the detriment of wild salmon migrating nearby, and the escape of fish that can migrate up rivers and compete for space and mates with wild salmon. Disease has caused net-pens in Cobscook Bay to be dismantled and sterilized.

Only limited research on and monitoring of the effects of salmon farms on Maine salmon have been carried out. Adverse effects of farms on wild fish have not been documented in Maine, but they have been elsewhere. There is no reason to believe that the harm to wild fish that has been documented elsewhere could not occur in Maine.

Acid Deposition

Deposition of sulfates and other chemicals from the atmosphere has acidified many lakes and streams in northeastern North America. In nearby Nova Scotia, acidification has led to the extirpation of salmon from more than a dozen rivers. Acid deposition has decreased in the past 25 years, but not all rivers and streams in Maine have become less acidified. The altered water chemistry of acidified streams especially affects the younger life stages of salmon and can be accompanied by a high mortality of smolts making the transition from freshwater to seawater. Although acidification has not been conclusively identified as a source of death for Atlantic salmon in Maine, recent information on poor survival of smolts and on water chemistry in Maine makes it appear that acidification could be a serious problem.

Fishing

Fishing has affected Maine salmon until very recently. At first, fishing was for subsistence, and its intensity is not well quantified. Commercial and recreational fishing were well established in the nineteenth century. Recreationally caught salmon were almost all killed before about 1985; but since 1994, most salmon caught have been released. High-seas fishing for salmon differs from fishing in rivers in that specific stocks cannot be targeted, so the number of Maine salmon caught by commercial ocean fishing is not easy to quantify. By 2000, all recreational angling for anadromous Atlantic salmon, even catch-and-release, was prohibited in Maine. Directed commercial fishing was eliminated by 1948 in Maine and almost completely eliminated at sea in 2002. Some poaching, accidental catch (bycatch), and take because of mistaken identity (anadromous Atlantic salmon resemble landlocked Atlantic salmon and brown trout) occur, but their magnitude is not known.

Change in Atmospheric and Ocean Climate

Atmospheric climate and oceanic conditions on earth have been changing for at least as long as life has existed; they will continue to change. Maine's climate has warmed over the past three decades, and ocean conditions have changed as well. Continued warming would make it more difficult to rehabilitate wild salmon populations in Maine. Change in precipitation patterns and related phenomena, such as ice cover and timing of snowmelt, probably also would make things more difficult.

Predation and Food Supply

Predation has always been a feature of the lives of salmon, but human activities have probably increased its severity. Salmon predators include birds, mammals, other fish, and—at some life stages—invertebrates. Many rivers now contain nonnative species of fish, some of which are strongly piscivorous. Ocean fishing has changed the composition and food supply of potential salmon predators. In addition, protection afforded to marine mammals under the Marine Mammal Protection Act has resulted in increases in species that prey on salmon. Finally, the human depletion of salmon populations might have made them more vulnerable to other predators. These changes have probably also affected the kinds and amount of food available to salmon at various life stages.

Research and Monitoring

Research and monitoring are essential for understanding the dynamics, status, and trends of Atlantic salmon in Maine and for assessing the effects and effectiveness of management actions. However, the trauma associated with capturing, handling, anesthetizing, and sampling fluids and tissues from fish—especially young fish—can result in some deaths. When populations are very small, as they now are in most Maine rivers, it is essential to weigh the value of new information against the possibility of the harm to wild fish caused by handling.

Governance

Governance institutions have a strong influence on the success or failure of management of natural resources in general, as they do for anadromous Atlantic salmon in Maine. Although a considerable amount is known about relationships between governance structure and resource management, each case is unique, and much basic information is needed before governance structures can be fully adapted to improve resource management. In Maine (as in most other places), much of the required information has not been collected or analyzed. In addition, most governance structures have much broader mandates than only resource management, and that can make resource management more difficult. It would be helpful to increase coordination of efforts across local, state, federal, and international levels of organization; adapt governance structures to more closely match the biology and geography of salmon populations; include stakeholders in the risk-assessment and risk-management processes; and develop and improve of adaptive-management approaches that allow people to test the efficacy of various governance structures.

RANKING THE THREATS

The committee's risk assessment led it to conclude that the greatest impediment to the increase of salmon populations in Maine is the obstruction of their passage up and down streams and degradation of their habitat caused by dams. This finding applies more to the non-DPS than to the DPS rivers, because the potential salmon habitat in the non-DPS rivers is so great.

The mortality of salmon—especially smolts and post-smolts—in estuaries and at sea appears to be a very serious problem. Despite some uncertainty about the causes of the excess mortality, the committee concludes that acidification of streams has the potential to be a major impediment to the increase of salmon populations in Maine by contributing to that mortality.

At the next level of importance, salmon farming has the potential to adversely affect salmon populations in Maine genetically and ecologically and might already have done so. Over the long term, hatchery supplementation of salmon populations in Maine is also likely to have deleterious genetic and possibly ecological effects. Predation and changes in oceanic conditions could be serious problems for salmon. Because populations of wild salmon in Maine are so low, the mortality associated with research and monitoring could be problematic.

Current agricultural practices, including forestry, do not appear to be an important problem for Atlantic salmon in Maine, although their effects should be monitored, especially for erosion, reduction of vegetation cover, and water withdrawals. Fishing is currently prohibited; therefore, it is not an important problem for Maine salmon. A rich and complex network of governance institutions in Maine influences how humans affect salmon. As is often the case with complex environmental problems, more information is needed on how well governance institutions are working together, and whether the government authority is sufficient to develop and implement effective recovery programs.

RECOMMENDATIONS

Many recommendations have been made for the rehabilitation of Atlantic salmon populations in Maine. Most of them are sound, but there are too many recommended actions to take at once. Moreover, not all of them are equally urgent. Most of the actions discussed below also have been recommended by others, such as the Maine Atlantic Salmon Task Force, but here an attempt is made to set priorities for them and to recommend those actions most likely to be effective.

Urgently Needed Actions

There is an urgent need to reverse the decline of salmon populations in Maine if they are to be saved. Other than the Penobscot River, only 80 adult salmon were recorded to have returned to Maine's rivers in 2002. The serious depletion of salmon populations in Maine underscores the need to expand rehabilitation efforts to as many of Maine's rivers as possible. Since most Maine salmon are now in the Penobscot River, that population should be a primary focus for rehabilitating the species in Maine. The committee recommends the following urgent actions:

- A program of dam removal should be started. Priority should be given to dams whose removal would make the greatest amount of spawning and rearing habitat available, which means that downstream dams generally should be considered for removal before dams upstream of them. In some cases, habitat restoration will likely be required to reverse or mitigate some habitat changes caused by the dam, especially if the dam is many decades old. A recent agreement to remove two Penobscot River dams is encouraging.
- The problem of early mortality as smolts transition from freshwater to the ocean and take up residence as post-smolts needs to be solved. If, as seems likely, early mortality in estuaries and the ocean is due in part to water chemistry, particularly acidification in freshwater, the only methods of solving the problem are changing the water chemistry and finding a way for the smolts to bypass the dangerous water. Liming has had considerable success in counteracting acidification in many streams, and the techniques are well known. Examples of its application are in nearby Nova Scotia. Liming should be tried experimentally on some Maine streams as soon as possible. Bypassing the dangerous water is best achieved by rearing smolts and acclimating them to seawater in controlled conditions. This approach is not appealing because of the degree of human intervention required and because of the adverse selection that must result from it. Given the extreme depletion of salmon populations, however, desperate measures are needed.

• Hatcheries need to continue to be used, at least in the short term, to supplement wild populations and to serve as a storehouse of fish from the various rivers. There is an urgent need to understand the relative efficiency of stocking of different life stages in the rivers in terms of adult returns per brood-stock fish and their reproductive success. Additional research on hatcheries and scientific guidance for their use is needed, because hatchery-based restoration of wild salmon populations remains an unproven technology.

The committee was asked to provide estimates of the approximate relative costs of the various options. Although it has not been able to provide detailed cost estimates for all of its recommendation, it does estimate that dam removal would cost between $300,000 and $15 million per year; and liming would cost on the order of $100,000 per stream initially plus $50,000–$100,000 per year. The cost of changing hatchery operations as recommended would not require major additional expenditures beyond what is currently spent on federal hatchery operations for Atlantic salmon in Maine. Although the costs of changes to salmon farming cannot be reliably estimated, it is clear that most of the modifications would likely cost enough to eliminate the profitability of salmon farm operations.

Actions Important over the Longer Term

• Over the longer term, the committee recommends a comprehensive decision-analysis approach to the rehabilitation of Atlantic salmon populations in Maine. The analysis should be conducted along the lines of the examples in Chapter 5 of this report but in more detail and with all major groups of stakeholders involved. Taking a Maine-wide view is more likely to be successful than focusing only on some rivers.

• No anadromous Atlantic salmon of any life stage should be stocked in rivers that have populations of wild Atlantic salmon unless those rivers are specifically identified as part of a hatchery-recovery program that uses river-specific stocks (that is, a program that takes brood stock from the river to be stocked with the aim of retaining any local genetic differentiation). Stocking of nonnative fish species and landlocked salmon also should be avoided in those rivers. Other rivers that once supported wild Atlantic salmon runs, but which lack them now, will probably become repopulated by strays from nearby streams if populations in those nearby streams recover. The advantages of such natural repopulation, which would be more likely than stocking to lead to local genetic adaptation, should be given serious attention before any decision is made to stock streams that currently lack wild Atlantic salmon runs.

• The current prohibition of commercial and recreational fishing for salmon in Maine, including catch-and-release fishing, should be continued. Any further reduction in the take of Maine salmon at sea would be helpful. Maximum and minimum size limits for trout and landlocked salmon should be established in rivers that have anadromous Atlantic salmon. The minimum size for retention should be large enough to protect Atlantic salmon smolts, and the maximum size should be small enough to protect adult Atlantic salmon. Any fishing that might take a wild Atlantic salmon, even inadvertently, constitutes an additional risk to the species. This risk should be carefully evaluated for all Maine rivers with Atlantic salmon, and additional measures should be taken if the risk is judged to be important. Habitat zones most heavily used by Atlantic salmon young and adults should be closed to fishing for all species until salmon populations have recovered.

• Research that increases the risk of death to wild fish should be curtailed. The value of any information obtained needs to be weighed against the likelihood of increased death of wild fish subjected to handling.

• Every effort should be made to further curtail the escape of salmon from farms. If accumulation of parasitic copepods (sea lice) or other pathogens is found to be a problem for wild salmon, the aquaculture facilities should be moved to a place where they will not adversely affect wild salmon.

• Hatchery practices should be evaluated in an adaptive-management context to further reduce adverse genetic and ecological effects and modified as needed.

• The monitoring of water quality and gauging of streams should be augmented. A network of metereological-monitoring, stream-gauging, water-quality-monitoring, and biological-monitoring sites should be linked to a geographic information system and an online database within 2 years.

• Government, industry, and private organizations and landowners should cooperate to evaluate forestry best-management practices and forest-road networks. Mitigation and pollution prevention should be organized to maximize the effectiveness of storm-water management and sediment control and the removal of barriers to fish passage.

• The State Planning Office should conduct a systematic governance assessment to see whether there are gaps in authority, overlapping authority, conflicts of goals and interests among agencies, and adequate cooperation among agencies.

• The State Planning Office, in cooperation with all other agencies, should implement adaptive management to monitor performance of governance activities related to Atlantic salmon, to experiment with alternative institutions for salmon recovery, and to systematically learn and adapt to the results of new information.

• The Maine Atlantic Salmon Commission should consider shaping governance structures so that they are consistent with salmon biology. That process could involve developing multistakeholder governance institutions for each drainage basin, each nested within larger scale governance bodies to address effects that are larger than individual basins, such as climate change and aquaculture.

• The suite of additional options with multiple environmental benefits outlined in Chapter 5 should be adopted. Those strategies are likely to help Atlantic salmon in Maine, and they will have other environmental benefits even if they do not help salmon. The energy and commitment of the members of many local watershed and river-specific groups focused on restoring salmon and their habitats is an important asset and should be included in any overall approach to rehabilitating Atlantic salmon and their habitats in Maine.

1

Introduction

BACKGROUND

Maine was once the home of abundant populations of wild Atlantic salmon (*Salmo salar*), but they have been declining since at least the middle of the nineteenth century (Baum 1997). Despite conservation efforts over the past 130 years or so, populations in Maine have continued to decline, and now they are seriously depleted in all the rivers that still retain natural runs. Only an estimated 862 adult salmon returned to Maine streams to spawn in 2002, down from 940 in 2001 (MASC 2002). Most of those fish returned to one river, the Penobscot (782 in 2002 and 786 in 2001). The declines led to the listing of Atlantic salmon in eight Maine rivers (Cove Brook, Dennys, Ducktrap, East Machias, Machias, Narraguagus, Pleasant, and Sheepscot) as an endangered distinct population segment (DPS) under the federal Endangered Species Act (ESA) by the Fish and Wildlife Service (FWS) and the National Marine Fisheries Service (NMFS) on November 17, 2000 (50 CFR 17, 224). Figure 1-1 and Table 1-1 show New England rivers with Atlantic salmon; the eight so-called DPS rivers are identified. Those eight rivers together had minimum estimates of 64 returning adults in 2000, 81 in 2001, and 33 in 2002 (MASC 2002, USASAC 2003). In 2002, no returning adults or redds were observed in Cove Brook, the Ducktrap River, and the Pleasant River.

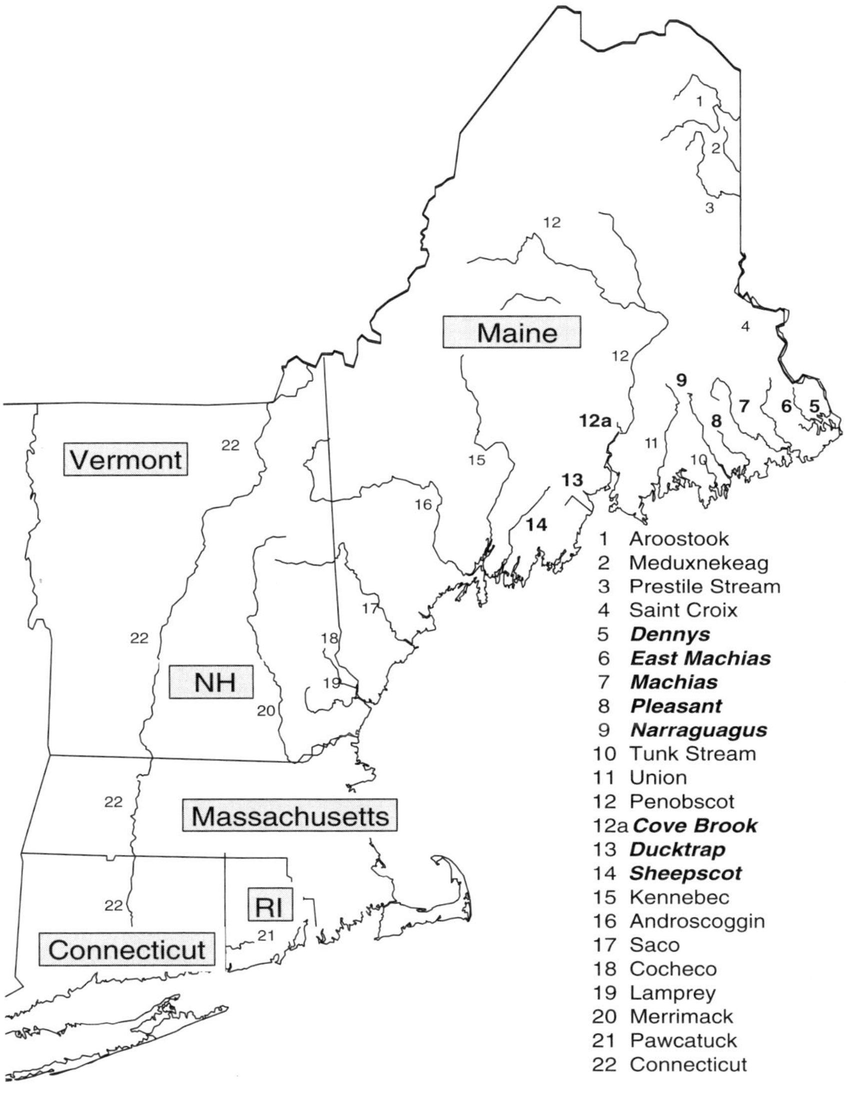

FIGURE 1-1 USA Atlantic salmon rivers with active restoration and recovery programs in New England. The eight DPS rivers in Maine with Atlantic salmon listed as endangered under the ESA are (5) Dennys, (6) East Machias, (7) Machias, (8) Pleasant, (9) Narraguagus, (12a) Cove Brook, (13) Ducktrap, and (14) Sheepscot. SOURCE: Baum et al. 2002. Reprinted with permission of the author.

TABLE 1-1 Inventory of Current U.S. Atlantic Salmon Rivers

No.	River	States	Length (km)	Drainage Area (sq ha)
1	Aroostook	ME	115	5,931
2	Prestile Stream	ME	39	562
3	Meduxnekeag	ME	65	1,287
4	Saint Croix	ME	50	6,475
5	Dennys[d]	ME	32	342
6	East Machias[d]	ME	59	650
7	Machias[d]	ME	98	1,191
8	Pleasant[d]	ME	45	220
9	Narraguagus[d]	ME	78	601
10	Tunk Stream	ME	27	104
11	Union	ME	100	1,295
12	Penobscot[e]	ME	267	22,196
13	Ducktrap[d]	ME	17	93
14	Sheepscot[d]	ME	55	591
15	Kennebec	ME	242	15,540
16	Androscoggin	ME	207	6,475
17	Saco	ME and NH	201	4,395
18	Cocheco	NH	70	479
19	Lamprey	NH	100	500
20	Merrimack	MA and NH	302	12,976
21	Pawcatuck	RI		
22	Connecticut	CT, MA, VT and NH	667	29,138
TOTAL			2,836	111,041

[a]Atlantic salmon habitat is defined as riffles and runs.
[b]Data based upon surveys in 1950s–1960s; a + indicates that some tributaries (mostly minor) have not been surveyed.
[c]North Atlantic Salmon Conservation Organization (NASCO) categories: L, lost; M, maintained; R, restored; T, threatened with loss; N, not threatened with loss. U indicates current population status unknown.
[d]Atlantic salmon populations in these rivers listed as endangered under the federal Endangered Species Act.

Atlantic Salmon Habitat Units[a] (unit = 100 sq mi)			
Surveyed Amount	Estimated Amount[b]	Total Amount (minimum)	Salmon Population Status[c]
30,000	30,775	60,775	L
–	835	835	U
–	10,000	10,000	U
29,260	+	29,260	L
2,414	+	2,414	T
3,006	+	3,006	T
6,156	+	6,156	T
1,220	+	1,220	T
6,014	+	6,014	T
–	627	627	L
–	8,370	8,370	L
–	125,000	125,000	T[f]
845	+	845	T
2,797	+	2,797	T
43,483	114,300	157,783	T[f]
–	47,900	47,900	L
12,540	15,000	27,540	L
	+	0	L
	+	0	L
	+	0	L
4,490	+	4,490	L
243,000	+	243,000	L
385,225	352,807	738,032	

[e]Cove Brook, a tributary to the lower Penobscot River, is one of the eight rivers identified in footnote 1 above.
[f]Designation applies to selected tributaries below the first hydrodam.
Abbreviations: sq mi, square mile; km, kilometer; sq ha, square hectare.
SOURCE: Adapted from NASCO Special Session on Salmon Habitat, Faroe Islands, June 2002.

No one disputes the general seriousness of the declines, but many people in Maine claim that the populations are not wild[1] and, therefore, oppose the ESA listing. They argue that the fish are derived mainly from hatchery stocking and aquaculture escapes. If so, then appropriate measures to increase the number of salmon in Maine's rivers could be quite different from appropriate measures to increase wild salmon runs in those rivers. The controversy led Congress to mandate a study of Atlantic salmon in Maine by the National Research Council (NRC). An interim report was to be prepared in time to help any recovery efforts (see Box S-1 for committee's statement of task).

The interim report (NRC 2002a) by the NRC Committee on Atlantic Salmon in Maine focused on the genetic characteristics of the wild populations in Maine, especially in the listed rivers. In this report, the committee focuses on the broader issues contributing to the decline of salmon in Maine and options for helping them to recover.

THE LISTING OF SALMON UNDER THE ENDANGERED SPECIES ACT

The ESA of 1973, as amended in 1988 (Public Law 100-478), defines species as including "any subspecies of fish or wildlife or plants, and any distinct population segment of any species o[f] vertebrate fish or wildlife which interbreeds when mature" (Section 3 {15}). A detailed description of the ESA's provisions as they affect Atlantic salmon in Maine is in Appendix C. The salmon in the eight Maine rivers, including "all naturally reproducing wild populations and those river-specific hatchery populations of Atlantic salmon having historical, river-specific characteristics found north of and including tributaries of the lower Kennebec River to, but not including, the mouth of the St. Croix River at the U.S.-Canada border," were listed as an endangered DPS by FWS and NMFS ("the Services") on November 17, 2000 (65 Fed. Reg. 69459 [2000]). The eight rivers are often referred to as the DPS rivers. The science that underlies the ESA; the concept of species, including subspecies and DPSs; and

[1]The term *wild* is used by the committee to mean populations of salmon that have been maintained by natural spawning for at least two full generations. This practical definition is used by the committee to distinguish salmon populations that are supported by human activities (hatchery fish) from those that have established themselves in the wild. The committee agrees with Baum (1997) that *pristine* salmon populations that have always been wild with no human influences on their genetic makeup almost surely do not exist in Maine. The term *natural* is used for salmon populations that are derived from parents' reproduction in streams rather than stocking.

the meaning of "endangered" under the ESA are discussed in considerable detail in two earlier NRC reports (NRC 1995, 1996a).

THE PRESENT STUDY AND REPORT ORGANIZATION

For this study, the committee met three times in Maine and heard presentations from representatives of the state government of Maine, including Governor Angus King; from the Services; from the Atlantic Salmon Commission; and from a variety of industrial, academic, environmental, and other private organizations and individuals. The committee also visited an Atlantic salmon farm and two blueberry farms in Washington County, a weir on the Pleasant River, the federal hatchery at Craig Brook, the site of the former Edwards Dam on the Kennebec River, road crossings over salmon streams, and other sites. Lists of the presenters and facilities visited are in the front matter. The committee met once each in Boston and Woods Hole, Massachusetts, and once in Washington, D.C., and considered an array of published literature and reports.

The committee has attempted to bring a new perspective to this much-studied problem. There is no lack of factors known to have adversely affected Atlantic salmon in Maine and throughout their historical range in eastern North America. Indeed, the problem is that too many such factors are known, and the difficulty is how to prioritize them and how consequently to prioritize potential actions to rehabilitate[2] the salmon populations and their environments. The committee has taken a risk-assessment and decision-analysis approach, in part for illustrative purposes but also to help it assess the importance of the factors affecting salmon and to prioritize potential rehabilitation options. The committee also considered the entire state as potentially available for rehabilitation efforts, rather than only the eight DPS rivers. As a result, our focus is much broader than only the requirements of the ESA. The committee's work along these lines is only a beginning, but we hope that by following the example and guidance in this report, decisions and actions taken by those responsible for and interested in rehabilitation of salmon populations in Maine will be made more productive and effective.

The report begins with a description of Atlantic salmon in Maine and the environments they inhabit (Chapter 2). Chapter 3 describes the most significant threats to salmon in Maine. The committee then describes risk

[2]Following the usage of the NRC Committee on Protection and Management of Pacific Northwest Anadromous Salmonids (NRC 1996a), the committee has chosen to consider *rehabilitation* as a practical and achievable strategy, rather than *restoration*, which implies return of ecosystems to some previous but unknowable pristine condition.

assessment and decision analysis and the committee's Atlantic salmon risk model in some detail (Chapter 4). Chapter 4 also discusses the committee's decision analyses for dams and for salmon farms; they are provided as examples of how to think about such issues systematically but not as a substitute for such analyses by the people who have to live with the results. Chapter 5 discusses methods of addressing the threats to Atlantic salmon in Maine, and Chapter 6 provides the committee's findings and recommendations.

2

Salmon Life History and Ecology

INTRODUCTION

Atlantic salmon (*Salmo salar*) once spawned in northern hemisphere rivers from Long Island Sound, New York, to arctic regions in the western and eastern North Atlantic, the Barents Sea, the Baltic Sea, and south to Spain and Portugal. The historical range has contracted northward in the past century (for the U.S. population, see Baum 1997; for current distribution, see also Collette and Klein-McPhee 2002). Climate warming may be a large factor in range contraction due to temperature and other weather-related effects on lacustrine and coastal marine conditions (see discussion by Dickson and Turrell 2000), but overexploitation of the salmon fishery and loss of habitat due to human activities (for example, dam construction, pollution, stream siltation, and introduction of nonnative species) must also be considered factors (Baum 1997). Natural runs of Atlantic salmon currently occur from Maine to northern Spain and Portugal (Figure 2-1), but spawning runs are at low or even endangered levels in most of those areas (Hutchinson and Mills 2000, O'Neil et al. 2000). The widespread hemispheric decline in salmon, even in streams with high-quality habitat where exploitation has been restricted or prohibited, points to a strong climatic impact in either the riverine or the oceanic portions of the salmon life history or both (Cairns 2001, Hutchinson and Mills 2000, Reddin et al. 2000). The strong coherence of declines for stocks from many varying areas implicates the marine part of the life cycle as a major factor (Reddin et al. 2000), although the spatial patterns of decline are complicated and

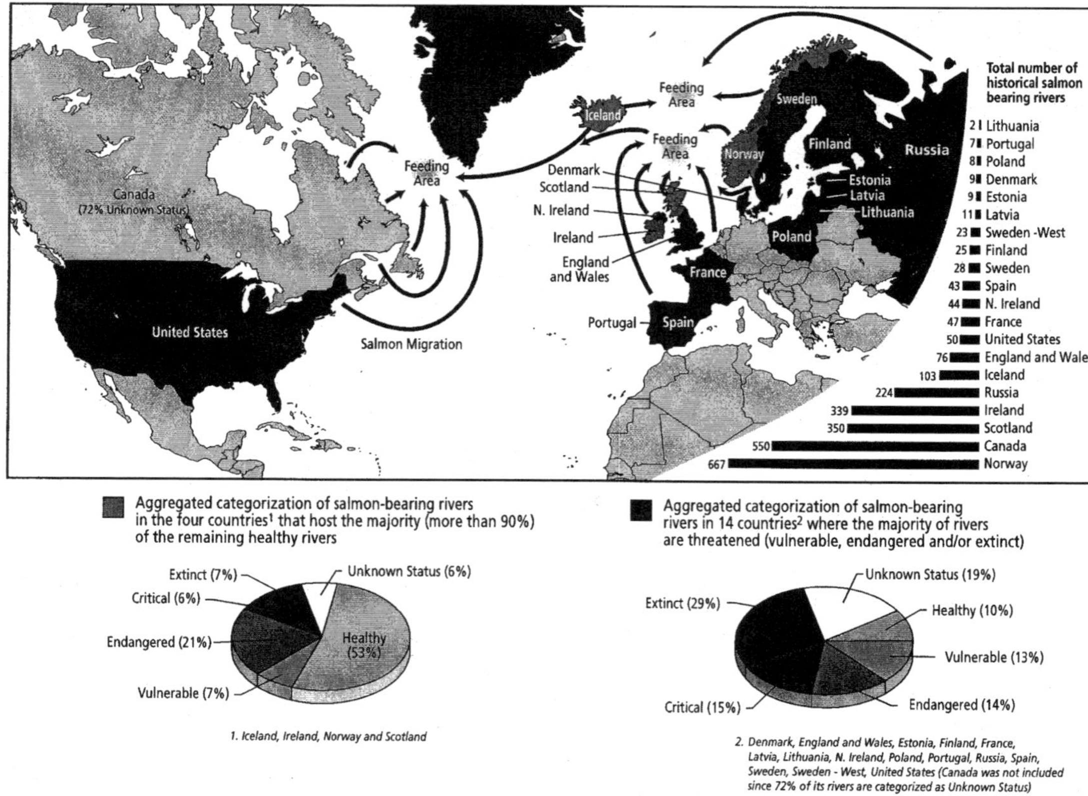

FIGURE 2-1 Map showing the wild Atlantic salmon's range in 2000 and its known migration routes.
SOURCE: WWF 2001. Reprinted with permission; copyright 2001, World Wildlife Fund-U.S.

suggest that a number of factors and adaptations (such as migration paths) may be at work.

Maine has the last of the wild Atlantic salmon populations in the United States. At one time, 300,000 to 500,000 adults probably entered U.S. rivers each year (Beland 1984, Stolte 1981). The Biological Review Team (1999) used zoogeographic information to construct ecological provinces, including aquatic ecological units (Bailey 1995, Maxwell et al. 1995), to analyze the distribution of Atlantic salmon in the United States. The results suggest that Atlantic salmon populations were divided into at least three distinct groups of populations: (1) those in Long Island Sound, in eight major rivers, including the Connecticut River; (2) those in Central New England, including the Merrimack River in the southern Gulf of Maine; and (3) those in the rest of the Gulf of Maine including the eight DPS rivers, where salmon are listed as endangered under the Endangered Species Act (ESA) (see Chapter 1). A map of Down East Maine showing Atlantic salmon habitat along these rivers is shown in Figure 2-2.

The Long Island Sound populations were gone by the early 1800s (Meyers 1994), followed by the central New England populations in the mid-1800s (Stolte 1981, 1994). The remaining U.S. populations might once have produced 100,000 adults per year, but those numbers have not been seen since the late 1800s. Fewer than 3,000 adults returned per year in the 1960s and 1970s (Figure 2-3). Large stocking efforts in the Penobscot River, especially of smolts, led to a brief period of annual returns numbering 3,000–5,000 fish, but returns to the Penobscot and the other DPS rivers have declined precipitously since the early 1990s (Figure 2-4 [returns versus time]). The decline has occurred despite sustained efforts at stocking and remediating anthropogenic impacts on the Penobscot, strict conservation measures on the DPS rivers, and general improvements in the way that riparian zones are managed. The total return of Penobscot fish for the cohort of smolts released in 1999 (now virtually all accounted for as 1, 2, or 3 sea-winter [SW] fish by 2002) was fewer than 700 adults; for the Gulf of Maine DPS rivers, a minimum estimate of 33 adults returned in 2002 (MASC 2002, USASAC 2003). (Spidle et al. [2003] provide estimates of returns modeled on redd and adult counts from a trap. These estimates include means and 95% confidence limits.) The population decline has been associated with lower return rates, which are now about 1% in the Narraguagus and about 0.2% in the Penobscot. These are below the 2–4% return rate published for many populations (e.g., Reddin et al. 2000). Recent electronic tagging studies in the Narraguagus indicate about half (range = 32–67%) of the total post-riverine mortality is experienced before smolts leave the coastal bay where the Narraguagus enters the sea, but it is not known whether this distribution of the total marine mortality is normal (J. Kocik, NMFS, unpublished data, 2001).

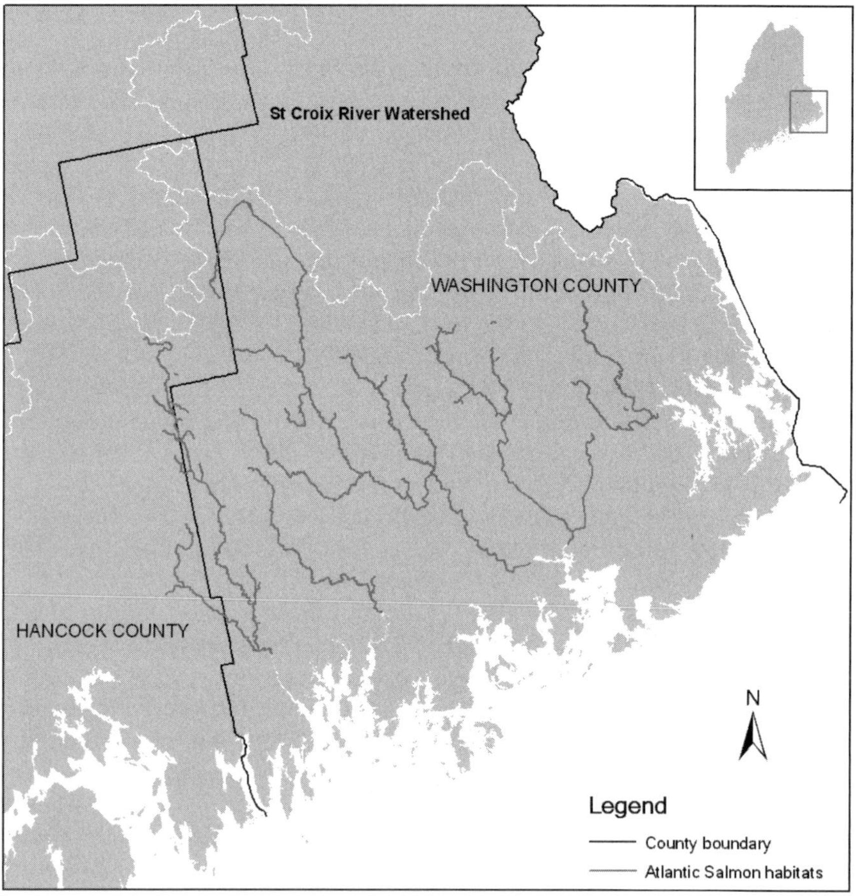

FIGURE 2-2 Atlantic salmon habitat in Down East Maine. SOURCE: Data from Maine Office of Geographic Information Systems. Drawing by Yanli Zhang, University of Massachusetts-Amherst.

The purpose of this chapter is to provide background on the biology and environment of Atlantic salmon specific to the task of understanding why the numbers of fish returning to Maine rivers are declining and recommending steps that would help ensure the survival of these populations. Topics covered in this section include salmon life history, historical and recent changes in abundance, and distribution and migrations. Then, the characteristics of environments that comprise salmon habitat are described: (1) geology and hydrology of soils and forests (including impacts

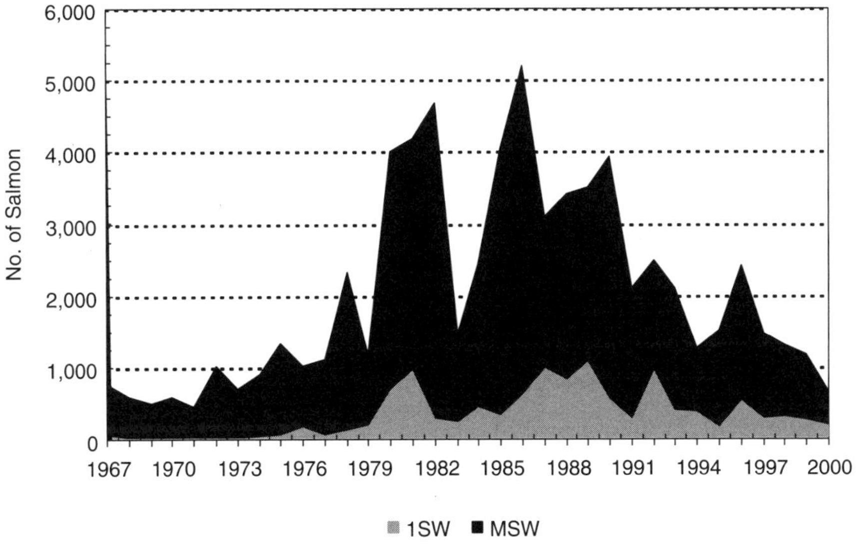

FIGURE 2-3 Documented adult Atlantic salmon returns to all Maine streams. These are rod and trap catches combined for the Penobscot River and primarily rod catches for other rivers. These numbers represent minimum estimates of adult returns. SOURCE: E. Baum, Atlantic Salmon Unlimited, unpublished material, 2001. Printed with permission of the author.

by human activities); (2) aquatic environments; (3) biological communities in the streams and estuaries and along the ocean migration routes; and (4) climate variability. Baum (1997) provides a readable and comprehensive history of salmon in Maine, including maps of individual salmon-producing rivers, detailed histories of stocking efforts, a map of historical fishing weirs, and tables of catch statistics. Those details are not repeated in this report. Bigg (2000) and Dickson and Turrell (2000) provide overviews of climate change and salmon, primarily from the perspective of European stocks. Drinkwater (2000) provides evidence of northern hemisphere climate impacts on North American fisheries. While there are strong suggestions of impact, the exact causal relationships remain unknown. Cairns (2001) provides a lengthy discussion of the many factors that affect salmon abundance and attempts to prioritize them on the basis of their *likely* and/or *potential* role in the recent declines. This is an assessment based on the experience and professional judgment of over 60 scientists throughout the range of Atlantic salmon. Finally, a combined report from U.S. National Marine Fisheries Service (NMFS) and U.S. Fish and Wildlife Service (Anadromous Atlantic Salmon Biological Review Team

a

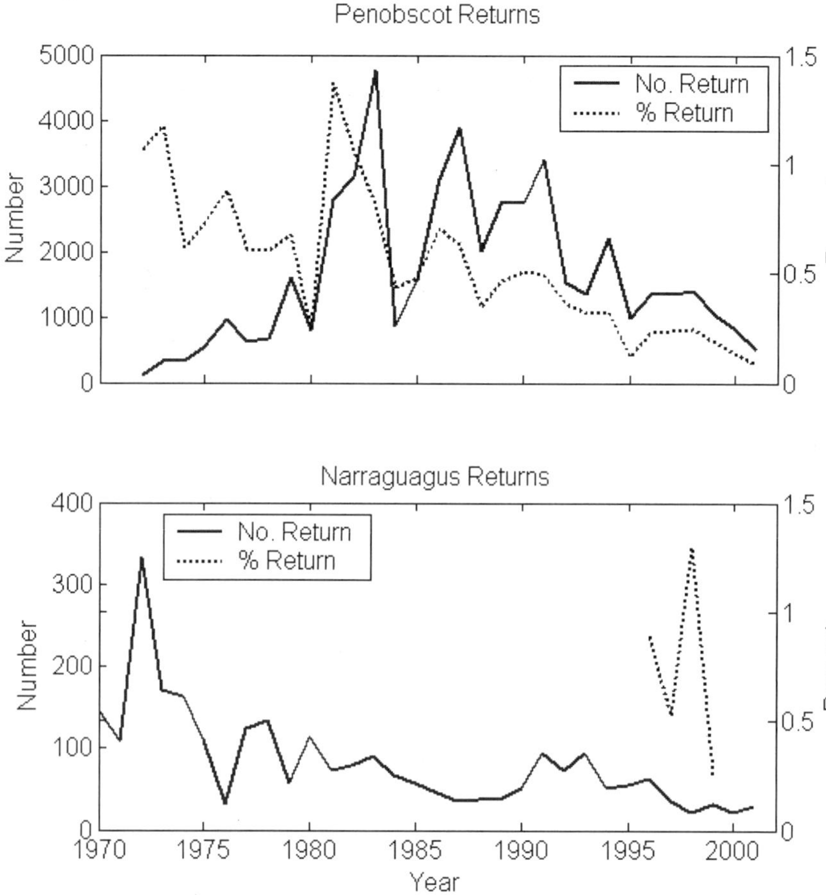

FIGURE 2-4 (a) Minimum estimates of number of fish returning to the Penobscot and Narraguagus rivers and the return rate (%) for that cohort. Penobscot return rate has been adjusted for all multi-sea-winter (SW) returns; the Narraguagus assumes all 2SW fish. SOURCES: Data from K.F. Beland, Maine Atlantic Salmon Commission, unpublished data, 2003; USASAC 1996–2004. (b) Minimum estimates of number of fish returning to seven of the DPS rivers. Solid lines are from fish counts based on rod catches. Broken lines are estimates based on redd counts. (The regression of returns to redds was done on the Narraguagus.) These are underestimates. SOURCES: Data from USASAC 1996–2004.

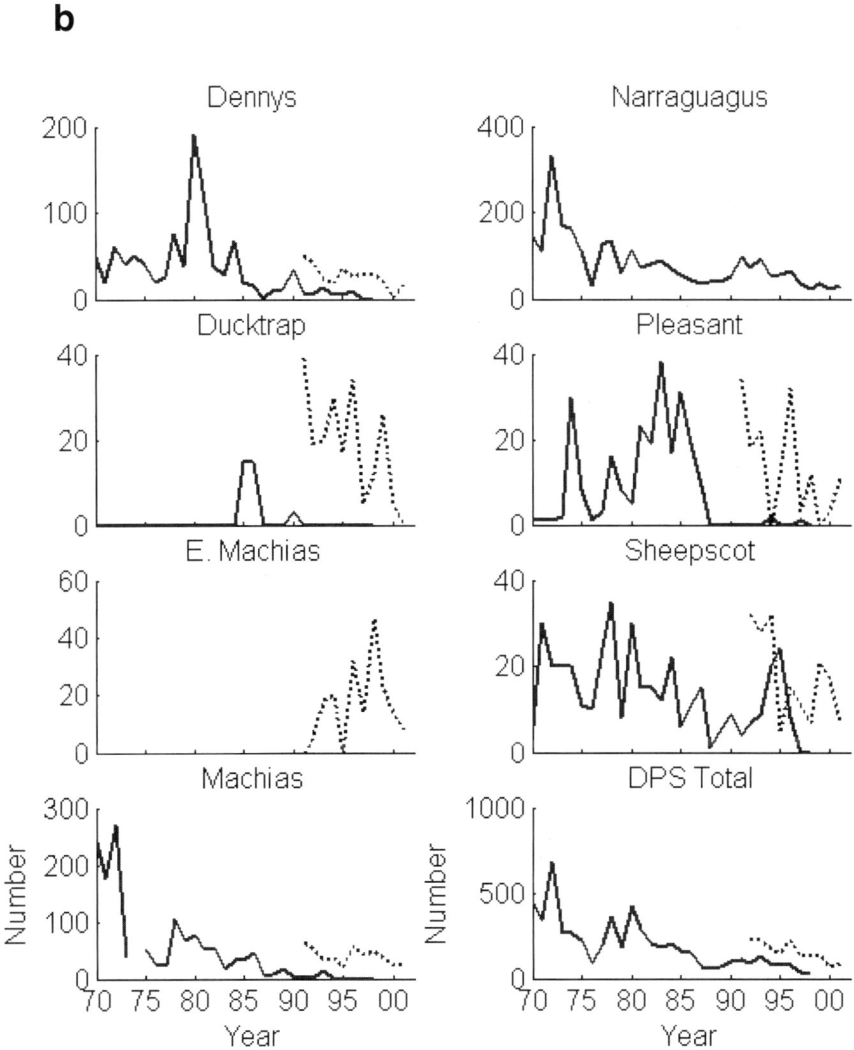
b
Dennys
Narraguagus
Ducktrap
Pleasant
E. Machias
Sheepscot
Machias
DPS Total
Number
Year
70 75 80 85 90 95 00

1999) provides an excellent summary of salmon biology and conservation issues.

SALMON LIFE HISTORY

This subsection provides basic information about the natural history of the Atlantic salmon necessary to understand problems facing their continued existence. The first topic is a description of the sequence of developmental stages and their timing in an individual Atlantic salmon. The second topic concerns key characteristics of the life history of Atlantic salmon as a species. These characteristics include alternative reproductive strategies and anadromy. Finally, recent and historical changes in the distribution and abundance of Atlantic salmon are explained.

Complex Life Cycle

Atlantic salmon are anadromous: they begin their lives in freshwater where the young grow to several inches in length, then migrate to the sea, where they grow more rapidly and become sexually mature after 1, 2, or 3 years[1] (Baum 1997). The complex life cycle of the Atlantic salmon consists of a series of morphologically, behaviorally, and physiologically distinct stages as the fish migrate from freshwater to the sea and back to freshwater again. This cycle differs from the simple life cycle of many fish species that complete the transition from juvenile to adult without migrating between different environments. The terms used to describe the Atlantic salmon's developmental stages are given in Table 2-1.

Key aspects of the stages and of transitions between them are as follows. The egg and alevin stages rely on the yolk deposited in the egg by the mother for nutrition. Hence the nutritional state of the adult spawning female affects the welfare of the offspring. Furthermore, any lipophilic pollutants (e.g., polychlorinated biphenyls) consumed by the adult female tend to be deposited into the lipid-rich yolk of the eggs. Embryonic development is especially susceptible to disruption by chemical agents. Successful transition to the feeding fry stage requires functional organ systems and appropriate behavioral responses. Timing of fry emergence from the gravel is important because of seasonal changes in prey availability. An important aspect of the parr stage is the rate of growth. Growth and size determine the timing of the parr-smolt transformation. Parr are young salmon with 8–11 vertical dark bands on their sides. Transforma-

[1]Fish that return after 1 year are termed 1SW (one sea-winter) fish; 2SW and 3SW mean fish that spend two and three winters at sea, respectively.

TABLE 2-1 Stages in the Life Cycle of the Atlantic Salmon[a]

Term for Stage	Begins	Duration	Appearance	Behavior
Egg	Mid-October to Mid-November	2–6 weeks	Amber, 6 mm diameter	Buried in gravel
Eyed egg	November	3–5 months	Two black eyes	Twitch
Alevin or sac-fry	At hatching in March or April	5–8 weeks	Mostly transparent with large yolk-sacs	Remain in gravel
Fry	Mid-May	1 summer	Pigmented	Emerge from gravel and feed
Parr	July 1		Stocky with black vertical parr marks	Territorial and solitary
0+ parr	July 1	6 months		
1 and 1+ parr	January 1	1 year		
2 and 2+ parr	January 1	1 year		
3 and 3+ parr	January 1	1 year		
Smolt	April	3 months	Streamlined and silvery	Schooling and migratory
1+ smolt	1 year old			
2+ smolt	2 years old			
3+ smolt	3 years old			
Post-smolt	July 1	6 months	Silvery	Marine and migratory
Salmon	January 1 of first year at sea	Variable	Subadult and adult	Feeding migration

[a]The terms used reflect an arbitrary hatching date of April 1 (equivalent to birth), and many stage increments are arbitrarily set at beginning and end dates of the calendar year (January 1 and December 31).
SOURCE: Adapted from Baum 1997.

tion to the smolt stage occurs in the winter and spring. Smolts are silvery, without parr marks and with a more streamlined body. As indicated in Table 2-1, most Atlantic salmon in Maine grow fast enough to transform to smolt in their second spring (they are called 2 parr), whereas slower-growing parr transform in their third spring (called 3 parr). The parr-smolt transformation is of key importance because the smolt faces the energetic challenge of seaward migration and the osmoregulatory challenge of the transition from freshwater to seawater. For the remainder of the calendar year in which smolts enter seawater they are called post-smolts. Beginning January 1, they are thereafter called salmon. The term *salmon* was used in English by the thirteenth century to describe the silvery salmon in the sea. It was not recognized then that the small parr in the streams were members of the same species.

Understanding the life cycle of Atlantic salmon is complicated by their alternative life-history strategies. For example, before reproductive maturity, these alternatives include variable durations in the stages before their seaward migration and variable numbers of years growing in the ocean (see Table 2-2). Reproductive alternatives include variable age and size at maturity, variable timing of homeward migrations to spawn, variable number of years of spawning, and variable fecundity between years. Maine's Atlantic salmon exhibit two run timings that are in part influenced by genetic factors. "Early run" adults enter freshwater between

TABLE 2-2 Life-History Strategies and Alternatives of the Atlantic Salmon in Rivers in Maine

Life History Factor	Description of Primary and Alternative Strategies
Development	
Duration of parr stage	Primary: 2 years (80%) Alternative: 3 years (20%) or 4 years (small %)
Anadromy	Primary: migrates to the sea for a growth period Alternative: landlocked populations Alternative: male parr become mature (precocious parr)
Time-at-sea	2 sea-winters (2SW, estimates of 84–94%) Alternative: 1SW, occurs in males, termed grilse (<0.3%); also estimates of <10% 1SW with >95% males Alternative: multiple sea winters (MSW, such 2SW, 3SW)
Reproduction	
Age at maturity	Primary: fifth fall of life Alternative: genetics and environment lead to alternatives
Timing of migration to natal streams	River dependent Early runs from May to mid-July Late runs from mid-July through September
Spawning frequency	Primary: Semelparity—spawn only once, then die Alternative: Iteroparity or repeated spawns Alternative: Precocious male parr constitute a large percent, which varies widely among rivers and years

May and mid-July, and "late-run" adults enter freshwater later in the summer. Most Atlantic salmon are semelparous, meaning they spawn once and die. However, 1–6% of anadromous spawning adults survive, return to the sea, and migrate home later to spawn again. Thus, a small percentage of anadromous fish is iteroparous. However, an unknown percentage of mature male parr survive to breed again, either as a parr or as an anadromous adult. The terms used to describe salmon with different reproductive alternatives are given in Table 2-2.

In addition to anadromous Atlantic salmon, Maine has populations of Atlantic salmon that complete their entire life cycle in freshwater. They are called landlocked salmon or ouananiche. They are the same species as the anadromous form. In Maine, they were originally found only in four drainages, but they have been widely stocked elsewhere in Maine. Although there is some small degree of genetic difference between landlocked and anadromous populations, it is not necessarily greater than the differences among anadromous populations, and it is not clear whether the difference in life history has a genetic basis (Tessier and Bernatchez 2000). Landlocked salmon are not endangered, but because they strongly resemble anadromous salmon and in some cases compete with them, they can complicate efforts to rehabilitate wild anadromous populations.

The salmon life-history pattern has major implications for the species' evolution and survival in different regions. Because the fish migrate upstream to spawn, they are particularly vulnerable to fishing. Because salmon migrate between ocean and freshwater environments, they are subjected to the vagaries of two ecosystems during different parts of their life history. This anadromous life history greatly increases the number of factors that could affect population size.

Salmon are known for their ability to return to the streams where they were hatched. Salmon return to their natal streams to spawn, a trait that segregates populations and leads to a variety of local adaptations, including the timing of spawning runs, growth rates, and other life-history features (e.g., Allendorf and Ryman 1987, Gharrett and Smoker 1993, Heggberget et al. 1986, Hutchings and Jones 1998, Kendall 1935, Kincaid et al. 1994, Nielsen 1998, for Atlantic salmon; NRC 1996a, Saunders 1981, Smoker et al. 1998 for Pacific salmon species; Taylor 1991 both Atlantic and Pacific salmon; Verspoor et al. 1991, Webb and McLay 1996). Straying to another stream occurs at low frequency. For example, Penobscot River salmon show over 98% fidelity to the home stream (Baum 1997).

The low frequency with which salmon stray to neighboring streams results in the development of a *metapopulation* structure—a set of local breeding populations connected by exchange of some individuals. This network of local populations provides a balance between local adaptation and the evolutionary flexibility that results from exchange of genetic material among local populations (NRC 1996a). That NRC report concluded that "maintaining a metapopulation structure with good geographic distribution should be a top management priority to sustain salmon populations over the long term." That conclusion was drawn for Pacific salmon, but it applies to Atlantic salmon as well.

Although strays probably have lower reproductive success than fish that are returning to their native streams, they provide a source of new genetic combinations—important for the salmon's evolutionary potential in the face of changing environments—and they may recolonize streams that have lost their own native runs. For Atlantic salmon populations to have colonized and survived for extensive periods near the southern limit of the species' range (currently Maine), they probably acquired adaptations to the distinct physical and environmental challenges of local waters. Local adaptations, established by strong homing and selection pressures, are a known property of salmon populations throughout the world (Allendorf and Ryman 1987, Taylor 1991).

The complex transition to saltwater at the smolt stage requires suites of behavioral adaptations for navigating, avoiding predators (including seals, cormorants, and striped bass), and finding marine invertebrate and fish prey. During the oceanic phase, juveniles from most river systems migrate to subpolar seas to feed for 2 or more years before returning to their native streams. A small number of fish, referred to as grilse, return after only 1SW. A known exception to this pattern occurs in rivers draining into the Bay of Fundy, Canada. Fish from these rivers remain within the Gulf of Maine and most return to their natal streams after only one winter at sea. For Maine salmon, maintenance of a stable population would require about 2% survival of smolt to 2SW stage (based on Baum's [1997] estimate of 90 smolts produced per female). A decrease in either freshwater or oceanic survival would cause a decline of Maine's wild salmon populations.

Adult salmon return to their natal streams from spring until fall. The peak migration time is a characteristic of individual populations and environments. Spawning occurs in autumn, and the eggs develop in gravel nests (redds) that are dug by the female. Because Maine's females are mostly large 2SW fish, they deposit more eggs, about 7,000 each. The fry emerge in mid-May and grow into parr during the summer. Vertical bars on parr provide camouflage protection from predators. Most parr remain in freshwater for 2 years before becoming smolts and migrating to the ocean. Some male parr mature in the stream and have some success in fertilizing eggs. Survival from the egg to the smolt stage is estimated to be 1.25% (Baum 1997, Bley and Moring 1988), and thus a rough calculation from Baum's data suggests that an average of 90 smolts are produced by a wild Maine 2SW female.

The anadromous pattern, with some repeat spawning, means that counting the fish returning to a stream gives information only on part of the population. The rest of the population is either in the river as fry, parr, or smolts or still at sea growing and maturing. In addition, salmon have overlapping rather than discrete generations as a result of precocious development and repeat spawners. The presence of early maturing males (precocious parr) tends to buffer the population somewhat against random variation in the return rate of anadromous (adult) male spawners (Garcia-Vazquez et al. 2001, Martinez et al. 2000). Repeat spawners are important because of the increased egg production of older females and their proven success in the face of natural selection. However, 3SW salmon and repeat spawners make up less than 1% of spawning adults (Baum 1997).

The use of freshwater habitats for reproduction and juvenile rearing improves the survival of early life stages because they are inaccessible to marine predators, although they are still susceptible to freshwater predators. Predation depends on the density of predators (Mills 1989), but it was recognized in early studies of juvenile salmon (Huntsman 1938) that precipitation, and thus stream flow and water depth, could affect predation rates and thus juvenile survival (Ghent and Hanna 1999).

When Atlantic salmon smolts enter the sea, they are entering that portion of their life that seems to have the largest variation in survival rate (Cairns 2001, Reddin 1988). At this point, they range in size from 13 to 23 cm fork length—most often 16 to 20 cm—and are 2 or 3 years old. Parr-smolt transformation is influenced by the size of the fish. Approximately 80% (range 70–90%) of the smolts are 2-year-olds that leave the river in spring (late April to mid-June). Most of the remaining fish leave the river as 3-year-old smolts in an outmigration the following spring. Despite the additional growing season, these smolts average only 1.1 cm longer than the 2-year-olds. A very small fraction of fish has been known to leave as 4- and 5-year-olds (Baum 1997). Initial feeding in the marine environment (estuaries) is on insects (at the surface), euphausiids, amphipods, and decapod crustaceans. (These groups may be found in the upper layer of the ocean, although deeper feeding cannot be ruled out.) Smolts soon begin feeding on herring, sand lance, capelin, and shrimp (Baum 1997). Smolts appear to spend most of their time in the upper part of the water column. Electronic tagging data near the mouth of the Bay of Fundy indicate most smolts are in the upper 10 m (G. Lacroix, unpublished data, 2001; see methods in Lacroix and McCurdy 1996). Norwegian studies show a rapid reduction in smolt catch rates when the upper portions of sampling trawls drop below the surface.

Salmon mortality is high during the rapid passage from river to Gulf. Studies conducted in the Narraguagus River from 1996 to 2000 (J. Kocik,

NMFS, unpublished data, 2001) indicate a loss of 38–63% of outmigrating smolts in this small bay (mean = 50%). This is nearly half of the total losses averaged over 1SW and 2SW fish from this river. The average survival of grilse plus 2SW fish from the Narraguagus is 1.1% (annual averages ranged from 0.87% to1.4% in this study), whereas the true "at-sea" survival over this period was >2% when corrected for the initial losses in the bays (J. Kocik, NMFS, unpublished data, 2001). While the near-shore loss is a large proportion of the total marine losses reported here, it must be remembered that the average return rate for this river and for all rivers in the Gulf of Maine and south for the period of record is low.

Salmon pass through the estuarine environment quickly. Electronic tagging reveals that smolts exiting the Narraguagus River pass out of Narraguagus Bay within a few days (J. Kocik, NMFS, unpublished data, 2001). In Penobscot Bay, where electronic tagging and detection are less practical, special trawling methods were used to follow the passage of elastomer-marked fish in 2001 (R. Brown, NMFS, unpublished data, 2001). Smolts passed through counting traps in the main stem of the Penobscot (Veazie Dam, north of Bangor) beginning in late April. By middle to late May, they were widely distributed throughout the bay (80% of tows were positive for smolts) and some had entered the shelf environment (more than 50% of tows outside the bay caught smolts).

Migration and Distribution

Oceanic migration affects growth, maturity schedules, availability to fisheries, and eventually recruitment of salmon populations (Friedland 1998, Narayanan et al. 1995). Migration routes in the Gulf of Maine are unknown. The migration patterns of European post-smolts appear to take advantage of prevailing strong residual currents, such as the Norwegian Coastal Current or the Slope Current along the margin of the shelf (Hansen and Quinn 1998, Holst et al. 2000). If post-smolts leaving the Down East rivers and Penobscot Bay exhibit similar behavior, a likely pathway would involve passage westward along the shelf to the central coast (Penobscot Bay region) and then across the Gulf following the prevailing circulation patterns around Jordan Basin, Georges Basin, and the northern edge of Georges Bank. Passive drift alone could cover this distance in a few weeks. The opposite choice for leaving the Gulf would involve migration eastward across the mouth of the Bay of Fundy and across the southern Scotian Shelf against a residual current that may average about 15–20 kilometers per day at this time of year. Some tagged post-smolts from Maine rivers have been recovered in the Bay of Fundy. Intermediary routes across the open Gulf are also possible and would result in intermediate advantages or disadvantages with respect to the influence of the residual circulation. A clockwise migration would keep the fish in colder water, and perhaps seawater temperature dominates the migratory behavior. Whichever route is taken, tag returns indicate that post-smolts arrive off northern Nova Scotia (Cape Breton Island) 45–50 days after leaving coastal Maine bays, and in southern Newfoundland shortly thereafter (mid-August, 60–65 days after leaving the Maine coast). In 100–110 days, many salmon have made it to the southern coast of Labrador (see review by Baum 1997; also see data from Friedland et al. 1998). Another factor that might influence post-smolt migration paths is the feeding environment, but this has not been studied in sufficient detail to resolve the relative advantages of the various routes. Bley and Moring (1988) and Friedland (1994a) compared return rates for rivers at different latitudes on both sides of the Atlantic Ocean and suggested that low rates might be characteristic of populations whose natal rivers are located near the limit of the species' range. The hypothesis was that mortality was higher due to the longer distances traveled between the natal rivers, winter feeding areas, and back. Despite the relatively large numbers of returning fish, the return rate for the Penobscot is less than half that for the Narraguagus, so other factors remain important. The highest return rates in the Gulf of Maine and south occur in the St. John River of New Brunswick and Maine. These higher rates might be due in part to the greater percentage of 1SW fish in the St. John (more than 90% vs. less than 10% in the Maine rivers).

Perhaps more important with respect to interpretation of mortality and return data, most fish from the inner Bay of Fundy do not leave the Gulf of Maine and therefore do not undertake the long migrations of Maine salmon (Ritter 1989). While the return rate for these fish is higher than that for salmon from Maine rivers, these rates also have been declining through the 1990s. Maine's salmon take part in extensive marine migrations, including movements to the waters off western Greenland (Friedland 1994a), where they become a small portion of a large mixed-stock complex of salmon from both European and North American sources. Unlike Atlantic salmon populations across the Canadian border from Maine, where 1SW fish are common among spawning adults, about 94% of adults returning to Maine are 2SW fish (USASAC 1999). Thus, the average body size of Maine adults is larger than Canadian adults. Because spawning populations of Maine salmon include several age groups (especially 2SW and 3SW adults but also precocious parr[2]), there is considerable exchange of genetic material across age classes (cohorts).

[2]Parr are young salmon actively feeding in freshwater. Even younger fish, with egg sacs, are called fry. Fish about to migrate to sea are called smolts. See Table 2-1.

The winter feeding grounds of Atlantic salmon are in the Labrador Sea (primarily North American stocks) and in the North Atlantic east of Greenland (mostly European and Icelandic stocks). These locations are associated with an apparent thermal preference of 4–8 °C (Reddin et al. 2000). There is a small amount of mixing of stocks from the two continents at this time of the life cycle, but straying of spawning fish from Europe to North America, or the reverse, is very unusual (Reddin et al. 1984).

PHYSIOLOGY

Physiology is the functioning of the individual, and it ties together genetics and ecology. There are three key concepts of particular significance to the discussion of Atlantic salmon. They concern homeostasis, temperature effects on rates, and the neuroendocrine transduction of environmental information (Figure 2-5). They are briefly explained, and their impacts on the timing of parr-smolt transformation and outmigration in the Atlantic salmon are discussed.

Homeostasis is the maintenance of a constant internal environment, which is necessary for life. The internal stability reflects a dynamic equilibrium and requires work. The internal environment differs from the external environment, whether the salmon is in a stream or the sea. The difference is created and maintained at the interfaces between the animal and its environment. These interfaces in the Atlantic salmon are the gills, gut, kidneys, and skin, and they are important for two reasons: First, it is these interfaces that are most susceptible to infection and insult; and, second, the roles of these sites change to meet the challenge imposed by the transitions between freshwater and seawater. At no site is this more obvious than in the gills. Gills regulate internal salts, gases, and nitrogenous wastes. The proxy used by salmon physiologists for indicating seawater readiness in Atlantic salmon during the parr-smolt transformation is an increased level of activity of the enzyme Na^+/K^+-ATPase in the gill. Gills are damaged by the environmental hazard of steam acidity, as discussed elsewhere.

Temperature affects all aspects of physiological functioning. The Atlantic salmon is ectothermic, meaning it has the temperature of its environment. Because the environmental temperature fluctuates, physiological functioning fluctuates as well. A general rule is that metabolic rate doubles with every 10 °C increase. Metabolism underlies development and growth. This means that hatchery, stream, reservoir, estuary, and ocean temperatures strongly affect rates of development and growth.

Neuroendocrine signals are specific chemical signals linking a salmon to its environment (Hoar 1965). A complex array of detectors receives information about the external and internal environments. This informa-

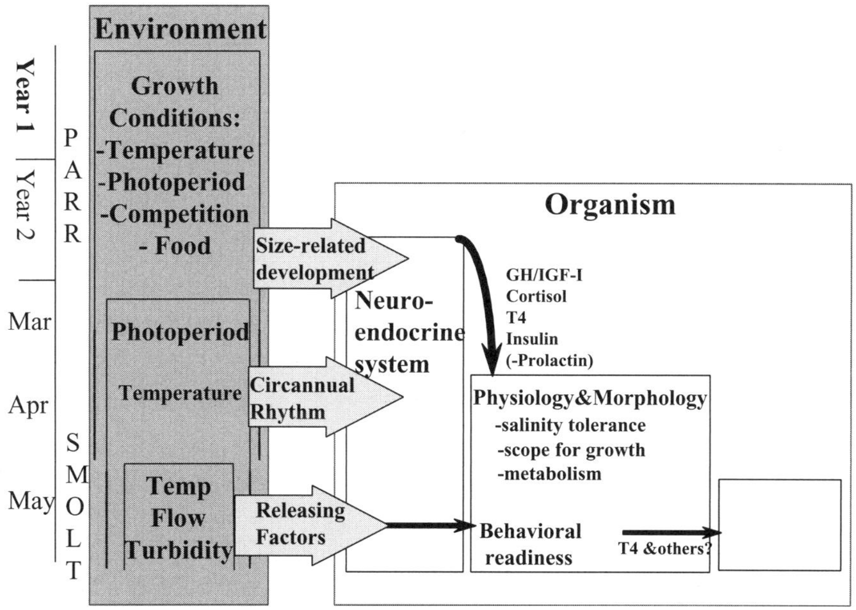

FIGURE 2-5 Multiple interactions between the environment and the organism that lead to smolt development. Growth conditions such as temperature, food, photoperiod, and competition determine growth of Atlantic salmon parr. A critical size (or size-related development stage) is necessary for smolting to proceed, and thus environmental conditions determine the age at which smolting occurs. Once this developmental stage has been reached, photoperiod and to a lesser degree temperature regulate neuroendocrine changes that bring about physiological changes in the spring. Releasing factors such as temperature, flow, and turbidity may have rapid effects (dashed arrows) to initiate downstream migration. Development of the smolt physiological condition (which presumably includes a behavioral readiness or a migration disposition), induced by prior development, photoperiod, and temperature, is necessary for releasing factors to initiate downstream migration (see Baggerman 1960). The possible neuroendocrine or physiological mediators of these rapid effects are not currently known. SOURCE: McCormick et al. 1998. Reprinted with permission; copyright 1998, NRC Research Press, Ottawa, Ontario.

tion includes daylength, sight, sound, odors, water flow, ambient and internal salinity, pH, and energy stores. Typically the central nervous system integrates the information and governs the effectors that regulate survival, reproduction, and behavior. All behavior requires neuromuscular activity and the expenditure of energy. In this way, behavior is shaped not only by genes but also by the interaction, mediated by the neuroendocrine system, between the salmon and the environment.

The requirement for homeostasis, the rate-setting role of environmental temperature, and the powerful role of the neuroendocrine system all interact to affect the timing and success of outmigration (see Figure 2-6). The Atlantic salmon prepares in advance for the transition to seawater

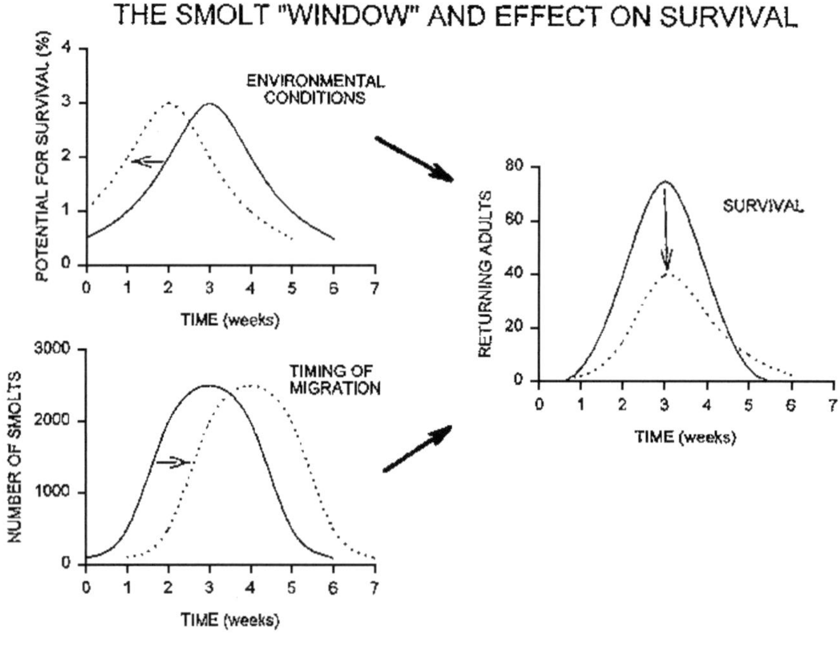

FIGURE 2-6 Simple mathematical model of the interaction of migration timing and environmental conditions and their effect on adult survival. Migration timing and survival estimates are typical for Atlantic salmon (e.g., Jonsson and Ruud-Hansen 1985, Hvidsten et al. 1995), but the temporal changes in environmental conditions are largely hypothetical. Values for migration timing are migrants per week. Adult returns are calculated from the weekly number of migrants and weekly survival rates. When migration timing and optimal environmental conditions coincide (solid lines), adult returns are high (total returns, 166). When migration timing and optimal environmental conditions are out of phase by 2 weeks (dashed lines), adult returns are lower (total returns, 94). Results of this simulation indicate that even when the magnitudes of migration and environmental conditions remain the same, alterations in their timing can have significant effects on adult returns. SOURCE: McCormick et al. 1998. Reprinted with permission; copyright 1998, NRC Research Press, Ottawa, Ontario.

rather than adjusting to it (Boeuf 1993); this is called *physiological preadaptation*. Preadaptation contrasts with acclimatization that occurs in many other species of coastal fishes that adjust to changes in salinity met while moving in and out of estuaries and rivers. The timing of preparedness has been an important issue because of the need to decide when to release hatchery-reared smolts or how to regulate water outflow from dams. It is also an important issue to be considered in our effort to understand the causes of low survival in the transition to the marine environment. The relevance to these issues is explained below.

The environmental cue or *zeitgeber* for the parr-smolt transformation is photoperiod, specifically the rate of day lengthening (Duston and Saunders 1990). The increase in daylength in the spring is transduced by the neuroendocrine system largely into increased output of pituitary growth hormone, which has the actions of elongating the stocky parr into a sleek and fast smolt and of coordinating preparation for osmoregulation in seawater (Björnsson 1997). Temperature is not a zeitgeber for smolting in Atlantic salmon, but rather temperature affects the rate of change in response to photoperiodic information (Johnston and Saunders 1981, McCormick et al. 2002). There is strong evidence for the significance of photoperiod and temperature in the timing of smolting (Sigholt et al. 1998, Solbakken et al. 1994). Smolting is a developmental phenomenon dependent on reaching a critical size of about 10 cm total length at the end of the previous growing season (Hoar 1988); this is an issue separate from the actions of growth hormone during smolting. This knowledge restricts solutions to the decline in Atlantic salmon to those including springtime seaward migration.

The environmental information used by smolting Atlantic salmon to time outmigration is complex and likely includes temperature, rainfall, and water flow, and the behavior of other smolts as "proximate cues" or "releasing factors" (Jonsson 1991, McCormick et al. 1998). Much of the endocrine system is highly activated for a prolonged period of weeks during smolting (Hoar 1988). The thyroid hormones and cortisol are linked with mobilization of energy stores, change in rheotactic behavior from an upstream to downstream orientation, and outmigration (Iwata 1995, Specker et al. 2000). Thyroid hormones and cortisol are not mediating photoperiodic information; rather they are mediating information about temperature, rainfall, and water flow, and possibly the behavior of other smolts.

The physiology of Atlantic salmon indicates that there is a "smolt window" that both opens and closes in the spring. Our current understanding of the impact of temperature on the window is that warmer temperatures accelerate the opening and closing of the window and can shorten the time during which salmon can successfully transition to the sea. Thus, dams and impoundments and other changes to the riparian environment can interfere with physiology and development. Three kinds of studies taken together illustrate the importance of the rate-setting role of environmental temperature in the timing and success of outmigration in Atlantic salmon. The first are studies providing evidence that increased temperature can accelerate the loss of smolt characteristics in Atlantic salmon in hatcheries (Duston et al. 1991). The second are studies concluding that hatchery-reared smolts released as smolt characteristics were declining had lower recapture rates, indicating reduced survival (Staurnes et al. 1993, Virtanen et al. 1991). The third study showed that migrating Atlantic salmon smolts in the Connecticut and Penobscot rivers lost their high salinity tolerance and gill Na^+/K^+-ATPase activity as the rivers warmed at the end of spring, whereas Atlantic salmon smolts in the more northern and colder Conne River and Catamaran Brook retained their smolt characteristics longer (McCormick et al. 1999). In the southern rivers, there was also year-to-year variation, supporting the conclusion that warm temperatures caused a more rapid decline of smolt characteristics.

AQUATIC ENVIRONMENTS

Salmon are a cold-water species. They spawn in streams characterized by clear, flowing water with gravel areas for egg deposition and embryonic development and productive, physically heterogeneous sections of river habitat for juvenile growth and survival. Stream size varies and ultimately affects the size of local runs through limitations of habitat space for spawning and the growth of parr.

The Pleasant, Narraguagus, Machias, and East Machias rivers empty into small coastal bays that develop seasonal stratification during warm months of the year. The Dennys River empties into a larger and more complex bay system (Cobscook and Passamaquoddy bays) with stronger tidal flows and less vertical stratification. The entire region is tidally energetic. All the Down East bays open to a coastal shelf (out to a 100-m isobath) that is dominated by the Eastern Maine Coastal Current (EMCC, see below). The Penobscot River empties at the head of Penobscot Bay, which is second in size on the East Coast only to Chesapeake Bay. The Ducktrap River also empties into Penobscot Bay, about halfway down its western shore (Figure 1-1). Penobscot Bay opens to the coastal shelf near the western end of the EMCC. The Kennebec and Androscoggin rivers empty into Merrymeeting Bay near Bath, and then the Kennebec River flows into the ocean just east of Casco Bay. The St. John River, more than twice as large as the Penobscot in drainage area and flow volume, is partly in Maine, partly shared with New Brunswick, and flows into St. John Harbour in New Brunswick.

The Gulf of Maine is characterized as a marginal sea because its connections with the North Atlantic are significantly constrained by large offshore banks (Georges Bank and Browns Bank). The Gulf has a general cyclonic (counterclockwise) residual circulation (the flow that is left after removing the tides). Surface water (upper 75 m) enters the Gulf of Maine across the southern Scotian Shelf, originating from the Labrador Sea and Gulf of Saint Lawrence to the north. Except for the occasional presence of a warm-core ring, the surface water is colder than the rest of the Gulf for most of the year. The temperature and biota associated with this circulation pattern provide a relative continuity of habitat between the Maine DPS rivers and the winter feeding grounds in the Labrador Sea. By contrast, rivers to the west of Penobscot Bay empty into an aquatic regime that is distinctly different during warm months of the year, approximately May to October. Most of the water exiting the Gulf does so via a narrow jet along the northern edge of Georges Bank, while a smaller volume leaves through the shallower Great South Channel.

CLIMATE CHANGE

Atlantic salmon distributions have been influenced by geological changes, including ice ages (MacCrimmon and Gots 1979). Populations in the United States probably date from the end of the last ice age 10,000 years ago. Atlantic salmon were probably present in all watersheds from the Hudson River in New York north to the St. Croix River at the U.S.-Canadian border (Kendall 1935). Atlantic salmon once occupied 34 rivers and streams in Maine (Beland 1984, Rounsefell and Bond 1949). Today, wild Atlantic salmon populations in the United States are found only in Maine,[3] from the lower Kennebec River in the southwest to the Canadian border, a range contraction that may in part be due to climate change.

Few fish species in the North Atlantic are as affected by climate variation over as wide a region as Atlantic salmon. The ocean migrations of Atlantic salmon rival those of the large pelagic species such as tuna, with documented returns of North American salmon from the eastern side of the Atlantic and European fish from the western side (Hansen and Jacobsen 2000, Tucker et al. 1999). The migrations themselves vary in response to currents and temperature distributions, among other factors. But the environment's effect on salmon is not limited to the conditions that adult

---

[3]Many Atlantic salmon have escaped from farms off the west coast of North America and concern has been expressed about their becoming established there (e.g., Volpe et al. 2001). Although adult Atlantic salmon have returned from the sea to spawn there, no population has yet become established.

salmon experience at sea. Freshwater mortality, hitherto considered less variable than marine mortality (Chadwick 1987), may be amplified by direct and indirect effects of changes in precipitation, seasonal ice formation, temporal patterns of stream flow and other local properties associated with global climate change (Cunjak et al. 1998). During their first year at sea, the migration cues, first feeding opportunities, and ocean nursery conditions affecting juveniles may be strongly affected by climate (Drinkwater 2000, Friedland et al. 1998, Montevecchi et al. 2002). To understand the relationship between salmon and climate variation, it is important to deal with three critical life-history stages of salmon: juveniles in freshwater, juveniles during their first year at sea, and maturing adults.

Climate can affect the dynamics of juvenile salmon populations in freshwater nurseries through modulation of growth rates, principally by the effect of temperature on growth. Habitat is a limiting factor in the production of juvenile salmon, and the factors that affect the pace at which cohorts move through rearing habitat the overall production of pre-recruits to the stocks (Bardonnet and Baglinière 2000). Juvenile rearing in freshwater may last for as long as 7 years in northern streams to as little as 1 year in southern habitats (Power 1981). Since migration from freshwater is growth-mediated, climate conditions that affect growth will determine the pace at which cohorts leave nursery streams. Smolt ages will probably decrease, and precocious maturation will probably become more frequent across much of the rearing habitat in North America if the anticipated increases in temperature associated with global climate change are realized (Juanes et al. 2000, Minns et al. 1995).

Concerns about parr mortality during the winter before they leave freshwater have increased recognition of the relationship between climate and the structure of the rearing habitat. Winter mortality is associated with the relationship between premigrant parr and their rearing habitat, which may be marginal in providing refuge during their last winter in freshwater. During their last winter in freshwater, premigrant parr are relatively large for their habitat and they often live beneath winter ice (Cunjak et al. 1998). The mortality of premigrant parr may be quite high for some populations and subject to climate variations that affect the stability of the ice cover (Whalen et al. 1999a,b). The smaller members of the nursery population may be better adapted to surviving these shifting conditions because their smaller size makes more specialized refuges available (Cunjak 1988). Changing climate conditions could destabilize ice cover and cause pre-migrant parr mortality to increase.

The next transition for salmon is the movement of smolts into the ocean, which is affected by climate conditions in many ways. At the outset, smolt migrations are cued by environmental signals, such as temperature in freshwater rearing areas (Jonsson and Ruud-Hansen 1985, Solomon

1978). In theory, smolts have adapted to environmental cues that deliver them to specific "migration windows" in the coastal ocean, where the fish are able to take advantage of prey, avoid predators, and find suitable habitat conditions. The fish are already under physiological stress since they are challenged by the transition of moving from freshwater to saltwater; timing the migration to optimize ecological conditions improves survival (Friedland and Haas 1988). If adaptations to initiate the migration to sea are not robust to climate variability, the consequence for regional stock groups may be profound, especially for stocks at the margins of salmon distribution that may already have low return rates.

Although many sources of mortality affect salmon throughout their marine life, the largest source is thought to be predation during their first few weeks at sea (Fisher and Pearcy 1988, Holtby et al. 1990). The size and variability of this source of mortality make it an important determinant of the return rate (Pearcy 1992, Salminen et al. 1995). Many attempts have been made to establish a link between salmon survival and climate (Friedland 1998, Friedland et al. 2000, and references therein). Although progress has been made with correlations, the causal mechanisms have remained elusive. Recent analyses (Friedland et al. 2003) provide the first indication for North Atlantic salmon that survival is negatively correlated with sea-surface temperature (SST) in June if SST exceeds the preferences of the local stocks of salmon. The importance of temperature during the first few months at sea is supported by data on salmon from Iceland and the Baltic (Salminen et al. 1995, Scarnecchia 1984).

The nursery zone for European post-smolts is located in the open ocean, whereas North American post-smolts appear to use inshore habitats. Holm et al. (2000) described the distribution of European post-smolts from surface-trawling operations in the northeastern Atlantic. The nursery is confined to a region within the Norwegian Sea, the northern extent of which appears to be defined by current transport. The post-smolts co-occur with surface schools of herring and mackerel and occupy a similar ecological niche (Jacobsen and Hansen 2000). In North America, post-smolts can be found in high numbers in the Labrador Sea during the fall of the year (Reddin and Short 1991). However, during the earlier part of the post-smolt period (i.e., through the spring and summer) fish are also found in the Gulf of St. Lawrence, along the coast of Nova Scotia, and elsewhere (Dutil and Coutu 1988, Friedland et al. 1999, Ritter 1989). Furthermore, North American stocks may not mix for many months after entering the ocean, while European stocks appear to be concentrated in a single, albeit large, ocean area (Friedland and Reddin 2000).

Age at maturation has important consequences for the total complement of eggs deposited during spawning. Younger fish do not produce as many eggs as multi-sea-winter salmon. Although the decision to mature has a strong genetic component (Gjerde 1984), environment also plays a significant role through effects on growth (Saunders et al. 1983a). Growth at various times during the post-smolt year may be important for achieving maturity (Duston and Saunders 1999, Gudjónsson et al. 1995, Scarnecchia et al. 1991). Alternatively, some investigators have suggested that climate variations that extend migrations beyond the normal return distance affect the proportion of grilse in the return (Martin and Mitchell 1985).

The North Atlantic Oscillation (NAO) is the dominant mode of atmospheric variation in the North Atlantic (Dickson et al. 2000) and has been associated with the effects of climate variation on the survival and maturation of Atlantic salmon. The spring thermal habitat areas associated with post-smolt survival of central European salmon stocks are derived from large ocean areas where the distribution of SST is affected by the NAO (Dickson and Turrell 2000). Likewise, the winter thermal habitat associated with the abundance of specific age components in the Northwest Atlantic are also derived from areas where SST distribution is correlated with the NAO (Friedland et al. 1993). However, it would be premature to suggest that the NAO is the only mode of climate forcing that affects salmon. For example, high-frequency fluctuation in currents in the Barents Sea appears to create a lagged linkage between Icelandic and Russian salmon stocks (Antonsson et al. 1996). Other atmospheric indices might be useful in developing hypotheses about transoceanic and global stock synchrony and in explaining salmon population trends (Klyashtorin 1998).

The unprecedented decline in Atlantic salmon abundances over the past few decades raises concerns about the effect that climate change may have on Atlantic salmon. With climate at the core of many of the factors contributing to the decline of stocks, the effect of further shifts, beyond the reactive norms to which salmon populations have adapted, now poses the threat of a range shift for the species. If climate changes are compounded by other anthropogenic factors affecting the health and size of the stock (e.g., through habitat effects on the freshwater part of the life cycle), local populations may be driven to extinction.

3

Threats to Atlantic Salmon in Maine

INTRODUCTION

The current state of Atlantic salmon in Maine appears to be the product of the cumulative effects of centuries of anthropogenic and environmental impacts. At present, the threats to Atlantic salmon in Maine are many and diverse. The challenge is not to identify them—that is relatively easy to do—the challenge is to make sense of all the threats and to rank them. The committee has attempted to do that in a risk-analysis model described in Chapter 4. In this chapter, we discuss the major factors that have adversely affected wild salmon in Maine since human contact.

Others have evaluated factors that adversely affect Atlantic salmon in eastern North America. For example, Cairns (2001) summarized a group effort to evaluate the possible factors contributing to the decline of salmon from 1984 to 1999. The document attempted to "catalogue all potential causes with any reasonable claim to credibility" and "systematically assess the plausibility of each hypothesized factor." Sixty-three factors ("hypotheses for the decline") were identified. They covered all stages of salmon life history and all aspects of their natural environments; they included human activities and structures, such as aquaculture, fishing, dams, and pollution. The conclusions were drawn from expert judgment, based on the literature and on a great deal of personal insight and experience. The plausibility analysis used a weighted scoring system and covered salmon originating from rivers in Quebec, the Canadian Maritime Provinces, and New England. Cairns's (2001) assessment was done before a workshop was held to develop research strategies. The deliberations and conclusions of that workshop were summarized by O'Neil et al. (2000). A separate report covers the potential causes of low salmon returns to Newfoundland and Labrador (Dempson and Reddin 2000).

The group of experts whose efforts were described by Cairns (2001) gave each factor a numerical score between 0 and 1 for its magnitude (proportion of habitat affected times degree to which the factor constrains survival or reproductive output) and its trend (positive numbers for increasing mortality or constraint on reproductive output and negative numbers for the reverse). Those two numbers were multiplied and the product was plotted. Five factors were ranked highest in the following order: (1) post-fishery marine mortality is higher than that assumed by fishery models (thus, the degree to which fishing reduces pre-fishery abundance is overstated); (2) smolt survival is reduced due to fish predation; (3) predation by birds and mammals is high at sea; (4) altered ocean conditions alter migration routes; and (5) bird and seal predation in rivers and estuaries affects smolts and adults. Limited spawning habitat ranked 57th and barriers to spawning migration ranked 60th out of the 63 factors. The low ranking does not mean that the factors are unimportant; it means only that their effects on salmon have not changed in a way that explains the recent declines in salmon populations. Two predictions arising from climate-change projections were listed but not scored.

The highest-ranked factor and two of the next three highest ranked were in the marine environment. The second highest-ranked factor overall was in the estuarine environment. The highest ranked factor in freshwater was ranked seventh overall. This analysis was done for all of eastern North America. Although most of the factors apply in Maine, they are not necessarily of the same rank there.

The primary causes cited by the U.S. Fish and Wildlife Service and the National Marine Fisheries Service (50 CFR 17, 224) to support listing Atlantic salmon as endangered under the Endangered Species Act (ESA) are (1) "Documented returns of adult Atlantic salmon within the DPS [distinct population segment] range are low relative to conservation escapement goals," and (2) "densities of young-of-the-year salmon and parr remain low relative to the potential carrying capacity. These depressed juvenile abundances, where not supplemented by stocking, are a direct result of low adult returns in recent years."

The services concluded that the threats contributing to the danger of extinction of Atlantic salmon in Maine posed by low adult return and depressed juvenile abundance are (1) predation or disease—potential for disease outbreaks in wild and in hatchery brood stocks; (2) inadequacy of existing protective mechanisms—insufficient protection against threat posed by agricultural water withdrawals, disease, and aquaculture; and (3) other natural or artificial factors affecting its continued existence—existing aquacultural practices and low marine survival rates.

This committee had somewhat different imperatives from those of the services because its charge leads it to take a broader focus than only the listed populations in the eight DPS rivers and the ESA's specific mandates. It is important to distinguish between those threats leading to endangerment of Atlantic salmon in the DPS rivers and the measures needed for recovery (in terms of regulations) of salmon throughout Maine. Following its charge, the committee considered the threats and evaluated recovery and restoration options for salmon in Maine rivers in general, not only in the DPS rivers. In general, threats on the listed rivers are similar to those on all Maine rivers, although there are some differences. For example, the complex problems associated with the presence of dams are not considered significant threats on the DPS rivers, yet the committee regards dams as a serious problem for successful restoration of salmon on a statewide scale because the larger drainages have greater potential to support large salmon populations.

As discussed above, the list of potential threats is broad, complicating the task of conservation planners. While a recovery plan called for under the ESA is being developed, conservation efforts are being carried forward under the Atlantic Salmon Conservation Plan for Seven Maine Rivers (Maine Atlantic Salmon Task Force 1997). The task force plan establishes conservation goals in terms of returning adults for seven of the DPS rivers (excluding Cove Brook), identifies threats, poses conservation measures, sets time tables and establishes responsibilities for implementation, and estimates implementation costs.

The factors judged by this committee to be the most important threats to the continued survival of Atlantic salmon in Maine are described below. Most of the threats identified by the committee are also considered by the Maine Atlantic Salmon Task Force (1997). The primary limitation of the existing plan is the lack of priority-setting for conservation actions. Following the recommendation in the final listing rule, the committee recommends that recovery planners develop a priority setting process for recovery actions with the use of information acquired after the adoption of the 1997 conservation plan. The recovery plan should focus resources and efforts to abate the most consequential threats. Because of different environmental conditions and land uses in the various watersheds affected, these actions will need to be adapted for specific watershed application.

A HISTORY OF THREATS TO ATLANTIC SALMON IN MAINE

Centuries of human activities and environmental change have in various ways influenced Atlantic salmon populations in Maine. Until the more recent population declines, the effects of these changes were different in the Kennebec, Penobscot, and Down East rivers. Tracing the patterns and trends of anthropogenic activities and environmental change in the region may provide insight into cumulative effects on Atlantic salmon and their habitat in Maine, helping to identify factors behind their pattern of persistent but regionally varied decline.

Geologic History

The advance and retreat of continental ice sheets during the Pleistocene epoch (10,000 to about 1.5 million years ago) had a dominant influence on the landforms, stream networks, and soils of Maine (Marvinney and Thompson 2000). Glaciers shaped mountains and valleys and the resulting stream and river networks; left sand and gravel deposits; and carved out hundreds of lakes, ponds, and depressions that are now wetlands. The dominant soil types are a direct result of glaciation; a cold, wet climate; and forest succession over the past 10,000 years. In general, soils are well drained, acidic, and relatively unfertile. The properties of the soils and watersheds generally yield high quality freshwater streams and rivers with good salmon habitat.

Changes in Climate and Ocean Conditions

For as long as information about the earth's and New England's climates has been available, the information tells a story of continual climate change. It is certain that climates will continue to change. The precise nature and magnitude of future changes is not predictable at present. However, as described in Chapter 2, there is evidence that Maine's climate has been warmer over the past half century than it was over the previous century. In addition, salmon in Maine seem to be at or near the southwestern limit of their range in North America. Thus, any prolonged or large warming of Maine's climate would probably make the survival of Atlantic salmon in Maine more difficult by warming the water in Maine's streams and changing their historical flow patterns. As an example, Table 3-1 shows that the number of ice days on the Narraguagus River has decreased in the past three decades, and the snow melt has occurred earlier. In addition, changes in the hydrologic regime not directly related

TABLE 3-1 Snow-Water Equivalent (SWE) (Amherst, Maine), Channel Ice Effects, and Median February and May Stream Flow (Narraguagus River), 1970–2000

Year	March 1 SWE (in.)	March 15 SWE (in.)	April 1 SWE (in.)	Ice Effect (no. of days)	Feb. 1 Median Q (ft^3/sec)	May 1 Median Q (ft^3/sec)
1970	5	5	4.5	60	300	600
1980	4	4.5	3.5	60	330	570
1990	3	3.5	2	55	350	520
2000	2.5	3	1.5	45	380	490

SOURCE: Dudley and Hodgkins 2002.

to temperature could also complicate the rehabilitation of wild Atlantic salmon populations.

The committee judges that some degree of climate warming or change in the hydrologic regime could be tolerated if most of the other problems affecting Maine's salmon are reduced. In addition, some methods are available to mitigate such climate changes. They include making sure that streams are protected by riparian vegetation and that their watersheds are managed so that flow volumes and seasonality are maintained. However, if climate warming is large and prolonged, eventually Maine's environment may not be within the natural range of Atlantic salmon.

Climate change also involves ocean conditions. The oceans represent a large black box into which many salmon venture and few return. The oceans are known to be highly variable, beginning with variations in atmospheric forcing from wind and temperature (see Dickson et al. [1996] and Dickson [1997] for a focus on the northwestern Atlantic and Dickson and [Turrell] 2000 for a discussion of the North Atlantic Oscillation [NAO] and European salmon). These forcings are linked to changes in the earth's climate system (Hurrell and van Loon 1997), which itself responds to feedback from the underlying ocean and to interactions between system components associated with the various ocean basins (Bigg 2000). Atmospheric forcing affects the large ocean current systems that transport heat and plankton, thereby affecting the physical and biological conditions experienced by fish (Colebrook 1991, Drinkwater 2000, Frank et al. 1996, Pickart et al. 1999, Reid and Planque 2000). The responses of fish populations to such changes are complicated, and the understanding of them is still small, especially in the high seas where biological data (in particular) are scarce. Still, the evidence for large-scale impacts that can be traced to population changes is strong (Hare et al. 1999, Mantua et al. 1997) even if the mechanisms remain elusive.

Variation in the ocean environment has emerged as a primary explanation for the changing abundance of salmon, because data on return rates permit an accounting of losses between freshwater and the ocean (Cairns 2001). Return rates clearly have been declining in many areas, including in all of Maine's rivers (Reddin et al. 2000). However, most return-rate data do not distinguish between losses occurring shortly after emigration to the sea and those occurring on the high seas. That makes it difficult to evaluate causes: those near land are easier to identify, and those at sea operate over a much longer period and may be harder to detect. Quantification is difficult in either case. The strong similarity of patterns along both sides of the Atlantic suggests a common cause of salmon losses in the ocean, probably modified by local processes. That idea is based on the improbability of different river and estuarine conditions co-varying to the degree needed to produce the coherent population responses observed if the dominant causes were continental or coastal in origin (see Friedland 1998). Among North American populations, salmon abundance patterns in Labrador and Newfoundland correlate with each other and not with patterns to the south, and those to the south (Quebec, Gulf of St. Lawrence, Nova Scotia, Bay of Fundy, and Maine) correlate with each other (Reddin et al. 2000). Although justifying and promoting the need to investigate ocean conditions, the authors are cautious about using the same arguments to deny other influences.

Identifying the causes of salmon losses in the ocean is difficult, especially since the international closure of the high-seas fisheries has eliminated a major source of data on the movements of salmon that might be correlated with remotely sensed data and augmented with increased research measurements. Friedland et al. (1993) showed that warmer temperatures in spring favored post-smolt survival of salmon in the northeastern Atlantic. They subsequently defined a spring habitat index (area of habitat between 7 and 13 °C) for two stock complexes (from Norway and Scotland) and showed a close correlation between the first principal component of that habitat and landings. The relationship is consistent with what is known about the migration of post-smolts in these stocks; therefore, its insight, although untested for its predictive ability, is promising. However, both data sets occupy a single cycle with a well-defined peak, and other modes of influence would not be surprising.

In addition, marine biotic assemblages have changed, partly in response to human exploitation of them and perhaps partly as a result of natural environmental changes. These changes mean that salmon in the ocean experience changing kinds and amount of food as well as changing kinds and degree of predation.

NATURAL PREDATION AND COMPETITION

Maine's Atlantic salmon confront documented predation and competition from other species both in Maine's rivers and in the estuarine environment. Natural predation and competition may also be factors in the natural mortality of migrating and overwintering salmon in the ocean environment, but that has not been well studied. Nonnative species that prey on salmon and compete with them are a potentially important anthropogenic threat to Atlantic salmon in Maine's rivers and estuaries.

Fish Predators and Competitors in Maine's Rivers

In addition to Atlantic salmon, Maine's rivers support populations of many other fish species. Some are prey of salmon, but others are competitors and predators. The list of fishes in Table 3-2 is for the Sheepscot River (Meister 1982), but it is fairly representative of other Maine coastal rivers, with a few notable exceptions.

Meister did not provide information on the relative abundances of those fishes, but it is clear from the table that the river supports a diverse fish assemblage, many of whose members are strongly piscivorous. In particular, the introduced brown trout, and largemouth and smallmouth bass and the native striped bass, chain pickerel, and lake trout are voracious fish eaters. Many of the other species also take fish, especially the larger individuals of the species. Other coastal rivers have similar assemblages. For example, the Machias River (Fletcher et al. 1982) lacks brown and lake trout and largemouth bass but supports rainbow trout (*Oncorhynchus mykiss*), bluefish (*Pomatomus saltatrix*), and Atlantic mackerel (*Scomber scombrus*), the latter two being in estuaries. In the Narraguagus and Pleasant rivers, non-anadromous Atlantic salmon also are listed among the fauna (Baum and Jordan 1982). Changes have probably occurred in these assemblages over the past 20 years, especially in regard to nonnative species.

In addition to preying on young salmon, many of the species compete with them, and many eat their eggs. Salmon evolved in environments that had predators and competitors but not the introduced species and not under today's conditions, when salmon populations are seriously depleted.

Compounding the problems faced by young Atlantic salmon in Maine rivers is the stocking of streams with various competitive and predatory

TABLE 3-2 The Fishes of the Sheepscot River

Sea lamprey (*Petromyzon marinus*)
Shortnose sturgeon (*Acipenser brevirostrum*)
Atlantic sturgeon (*Acipenser oxyrhynchus*)
American eel (*Anguilla rostrata*)
Blueback herring (*Alosa aestivalis*)
Hickory shad (*Alosa mediocris*)
Alewife (*Alosa pseudoharengus*)
American shad (*Alosa sapidissima*)
Atlantic salmon (*Salmo salar*)
Brown trout (*Salmo trutta*)[a]
Brook trout (*Salvelinus fontinalis*)
Lake trout (*Salvelinus namaycush*)
Rainbow smelt (*Osmerus mordax*)
Chain pickerel (*Esox niger*)
Golden shiner (*Notemigonus crysoleucas*)
Common shiner (*Notropis cornutus*)
Blacknose dace (*Rhinichthys atratulus*)
Fallfish (*Semotilus corporalis*)
White sucker (*Catastomus commersoni*)
Brown bullhead (*Ictalurus nebulosus*)
Atlantic tomcod (*Microgadus tomcod*)
Banded killifish (*Fundulus diaphanus*)
Mummichog (*Fundulus heteroclitus*)
Brook stickleback (*Culaea inconstans*)
Threespine stickleback (*Gasterosteus oculeatus*)
Ninespine stickleback (*Pungitius pungitius*)
White perch (*Morone americana*)
Striped bass (*Morone saxatilis*)
Pumpkinseed (*Lepomis gibbosus*)
Smallmouth bass (*Micropterus dolomieui*)[a]
Largemouth bass (*Micropterus salmoides*)[a]
Yellow perch (*Perca flavescens*)

[a]Not native to Maine.
SOURCE: Adapted from Meister 1982.

species, native and nonnative, that has been and is occurring. Among the species stocked are such predators as striped bass, smallmouth bass, and various species of trout, including brown trout. At least three agencies in Maine are stocking fish (most of which are piscivorous): The Maine Atlantic Salmon Commission, the Maine Department of Inland Fisheries and Wildlife, and the U.S. Fish and Wildlife Service.

Predators at Sea and in Estuaries

Due to the protection now afforded certain predator groups (birds and seals), predation on Atlantic salmon in estuarine areas in Maine is probably higher than it was during the period of higher return rates in the 1970s. After a period of virtual elimination by people, cormorants became reestablished on the Maine coast in the 1920s. Since then, their numbers have increased, and attempts have been made to control the population. Cormorants were added to federal bird protection laws in 1972, and the number of breeding pairs in Maine increased more than 80% but may now be relatively stable (Krohn et al. 1995). Double-crested cormorants (*Phalcrocorax auritus*) are a significant predator on smolts at the time they are leaving the rivers (Baum 1997). Studies conducted in the 1960s and 1970s showed high rates of predation (e.g., 55 Carlin tags from salmon smolts in the stomach of a single bird [Baum 1997]). These rates are attributed partly to less-adept predator-avoidance skills on the part of hatchery-bred fish (Hockett 1994). Despite this conspicuous threat to smolts, the overall loss of hatchery-reared fish to cormorants in the Penobscot River was estimated at less than 7% by Blackwell (1996), and the rate seems to be much lower for wild smolts (for which there are few documented instances of consumption by cormorants [Baum 1997]). The loss might be higher in the smaller salmon rivers with shallow water and pools closer to the coastal rookeries, but the committee has seen no evidence that the overall return rate of salmon to those various rivers is significantly less than the return rate to the Penobscot.

Similar facts and arguments can be developed for another conspicuous predator in the coastal marine environment: seals (mainly harbor seals, *Phoca vitulina*, and the larger gray seals, *Halichoerus grypus*). Seals are protected by the Marine Mammal Protection Act of 1972, and their populations in Maine have increased since the law was enacted. The frequency of seal bites on surviving salmon returning to the Penobscot River on spawning runs increased from less than 0.5% to greater than 3% from the early 1980s to the mid-1990s. (Data extend back earlier than 1980, but with much smaller sample sizes and perhaps less focus on this question. The 3% figure is lower than that shown by Baum [1997] and is meant to reflect questions raised by that author about possible observer bias in the data.) There are no data with which to estimate the number of salmon consumed. One would need to know the relative rates of encounters, unsuccessful pursuits, nonfatal "near-misses" (bite marks detected on the survivors), and successful pursuits. From such a model, one might propose that the rate of encounters in the smaller estuaries in Down East Maine is higher than that in the Penobscot due to more confined spaces, possibly denser concentrations of seals, and possibly lower concentrations of alternative natural prey in the smaller systems. The possibility that seals have a significant impact on returning salmon cannot be dismissed. However, as with bird predation, the committee has seen no indication that salmon in the smaller rivers experience a higher predation rate than in the Penobscot.

It remains unclear whether seals significantly affect the abundance of outmigrating smolts. Seals are opportunistic feeders, so they could be a serious threat under certain circumstances. A beach-seine study in the Narraguagus estuary showed that smolts composed less than 1% of the similarly sized small pelagic fish (J. Kocik, NMFS, unpublished material, 2001). Some of these fish are known prey of harbor seals, the most abundant pinniped species along the Maine coast. The abundance of other forage species might make it less likely that seal predation has a significant impact on smolts.

Predation is a major factor determining the abundance of many animals in the sea. For salmon, this seems to occur both in a focused time and area (as in the case of an estuary at the time of outmigration) or as a gradual process over the 1–2 years of at-sea migration and growth. The transition from freshwater to saltwater imposes additional physiological challenges for anadromous fishes, and some of the mortality in the marine environment may be the result of additional stresses experienced during the riverine phase. It is not clear in the Kocik study how much of the estuarine mortality is due to predation, but no single source emerges as a likely candidate. When salmon populations are low, perhaps the impact is significant. The question is important for distinguishing between factors that might threaten the populations when they are small and those, if any, that might be responsible for the populations' current condition.

ABORIGINAL, COMMERCIAL, AND RECREATIONAL SALMON FISHERIES IN MAINE

Atlantic salmon have long been valued for sport and for food. Native Americans used them for subsistence, at least to some degree, as did early European settlers. They have been commercially fished by the United States, Canada, and Greenland. Sport fishing for salmon has been important in Canada and New England since the mid-nineteenth century. Fishing was by hook and line and nets both in rivers and at sea (Baum 1997). Commercial fishing for salmon in Maine was eliminated in 1948. All directed fishing—including catch-and-release angling—for anadromous Atlantic salmon in Maine and its offshore waters was prohibited by 2000. Some Atlantic salmon were caught in the Greenland fishery, but that was eliminated or very nearly eliminated in 2002.

Fishing in the Past: Pre-1800s

Although the history of human use of salmon prior to European settlement is murky and poorly documented, salmon—particularly adult salmon—may have been targeted by humans since the first aboriginal occupations of Maine several thousand years ago. The history of the subsistence, cultural, and commercial importance of salmon to Native Americans in Maine appears to be relatively poorly documented and subject to dispute. The archaeological record shows a succession of aboriginal occupations of the Maine area following the Ice Age, starting with Paleoindians between 11 and 10 thousand years ago. Archaic Indians came second and, by the time of the Middle Archaic period (7500–6000 BP), Maine had a substantial Indian population that is thought to have hunted white-tailed deer and to have fished for a variety of species along rivers and stream and lake inlets or outlets. Bourque (1995) suggests that they fished seasonal runs of shad, alewives, salmon, and eels.

Late Archaic (6000–3000 B.P.) human populations were larger than earlier and more dispersed. Late Archaic coastal archaeological sites have shell middens containing animal and fish remains. These remains have been protected from acid soils by mollusk shells that render middens slightly alkaline (Bourque 1995). Late Archaic peoples have been divided into three somewhat distinct cultures: Laurentian Tradition, Small Stemmed Point Tradition (SSPT), and the Moorehead Phase. Analysis of the contents of a refuse pit in Penobscot Bay has produced clam, sea urchin, cod, swordfish, deer, and duck remains. Cod and deer bones were found in the tidal falls on the Sheepscot River estuary. Cod and swordfish appear to have been important in the diets of people in the Moorehead Phase. The Susquehanna Tradition replaced the Moorehead Phase around 3800 B.P., occupying the same coastal sites but having a more shore-based diet consisting of deer, moose, shallow-water fish, shellfish, and seals. Around 2500 B.P., Maine Indians occupied most coastal shell middens and showed a renewed dependence on fish and marine mammals, such as gray and harbor seals; moose and deer; and shallow-water fishes such as flounder, sturgeon, and cod (Bourque 1995).

Historical accounts of Maine Indians may reflect earlier contacts with Europeans, including effects of the devastating diseases introduced as early as the fifteenth century. There is some disagreement as to whether early descriptions by Champlain and others represent traditional cultures (Bourque 1995) or whether archaeological reconstruction is the more reliable source for assessing aboriginal use of salmon in New England during pre- and post-contact periods (Carlson 1993).

Common folklore and some historical accounts suggest that Atlantic salmon were abundant at the time of European colonization and that salmon runs were a valuable resource for Native Americans in New England (Carlson 1993). In historical records and accounts from the seventeenth and eighteenth centuries, Carlson finds ambiguities in the use of the term *salmon* (e.g., salmon could be used to refer to shad); salmon is less prominent in descriptive accounts, possibly indicating that salmon were relatively less abundant than cod, shad, bass, and some other species. In addition, salmon may have been difficult to catch at variable run times of short duration (Carlson 1988). The authors may have been encouraged to "put a brighter picture on life in New England to folks back in the old country than was necessarily the case" (Carlson 1988), implying that they may have inflated statements about salmon abundance.

Archaeologists have identified a critical role for anadromous salmon in the development of Pacific Northwest aboriginal cultures on Canada's west coast. There may be a tendency to extrapolate this finding to New England, according to Carlson (1988). However, her archaeological research on bone remains from over 75 sites in New England found only four possible reports of salmon vertebrae, all of which could have been from trout (Carlson 1993). The absence of salmon bones in the archaeological record could reflect a scarcity of fish or difficulty in catching them. Alternatively, salmon bones may not be preserved in the archaeological record. Carlson concludes that there is no basis for the loss of salmon remains, therefore salmon were probably not fished either for cultural or biological reasons. She considers the biological explanation (that salmon were relatively rare) to be the most probable because the archaeological record contains ample evidence that Native Americans had the capacity to harvest salmon (Carlson 1988, 1993).

Carlson (1988) argues further that "the generally disappointing results of the modern salmon enhancement programs in New England may be due more to the fact that salmon is not naturally abundant in these waters than to historical and modern dams and pollution." According to Carlson's hypothesis, salmon did not migrate from Europe to North America until relatively recently (A.D. 900–1300). The presence of salmon in New England's rivers was a consequence of the Little Ice Age between 1550 and 1800 when cooler water may have temporarily extended the southern range of salmon, a pattern that reversed after 1800 (Carlson 1993).

Carlson acknowledges that cultural factors may have influenced the consumption of salmon by Native Americans. Archaeological remains suggest that aboriginal culture in New England, unlike the Pacific Northwest, was based on marine fish exploitation rather than anadromous fish exploitation (Carlson 1988). Salmon runs in New England occur in the spring and summer when other resources are abundant, and "there is little evidence in the ethnohistorical accounts for New England of extensive fish storage and preservation technology" (Carlson 1988). In short, there may well have been salmon in New England's rivers that were not targeted by Native people.

There may be a problem with using the archaeological record as the basis for assessing the use and abundance of Atlantic salmon in Maine in the past. Faunal remains survive best in New England soils when shellfish are present to neutralize soil acidity. Shellfish would be primarily associated with marine and estuarine sites. Given that most salmon interceptions would have happened in riverine environments, it is possible that salmon remains have been particularly poorly preserved in the archaeological record.

1800s to Present

Most of the commercial landings in Maine came from upper Penobscot Bay and the tidal mouth of the Penobscot River. The majority of the catch was clearly of Penobscot River origin, although a cluster of weirs at the mouth of the Ducktrap River in the late 1800s indicates a sizeable run in that (DPS) river as well (Baum 1997). Anadromous fish were plentiful in the Penobscot, but there are only fragmentary data specific to salmon prior to 1867. Although there are gaps in the record, the period from 1867 to 1890 seems to have sustained catches of more than 75,000(lb) per year. The reported harvest in 1880 was 110,016 lb (10,016 fish).

Although the salmon landings of the late nineteenth century appear to have been high, legislative actions in the mid-1800s directed at protecting and restoring the runs of fish in inland waters demonstrate that the stocks were already in an obvious decline. A 3-year period from 1888 to 1890 recorded harvests of over 145,000 lbs per year. Whether this was the result of extraordinary runs or of large runs coupled with extraordinary fishing effort is not clear, but the following 5 years witnessed a decline in landings, suggesting a decrease in the stock. By 1895, commercial fishing effort declined by about 20%, but the catch declined by 50% and never fully recovered to pre-1888 levels. From 1895 to 1914, the harvest averaged about 50,000 lb per year, with a noticeable dip from 1907 through 1909. Except for a few years surrounding the Great Depression, harvests never rebounded. The two world wars, and declining stocks, water quality, and human interest probably all contributed to the variable but inexorably downward trend in salmon landings after 1910. The commercial fishery in the region was closed after 1948, when fewer than 500 lb were landed.

The long-distance migrations of Maine salmon to their overwintering areas was discovered relatively recently, dating to the capture of a tagged

Narraguagus fish at approximately 67 °N off the west coast of Greenland in 1963 (Baum 1997). Before 1963, the known extent of migration was the Atlantic coast of Nova Scotia. Tag returns from the escalating high-seas fisheries in the 1960s through early 1980s identified the northern extent of overwintering areas and provided estimates of the take of Maine-origin salmon in distant waters. The estimates vary widely, from 1,534 fish per year (1967–1989) to over 7,000 per year (1980–1992). The best estimate, made for the 1987–1992 seasons, suggests a catch of 2,896 Maine salmon/yr (Baum 1997). As late as 1997, commercial fishing off Canada and Greenland took 144 metric tons of adult salmon, equivalent to about 27,000 multi-sea-winter (MSW) fish (Nightingale 2000).

Despite uncertainty in the estimates, high-seas landings are large compared with the Penobscot returns of the periods (about 3,000 fish per year), even after adjustment for age. High-seas landings were mostly 1SW fish (about 95%), whereas most river returns are 2SW fish (70–90%). Baum (1997) estimated that the 1SW take should be reduced by 12% to estimate the impact on returning fish. Empirical evidence of the declining stocks in the overwintering areas is evident based on landings data since 1986. The high-seas fisheries for salmon were gradually reduced through regulations and international treaties, beginning with partial closures in Canada in 1985 and culminating in virtual elimination of sanctioned fisheries in regions affecting North American stocks after 1992. Despite this ban, returning salmon (and return rates) have continued to decline in Maine (see Figure 2-3) and in most North American rivers (MASC 2002, WWF 2001), as well as in many of the rivers of Europe and Scandinavia (e.g., Hutchinson and Mills 2000, Reddin et al. 2000, WWF 2001). Possible explanations are discussed in sections that follow.

Recreational angling for salmon has also taken its toll. Baum (1997) reports 16,864 salmon caught and killed by recreational anglers between 1935 and 1994, the latest year for which any kill was reported. Recreational kills peaked at 1,396 in 1980. Many salmon were caught before then, but Baum (1997) estimates that 80% of all recreational catches of salmon occurred after 1950. Until 1985, very few fish were released alive. Beginning in that year, 392 were released, and the number of fish released exceeded the number retained every year from 1989 until 1995, when no angled fish were reported kept (Baum 1997). Since 2000, all recreational angling for salmon in Maine, even catch-and-release angling, has been prohibited.

Fishing Today

Even though catch-and-release angling for Atlantic salmon is now prohibited in Maine, some unknown number of salmon are killed by anglers each year, a consequence of catching them (mainly as parr or smolts) by accident while fishing for other species; retaining them by mistake, thinking they are something else (mainly brown trout or landlocked Atlantic salmon); or illegally targeting them (poaching). Not all fish caught and released, either by commercial or recreational fishing, survive (Muoneke and Childress 1994, Policansky 2002). Wilkie et al. (1996, 1997) reported hooking mortalities as high as 40% for Atlantic salmon at water temperatures of 22°C (but zero at 6°C).

Recreational angling for salmon continues in some areas of Canada today, although no-take (catch-and-release) angling is much more widespread than it was even when New Brunswick instituted a no-take policy in 1984 (Nightingale 2000). Take of salmon by First Nation peoples in Canada does continue, although much less than formerly (Nightingale 2000). Canadian recreational angling probably involves few if any Maine salmon.

Almost all commercial fishing for Atlantic salmon in the waters off North America has ceased, but some continuing catches likely take some Maine salmon. Approximately 2 metric tons (t) per year are taken by the French islands of St. Pierre and Miquelon off the coast of the Canadian province of Newfoundland (Chase n.d.), although NASCO (2003a) reported a catch of 3.6 t in 2002, with almost one-third of that total being taken by recreational anglers. In Greenland, allowable commercial catches of salmon were as high as 924 t per year in the late 1980s and early 1990s, but decreased thereafter (NASCO 2003b). Reported landings declined from 966 t in 1987 to 237 t in 1992 and less than 100 t per year thereafter (ICES 2002). In 2002, an agreement was negotiated between the North Atlantic Salmon Fund and its partners, and the Greenland Association of Hunters and Fishers (KNAPK), to suspend the commercial part of the salmon fishery, similar to the agreement that covered the years 1998–2000. Fishing for internal subsistence consumption is allowed. The total taken for that purpose has been estimated at 20 t per year (NASCO 2003a, b). The agreement is for a total of five years, and is automatically renewed annually unless one of the parties gives notice in advance of the fishing season of their intention to withdraw. In addition to the foregoing, bycatch of Atlantic salmon occurs in other fisheries; its extent is not fully known. Ocean fishing for other species probably affects the availability of food for salmon and the amount and kind of predation on them.

FORESTRY, FARMING, AND FRESHWATER HABITAT QUALITY

Anthropogenic disturbance has occurred for centuries in New England's forests. Before European settlement, Native Americans used fire to alter wildlife habitat and enhance or maintain the productivity of wild foods and medicinal plants (Cronon 1983, Russell 1980). The commercial exploitation of Maine's land-based natural resources has taken place over the past three centuries. European settlers and their descendents made sweeping changes to forests, wetlands, streams, rivers, and the atmosphere. Since the mid-1700s, Maine's environment has been altered by timber harvesting, clearing for agriculture, farm abandonment, industrial development, and more recently, residential land use. These changes can affect water quality and hence interact with aspects of salmon physiology described in Chapter 2.

Estimates of Maine's forest area between 1600 and 1995 were recently compiled and analyzed by Irland (1998). He estimates that Maine (land area of 19,253,300 acres) was 92.1% forest in 1600. The forested area decreased dramatically when the combined effects of forest clearing for agriculture, industrial logging and milling, and subsequent forest fires reduced coverage to 53.2% by 1872. Forests regenerated on abandoned agricultural land and cutover areas, reversing this trend. The most recent (1995) U.S. Department of Agriculture Forest Service estimate places Maine's forest cover at 17,689,100 acres or 89.6% (Griffith and Alerich 1996), but the composition is much different from that of a few centuries ago.

In Maine, virgin white pine forests were the first to be cut, followed by an increasing proportion of red spruce. Maine led the nation in lumber production in 1850 (Irland 1999). After that, a suite of factors influenced the industrial use of Maine's forests. They include but are not limited to migration of the industry to the Adirondacks (New York), the Alleghany Plateau (Pennsylvania), and northern Lake States (Michigan, Minnesota, Wisconsin); railroad links between the Midwest and East Coast; industrialization during and after the Civil War; expanding markets in the Midwest; technological change (steam mills and logging railroads); the California Gold Rush; and the steady depletion of Maine's forests relative to other areas of the United States.

The declining fortunes of Maine's timber barons changed dramatically when "the development of wood pulp paper in the 1880s produced a spectacular change in the region's paper industry, and the industry moved north to find wood" (Irland 1999, p. 278). At a time when much of New England was cleared for agriculture, only Maine had abundant supplies of small diameter softwood pulp (Whitney 1994) close to major urban markets, such as Boston and New York. The Maine forest industry readily transitioned from large, high-value saw timber to smaller, low-value pulpwood used for the manufacture of paper (Irland 1999, Whitney 1994). Between the 1890s and World War I, the ownership of industrial forests in Maine was radically reshuffled as major firms, such as International Paper Company, St. Regis, Great Northern, Champion, and others, were formed (Irland 1999, p. 79). Large conglomerates, such as the International Paper Company (established 1898) and Great Northern Paper Company (established 1899), located in Maine so that they could simultaneously obtain an enormous supply of high quality raw material (red spruce) and access large, lucrative markets. A second wave of logging then began to supply pulp mills as well as sawmills capable of efficiently using smaller logs. Even before major companies began operations, Maine led the nation in the production of wood pulp by the 1890s. Maine's lumber production peaked in 1909, exceeding even the enormous volumes of the mid-1800s. Logging and related activities were widespread in Maine through the nineteenth and much of the twentieth centuries.

Forests in Maine: 1900–1990

At this point along the timeline for Maine's forests, it is important to make a clear distinction between exploitive logging and sustainable forestry. Simply put, exploitive logging operations "cut the best and leave the rest," with the best being defined by species, size, quality, accessibility, and market demand at a given place and time. In this case, the landowner or mill is only interested in maximizing the short-term profits from cutting. This is not necessarily done by clearcutting large areas, although, again, they are often perceived to be the same thing. More often, exploitive logging is referred to by foresters as "high grading" wherein only the largest, most valuable trees are cut. Smaller, poorly formed, damaged, or diseased trees are left.

The principles and practices of forestry were transplanted from Europe to North American beginning in about 1900 as the antidote to exploitive logging. Forestry is the art and science of managing forests for multiple benefits and values (e.g., wood, water, biological diversity, wildlife, fisheries, recreation, and aesthetics) over the long term. Foresters usually face the complex task of balancing multiple conflicting demands for natural resources in a financially (in relation to the firm) and economically (in relation to societal values) sound manner. Although they are often used interchangeably, logging (also lumbering and timbering) and forestry, far from being synonymous, define a broad spectrum of motives, standards, and effects. Like most of the history of forests in the United States and Canada, the history of forests and forestry in Maine tracks the gradual transition from exploitive logging to sustainable forestry during the twentieth century. This is important because the overall condition of a forest ecosystem (e.g., water quality, and aquatic habitat) is directly affected by when, where, and how trees are cut.

Maine is unique in the region for the proportional area (about 85%) and sheer size of its forest, the dominance of spruce and fir, land ownership patterns, and low population density. Relative to other parts of New England, Maine was least affected by the conversion of forests to agricultural land in the 1800s. Before and after World War I, and even during the Great Depression, forest products companies assembled large landholdings through purchases from families long engaged in logging and milling, tanning, and iron production. Even with more land, the legacy of repeated logging (young forests with small trees) meant that the 1920s were "years of hard scratching for wood to keep mills turning" in many areas (Irland 1999, p. 80). The Depression reduced demand for wood and other manufactured goods and allowed more time for forests to recover. The mobilization and supply efforts for World War II caused many foresters, firms, and public agencies to relax or abandon standards in order to "get the wood out." Many areas were damaged as severely as they were during the 1800s. The pulp and paper industry grew dramatically between 1940 and 1970 across the United States (primarily in the Southeast). Maine lagged behind other regions until corporations, such as Georgia-Pacific and Scott, purchased land, refitted and expanded mills, and, along with companies like International Paper, changed the nature of field operations in the 1960s and 1970s.

The overall changes in forestry operations, standards of practice, and associated environmental impacts reflect and will continue to reflect changes in science and technology, population and markets, and competition (regional, national, and global). Little changed in the forests until hand tools, horses, and log drives were supplanted by chainsaws, bulldozers, skidders, and trucks after World War II. (Logging railroads were used in some parts of Maine but not as extensively as in the Adirondacks, Lake States, and Pacific Northwest.) Mechanized logging equipment (feller-bunchers, forwarders, and cut-to-length systems mounted on tracks or low ground pressure tires) and large trucks (up to 80 tons when loaded) have replaced chainsaws and skidders in many areas since the mid-1980s. Forest-cutting practice acts and increased enforcement efforts substantially reduced logging and road construction impacts.

Milling technologies and water and air pollution control measures changed even more dramatically during the twentieth century. The unregulated discharge of noxious and toxic compounds was first curtailed in the 1970s with the passage of the Clean Water and Clean Air Acts. Further improvements in pollution prevention (during storage, processing, manufacturing, and transport) and pollution control along with more stringent environmental laws and regulations have dramatically reduced total pollutant discharge and toxicity in recent years. Waste materials, such as sawmill slabs, edgings, chips, and bark, are being converted to such products as landscape mulch or used to generate steam and electricity instead of being burned, pushed into mountainous piles, or dumped directly into rivers. Similarly, pulping chemicals are being recycled or converted to other products rather than being discharged into rivers.

It is beyond the scope of this study to quantify the net effect of a century of changes in logging, transportation, milling, and environmental regulation on aquatic ecosystems and Atlantic salmon in Maine. However, by all accounts, acute disturbance from log drives and the toxic effects of point-source discharges have been replaced by the chronic effects of road networks (sedimentation and barriers to fish passage), other forms of non-point-source pollution (e.g., fuel spills), and regional air pollution. When evaluated with general metrics, such as biochemical oxygen demand, temperature, dissolved oxygen, turbidity, and specific conductance, water quality has improved.

Contemporary Forestry

Foresters use single-tree or small-group selection or small-patch cuts (less than 1 acre) to mimic canopy gaps or small openings during timber harvesting operations (Barten et al. 1998, Smith et al. 1997). Clearcutting, sometimes with prescribed fire, complete overstory removal (so named when regeneration is already present in the understory), or shelterwood cuts (two or three stages about 5 to 15 years apart to prepare seed trees, establish regeneration, then remove the seed trees) are used to mimic "stand replacement events," such as hurricanes or fires (Oliver and Larson 1990, Smith et al. 1997). Diverse forest ecosystems are more resistant to rapid, undesirable changes and are more compatible with other forest uses than industrial tree farms.

Of Maine's 17.7 million acres of forest, approximately 7.3 million acres are owned by forest-products companies (Irland 1999, MFS 1999). During the 1990s, timber harvesting increased from about 400,000 to more than 500,000 acres per year. The increase in area harvested reflects a shift away from clearcutting toward selection and shelterwood systems (MFS 1999, 2000), because a larger area must be selectively cut to yield the same amount of timber as one from a clearcut. In 1999, clearcutting was used on only 3.5% of the area harvested (18,754 acres). Virtually all clearcuts (99%) were less than the 75-acre limit mandated by the Forest Practices Act of 1989; 83% were prescribed by landowners with more than 100,000 acres.

Farming

During the late nineteenth century, large areas of forest were converted to farms. By 1920, a wide swath from York to Hancock County was a "hay and dairy region," while most of Washington County remained "forest and hay." (See Figure 3-1 for a map of the counties of Maine and

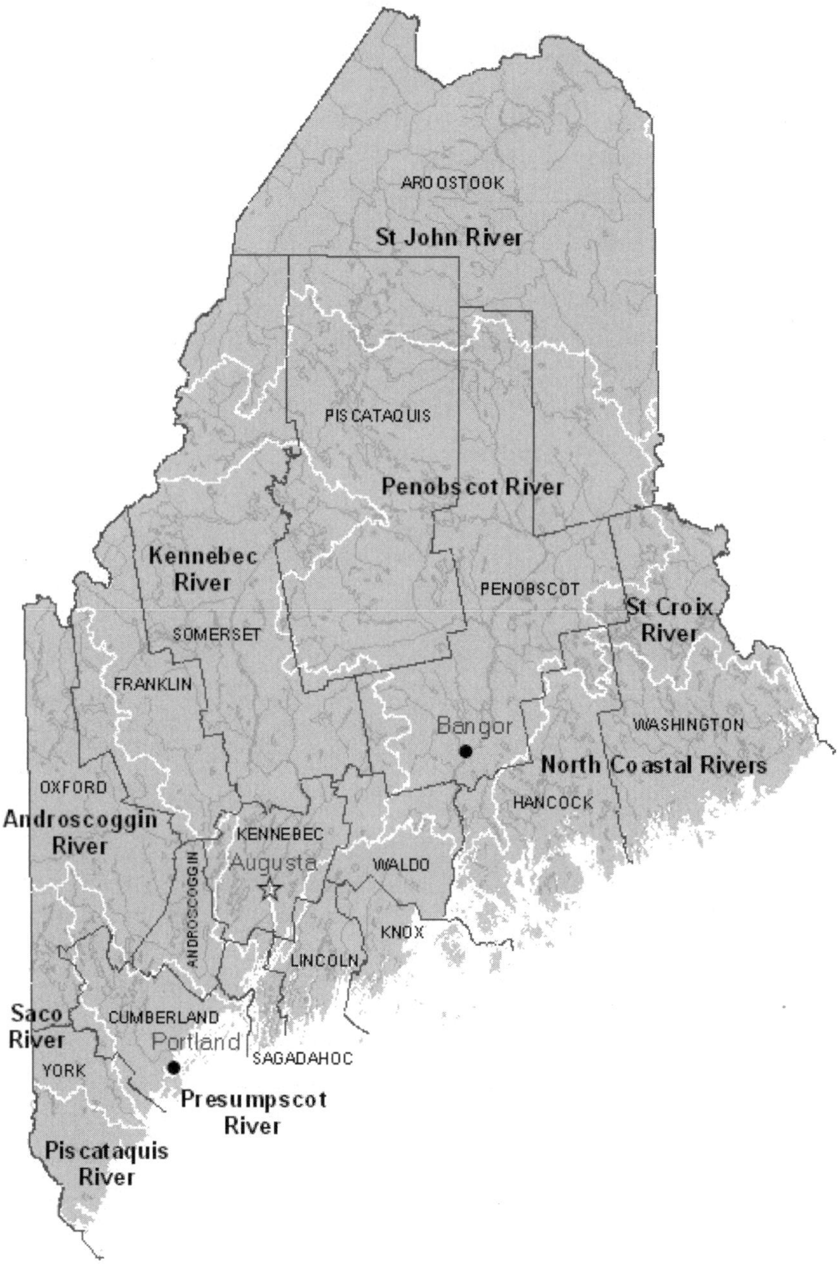

FIGURE 3-1 Boundaries of major watersheds and counties in Maine. SOURCE: Data from Maine Office of Geographic Information Systems. Drawing by Yanli Zhang, University of Massachusetts-Amherst.

the boundaries of watershed areas.) Table 3-3 highlights the differences in forest clearing for agriculture between Washington County and the other coastal counties between 1880 and 1995 (Irland 1998). Some of the land in the 1880 survey was recently cut and, therefore, classified as open. Most, however, was agricultural land that was abandoned for a variety of reasons.

Legacies of Logging, Milling, and Farming

About 22% of Maine is secondary forest land that regenerated after farm abandonment (Irland 1998). Almost all of the remaining 78% is primary forest (cut, perhaps repeatedly, but never cleared for farming). A very small area of virgin forest (never cut) might still survive in inaccessible areas (Foster 1999, Foster and O'Keefe 2000). It is reasonable to assume that most forested land in Maine has been subject to one or more cycles of logging. By 1920, most of the forest left in the Penobscot, Kennebec, and Androscoggin watersheds had been altered by one or more cycles of logging. By contrast, a larger proportion of the Down East region still had areas of virgin timber greater than 25,000 acres (Whitney 1994). A suite of factors related to the lower impacts of farming and logging in

TABLE 3-3 Forest Area for Selected Counties in Maine in 1880 and 1995

County	Total Area (km^2)	1880 Forest Area (km^2)	1880 % Forest	1995 Forest Area (km^2)	1995 % Forest	% Change 1880 to 1995
Androscoggin, Kennebec, and Penobscot watersheds and estuaries						
Androscoggin	1,218	425	35	850	70	100
Kennebec	2,247	661	29	1,653	74	150
Knox	947	353	37	706	75	100
Oxford	5,382	2,949	55	4,915	91	67
Penobscot	8,796	7,013	80	7,793	89	11
Piscataquis	10,273	8,477	83	9,972	97	18
Sagadahoc	658	250	38	499	76	100
Somerset	10,171	5,800	57	9,668	95	67
Waldo	1,890	538	28	1,537	81	186
Down East watersheds and estuaries						
Washington	6,653	5,110	77	6,012	90	18

SOURCE: Data from Irland 1998.

Down East Maine may have contributed to the continued survival of wild Atlantic salmon in the rivers such as the Narraguagus, Pleasant, Machias, East Machias, and Dennys.

Early logging operations were associated with a variety of environmental impacts. Large, stream-side trees were the first to be felled by loggers, removing trees whose roots supported stream banks and that would have eventually become large woody debris. The loss of both functions inevitably reduced stream channel stability and increased bed and bank erosion. During and after spring ice breakup, log drives on streams swollen with melting snow and early season rains carried enormous volumes of wood to downstream mills. Dams were used on many headwater lakes to store water, raise levels, and regulate outflow. On smaller streams, "splash" dams were built to store water (and energy) for the drive. These splash dams were deliberately breeched by releasing blocks, removing a key log, or setting off a well-placed charge of black powder, sending a torrent of water and logs downstream (Irland 1999, Verry 1986, Williams 1976). The log and pulpwood drives must have had a devastating impact on stream-channel stability and aquatic habitat quality in some stream and river reaches. At the mills, booms that were used to capture and store logs also fouled the water and riverbeds with tannins, loose bark, and "sinkers." In addition, mill waste and sawdust were commonly discarded directly into rivers. Before conversion to steam power, all the mill equipment was powered by water. Eventually, many large mills with high dams also generated hydroelectric power. Augusta, Bangor, Bath, Ellsworth, Orono, Old Town, Skowhegan, and Waterville all had large mill complexes in the 1800s. Bangor alone had 410 saws (Holbrook 1938). The huge salmon runs in 1888–1891 may have been related to short-term reductions in logging, log drives, and milling and the corresponding improvements in water quality and habitat conditions through the 1880s. Beginning in the 1700s, large, high-quality white pine and red spruce logs close to streams and rivers were cut for the manufacture of lumber. Smaller, inferior trees were left behind and species such as balsam fir and red spruce filled openings in the forest. In the 1890s and early 1900s, these trees would be exploited once again. For more than a century, water quality had been degraded by waste products (principally sawdust) from mills and residues from log drives and booms. Small dams constructed for log drives and large dams for booms (log storage in streams and rivers) and water power at mills blocked and degraded salmon habitat (Judd 1997). Water pollution from logging and milling, barriers to fish passage, and degradation of aquatic habitat increased in direct proportion to soaring industrial production and population growth. The brief window of ecological opportunity for Atlantic salmon in Maine's streams and rivers of the late 1800s was closed.

The statistics summarized in Table 3-3 describe the state of the land, and they indicate the biophysical conditions encountered by Atlantic salmon for more than a century in the Kennebec, Penobscot, and Down East watersheds. Timber harvesting changes the water balance, energy balance, and rates of soil erosion and biogeochemical cycling in a watershed (Likens et al. 1977). The magnitude and persistence of changes in the quantity, quality, and timing of stream flow depends on the proportion of the watershed that is treated and the proportion of the biomass removed (Bosch and Hewlett 1982; Hornbeck et al. 1993, 1997; Reinhart et al. 1963; Verry 1986). Even when every tree in the watershed is logged, the treatment effect decreases rapidly from the first-year maximum back to an equilibrium condition when the leaf area of the new vegetation approaches that of the mature stand. Although the tree seedlings and saplings are small, their high densities (e.g., up to 100,000 white pine seedlings per acre) and rapid growth rates usually restore watershed functioning in 5 to 10 years. A light thinning or timber stand improvement cut that removes a small percentage of the biomass may have no measurable effect on the quantity, timing, or quality of stream flow. The residual trees quickly and completely make use of the temporary surplus of light, water, and nutrients.

Changes in soil erosion and biogeochemical cycling rates, and attendant degradation of water quality, are closely linked to water and energy balance changes. In most forest soils, water moves through the soil surface at a rapid rate (known as infiltration capacity) because the soil is rich in organic matter, contains large pores, and is protected by leaf litter. Overland flow does not occur unless the soil mantle is saturated. Unless a logging operation exposes and compacts the soil surface, initiating raindrop splash and overland flow, detached soil particles and organic matter (now sediment) will not be lifted and carried to streams, lakes, or wetlands. When overland flow and soil erosion occurs, nutrients that are adsorbed to the surface of sediment particles (especially clay and silt particles with charged surfaces) will be carried downstream.

Forest soils are unsaturated most of the time because of their high permeability. When some or all of the forest vegetation is removed, soil-water content increases with a consequent increase in the rate of subsurface flow. Tree removal also reduces nutrient uptake, increases dry deposition (dust and aerosols from the atmosphere that would have been deposited on the forest canopy), and stimulates microbial decomposition of organic matter due to higher soil temperature and water content. This increases the concentrations of nutrients, dissolved organic carbon, and trace metals that, when combined with the increased subsurface flow, results in greater loading to streams, lakes, and wetlands.

Recent reviews of paired watershed experiments show that vegetation must be removed from about 25% of the watershed to produce a significant increase in stream flow. The soil water content and stream flow increases are the necessary precondition for increased sediment yield and nutrient loading (non-point-source pollution) in receiving waters. Because logging was dispersed over large areas, it is unlikely that Washington County reached this threshold at the peak of forest clearing around 1880 (Table 3-3). Except for the two large northern counties, Piscataquis and Penobscot, the forest cover in other areas of the Kennebec and Penobscot drainages ranged from 28% to 55%, with a mean of about 40%. Because a large proportion of the land was converted to agriculture rather than naturally regenerating as forest, the changes in stream flow, and associated increases in nutrient and sediment loading, were probably much more severe. Mean annual erosion rates from active agricultural land range from 2 to 5 tons/acre, while forests rarely generate more than 0.1 ton/acre (Patric 1976). So while aquatic ecosystems in the Down East watersheds, such as the Narraguagus, Pleasant, Machias, East Machias, and Dennys rivers, may have been somewhat affected by logging and log drives, the Penobscot and Kennebec watersheds were subject to significant and sustained changes.

DAMS

Dams are a major cause of salmon declines worldwide. Dams have two major effects on anadromous fishes, such as salmon. They prevent or impede fish passage up- and downriver, and they change or destroy habitat (American Rivers et al. 1999, Heinz Center 2002, NRC 1996a, NWPPC 2000). The first effect, especially the blocking of upstream migration of adults, has long been recognized, even in the writings of Atkins (1874) and Kendall (1935).

Although fish-passage facilities can alleviate the difficulties that adults have in upstream migration, the effects of dams on the downstream migration of smolts has been recognized only recently, and they are more difficult to reverse. The slow-moving pools behind dams confuse smolts during migration, increase the energetic costs of their movement, and can increase predation on them. The dams can injure smolts or block their passage. Athough smolts do swim, their travel time to the estuary also can be greatly increased as a result of dams, as has been shown on the Columbia River system in the Pacific Northwest (NMFS 2000b). Although the western dams are larger than those in Maine, effects documented in the West are likely to occur to some degree on dammed streams in Maine.

The second effect needs wider recognition. By creating pools behind them, dams change habitat by eliminating flowing water and riffles. They flood riparian habitats, and they change the patterns of sedimentation and erosion. Dams usually cause changes in water temperatures and chemistry, and reservoirs behind dams are often stratified, while undammed rivers usually are not (American Rivers et al. 1999, Heinz Center 2002). In addition, the large woody debris, gravel, and sediment that were formerly carried down the river and that provided spawning and rearing habitat, as well as cues that helped adults to return home to their natal streams, are now stopped by dams. As a result, these altered habitats are less suitable for spawning and juvenile rearing. Rivers behind dams become pools, more like lakes than rivers. Most anadromous salmonids are not adapted to such habitats. Other species of vertebrates and invertebrates that can thrive in lakes proliferate and thereby change the prey resources available to salmon, as well as the number and kinds of their competitors and predators.

Dams on Maine's Salmon Rivers and Their Legacy

Maine's rivers and streams have many hundreds of dams (Figure 3-2). Not all dams are necessarily large and completely impervious barriers to fish, especially in Maine. Even the relatively large wood and concrete Edwards Dam on the Kennebec River, which was removed in 1999, had previously been breached by high flows. Thus, the upstream habitat had been available (at least to the next dam) for adult salmon for periods up to 12 months. Other Maine dams are smaller, and many are made entirely of wood. Those often allow some passage during periods of moderate-to-high flow, thus allowing some downstream passage of small fish. Many are not maintained and have deteriorated to varying degrees. Many dams in Maine are breached, overwashed, or even washed out during periods of high flows. Therefore, simple inspection of maps that illustrate dam placement is not sufficient to assess the availability of habitat to migratory fishes or the quality of that habitat in Maine.

The effects of dams on salmon in New England rivers are sobering. Kendall (1935), citing Atkins (1874) but adding newer information, provided the summaries below, starting from the southwest. (Rivers with no mention of dams or records of salmon abundances are omitted except for the eight DPS rivers.)

- Housatonic River—Salmon disappeared from this river many years ago. There is a record of plenty about 1750; about 1868 one of seven or eight pounds was reported to have been caught below the dam at Stratford.
- Connecticut River—This magnificent stream was formerly one of the best of New England rivers in which salmon are said to have been

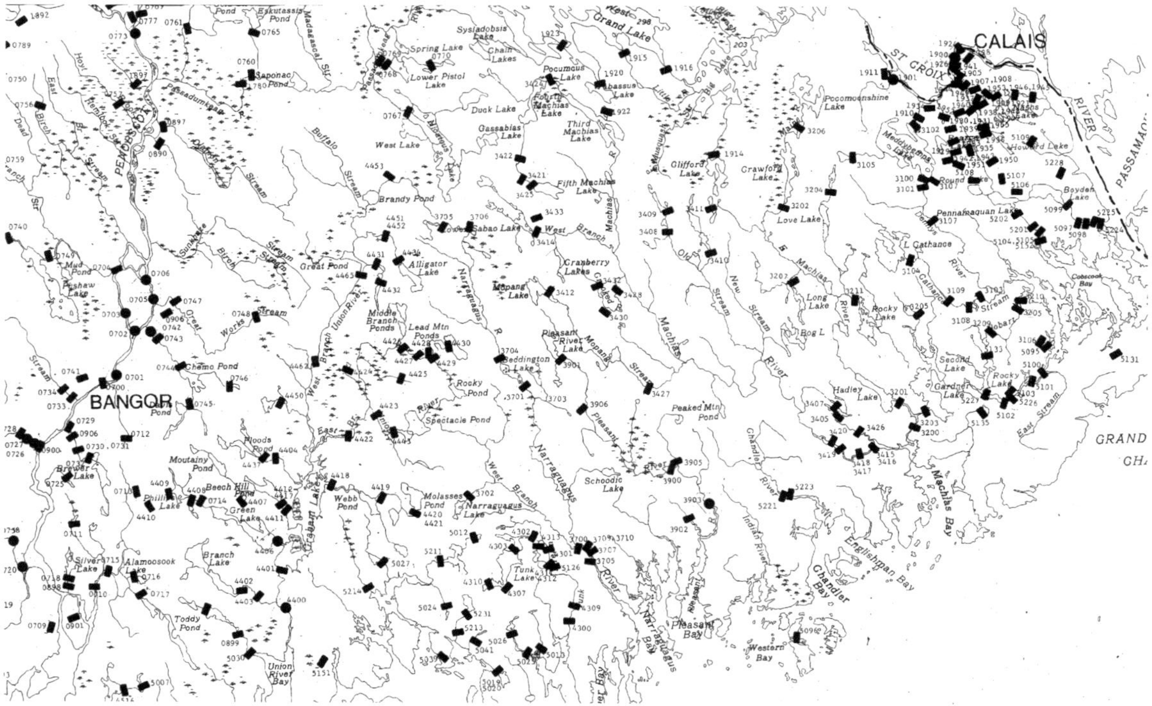

FIGURE 3-2 Existing and former dams in Down East Maine, including part of the Penobscot River. The rest of Maine looks similar. A black rectangle (■) indicates one nongenerating dam, and a black circle (●) indicates one hydroelectric/power dam. The numeral beside each dam site corresponds with data provided in the Inventory of Existing and Former Dams in Maine. SOURCE: Elder 1987a,b.

plentiful up to 1797, after which they disappeared, owing to a dam just below the mouth of Miller's River.

• Thames River—Salmon formerly inhabited the Thames and some of its tributaries until dams effectively prevented ascent. There are no records of salmon since 1822.

• Merrimack River—The Merrimack was once one of the best salmon rivers in the United States, but for years after the erection of dams at Lowell, Lawrence, and Manchester no salmon were able to pass them.

• Piscataqua River—Formerly salmon were very abundant, breeding in the Salmon Falls branch and to some extent in the Cocheco. The rivers have been obstructed by dams for over 200 years.

• Presumpscot River—This was once one of the finest salmon rivers for its size in the state of Maine, but was early obstructed by dams and only a few salmon have since been taken.

• Royal River—Salmon were common in the river up to 1800, and some occurred later. The last salmon seen here was taken in 1853. For years, owing to the dams at Yarmouth, no fish could ascend the river, and in later years besides the dams, excessive pollution has effected occlusion of fish of any kind in that vicinity.

• Androscoggin River—The Androscoggin and its tributaries were naturally adapted to salmon and were frequented by them until dams prevented ascent.

• Kennebec River—In its original condition, the Kennebec was scarcely surpassed by any salmon river in the country. The salmon fisheries of the Kennebec were in flourishing condition in 1873, when the dams at Augusta were completed. For a few years they continued plenty, and then rapidly declined until they almost disappeared.

• Sheepscot River—The Sheepscot was formerly frequented by salmon in great numbers, but the stream was obstructed many years ago. However, occasional salmon have been observed and taken in recent years below the dam at Alna.

• Medomac River—Obstructed for many years, the only salmon taken in recent years have been caught near the mouth of the river. It has been over 100 years since any considerable numbers were taken. In those early days they used to be dipped below the dam at the head of tidewater.

• Penobscot River—At the present time [1935] the Penobscot is the only New England river affording any extent of commercial salmon fishery. Atkins [1874] wrote that besides being the largest between the Saint John and the Connecticut, it is distinguished by the manner in which it discharges its waters into the sea, namely, through a large bay or estuary, narrow at its head, where it receives the waters of the river, but widening gradually to its junction with the open ocean. The works of man have interfered less with the migration of salmon in the Penobscot than in any other large river south of the Saint John. Owing to its great volume and other favorable circumstances, dams, quite impassable by salmon, have never been in existence many years at a time. The four points on the lower part of the river at which dams have been built are Veazie, Ayer's Falls, Great Works, and Oldtown.

- Union River—Once a productive salmon river, it has not yielded a single salmon for over seventy years. Formidable dams at Ellsworth, within three miles of tide water, effectually obstruct the ascent of fish.
- Narraguagus River—Salmon were plentiful here 90 or 100 years ago and the river afforded a productive salmon fishery. A few salmon even now appear at Cherryfield.
- Machias River—It is stated that in olden times salmon were extremely abundant in this river. Something over 80 years ago, it is said, a fisherman with a dip net could take 60 salmon in a day at the lower falls. As in other streams, dams have practically effected extermination so far as that river is concerned, although a few appear at times below the dam.
- East Machias River—While in former times Machias River was regarded as the better salmon river, at present and for a long time the East Machias is and has been the better stream. Salmon are now and then taken, and apparently they breed to some extent in Chace's Stream, the outlet of Gardner's Lake. Several salmon were caught with a dip net at East Machias in the latter part of June 1876 (S.B.H. 1876).
- Orange River—It does not appear that salmon ever very numerously frequented this stream, although before dams obstructed it, some entered it for breeding.
- Dennys River—Atkins [1874] wrote that in its primitive state salmon abounded in this river. In Notes from Dennysville, Robert T. Morris (1900), under the date of July 1, 1909 [sic], wrote: "As a salmon stream the name of the river is Dennys. Sawmillafecit[1]—Until very recently, the river was full of salmon. But these things are all spoken of in the past tense, because the lumber company has a sawmill at the head of tide water, and the artificial fishway will not allow breeding fish to pass."
- St. Croix River—The St. Croix by its eastern and western branches respectively discharges the waters of two extensive lake systems, and salmon, once abundant, ascended nearly to the headwaters of both branches. Obstruction and pollution, augmented by poaching, have practically eliminated salmon from the river, excepting the few which yearly, at least up to recent times, appeared in the pool at Calais or Milltown.

[1]Jocular latinization meaning "The sawmill did it."

The dams on the Penobscot were the subject of acrimonious debate beginning in 1825 (Judd 1997). The mill owners argued the dams provided a much greater community benefit than anadromous fish. On smaller rivers with smaller mills, dams, and communities, moral arguments and traditional rights held more sway than in the large cities where "Dams were an exercise of class prerogative, perpetrated by 'gentlemen lawyers' in league with the mill owners" (Judd 1997). On the Machias River in about 1850—"where fisheries were rebounding under patient care"—as well as the Pleasant and Narraguagus rivers, communities were concerned about poaching and other conditions that were detrimental to the alewives, shad, and Atlantic salmon runs. They urged state and local officials to be more diligent in protecting the resource (Judd 1997). A recent agreement has been reached to remove the Veazie Dam above Bangor and the Great Works Dam in Old Town, significantly improving Penobscot habitat and access to it (Richardson 2003).

HAZARDS OF CHEMICAL CONTAMINANTS IN RIVERS AND STREAMS

Synthetic chemicals that could cause detrimental effects on salmon originate from residential, industrial, and agricultural activities. An important question is how quickly sick and dying fish disappear in nature. The answer is probably quickly, so it is difficult to know the extent of potential damage. This section briefly reviews a few key concepts in toxicology, highlights some examples that specifically concern salmon, and comments on progress in this area. More extensive treatment can be found in publications by the National Research Council (NRC 1999, 2000) and the Society of Environmental Toxicology and Chemistry (SETAC; DiGiulio and Tillitt 1999); also see e-Hormone 2003.

Ecological toxicologists investigate impacts at ecosystem, population, individual, and suborganismal levels of organization. Basic mechanisms of toxicity at the suborganismal level include damage to cell activities and cell death. The distinguishing feature of endocrine-disrupting chemicals (EDCs), also known as hormonally active agents (HAAs), is that they tend to exert their actions by mimicking hormones or by blocking the action of hormones; that is, they operate through specific receptors. EDCs may also alter metabolism of hormones and receptors. Hormones of the neuroendocrine system coordinate growth and development, metabolism, physiological adaptation to a changing environment, reproduction, behavior, and, importantly to salmon, the parr-smolt transformation. A generalization with many exceptions is that the end point in toxicology tends to be mortality, whereas many actions of EDCs are sublethal.

There are important research findings that concern Atlantic salmon. For example, the lowest returns of Atlantic salmon to 16 rivers in eastern Canada over the period 1975–85 coincided with spraying Matacil 1.8D, an insecticide used in forestry (Fairchild et al. 1999). The culprit was 4-nonylphenol (4-NP), a nonionic detergent metabolite, and the spray was reformulated as Matacil 1.8F, without 4-NP. In retrospect, 4-NP was probably having estrogen-like effects in juvenile salmon. As an additional example, milt production and reproductive hormones of mature male parr are reduced by exposure to atrazine under experimental conditions (Moore and Waring 1998). Atrazine is an herbicide that inhibits photosynthesis in atrazine-sensitive plants used on food crops and in non-crop areas across the United States (EPA 2002a). Atrazine is persistent and mobile. As a final example of research on this topic, Atlantic salmon from a stream contaminated with polychlorinated biphenyls (PCBs) were found to have greater expression of the gene coding for the detoxifying enzyme cytochrome P4501A (CYP1A) than were salmon from a nearby stream with no known contamination (Rees et al. 2003). The gills and kidney, both interfaces involved in osmoregulation, showed induction levels of two and five orders of magnitude. Induction of CYP1A in fishes in remote ocean areas has been suggested as an indicator for chemical contaminants (Stegeman et al. 2001).

The Toxic Substances Hydrology Program of the U.S. Geological Survey used new analytical methods to measure pharmaceuticals, hormones, and other organic wastewater contaminants in 139 streams in the United States (Kolpin et al. 2002). 4-NP was found in half the samples at concentrations adequate to affect reproduction in mature male parr (discussed above). Hexazinone, found in the herbicide Velpar, is used for controlling weeds in blueberry stands. Hexazinone is toxic to juvenile Pacific salmonids, although at fairly high concentrations (276 liter, Wan et al, 1988). The potential for ecotoxicity of hexazinone to Atlantic salmon in Maine merits investigation.

There are a number of sites in Maine listed as Superfund sites by the U.S. Environmental Protection Agency, with 13 sites on the National Priorities List, and 58 sites on CERCLIS—a list of potential and confirmed hazardous waste sites. Six sites have been cleaned up. The sites were used by clothing mills, paper companies, and the military, for example. They tend to be near streams and rivers and some are being cleaned up, although funding was cut in 2003. For example, the Sebasticook is a tributary of the Kennebec River and the former site of the Eastland Woolen Mill, which declared bankruptcy in 1996 and is now a Superfund site. The groundwater in the area was heavily contaminated with chlorobenzene compounds used in the dyeing of wool cloth. The mill dumped the chemicals directly into the Sebasticook or into a tail-race that led to the Sebasticook. Another example is the Eastern Surplus site in Meddybemps, near Meddybemps Lake and the Dennys River, formerly used for storing military surplus from 1946 through 1976 and placed on the National Priorities List in 1996. The Agency for Toxic Substances and Disease Registry (ATSDR) began investigations in the 1980s that found chemical contaminants, such as organic compounds (e.g., benzene, PCBs, DDE) and metals (e.g., mercury, chromium, arsenic, and lead). Fish collected in 1997 included brook trout; and pesticides, PCBs, and metals were found above comparison values in the fish fillets. The Passamaquoddy Tribe lives downstream of the site; however, because of lack of data, ATSDR had to classify "the current and future exposures at the site as posing indeterminate public health hazard" (ATSDR 2003). No additional monitoring is planned. The Penobscot Nation is advocating the cleanup of the sites owned or previously owned by paper companies. The Great Northern Paper Company filed for bankruptcy in January 2003. Some polluted sites it owned are near the Millinocket Stream, which empties into the West Branch of the Penobscot River upstream of the Penobscot Indian Reservation. Monitoring water quality in streams throughout Maine would contribute substantially to habitat assessment and management. Efforts like those of Maine's Board of Pesticides Control to monitor pesticides (Jackson 2002, 2003) should be continued and strengthened. Where possible, they should be coordinated with streamflow, water-quality, and biological assessments conducted by the U.S. Geological Survey, the Maine Atlantic Salmon Commission, the University of Maine, and others.

HATCHERIES

Stocking of hatchery fish has long been a major component of fishery management programs. With the increasingly widespread decline of fish populations, managers have turned to hatcheries to rehabilitate depressed populations. A fundamental premise of these programs is that use of hatchery programs, when properly designed and implemented, provides one tool for rebuilding wild populations of salmon. Although the evidence available in Maine does not allow an evaluation of that premise—Atlantic salmon populations there had not been rebuilt as of 2002—it seems clear that success will not be achieved without the use of the best available techniques, if then. In addition, careful research and monitoring are needed to increase the likelihood that hatchery stocking will help to achieve the goal of recovering wild fish populations. When salmon populations are as low as they are now in many of Maine's streams, hatcheries might offer the only possibility of avoiding extinctions in the short term while longer-term solutions are implemented.

Three caveats are important (Miller and Kapuscinski 2002). First, without proper adherence to genetic, evolutionary, and ecological principles, integration of hatchery and naturally reproducing salmon could lead to adverse consequences for naturally reproducing fish, thus undermining other rehabilitation efforts. Second, the use of hatcheries to rebuild depressed populations is still an unproven technology; therefore, it should be conducted with a commitment to the concept of adaptive management. Adaptive management requires explicit design and implementation of actions or programs as experiments, regular monitoring to obtain reliable data and track progress toward program goals and objectives, systematic evaluation of outcomes of actions, and most crucially, adoption of adaptive changes (mid-course corrections), on the basis of conclusions drawn from such evaluations (Lee 1995, NRC 1996a, Walters 1986). Finally, hatcheries should be viewed as only one part of a more comprehensive strategy to remedy factors, such as lack of habitat and poor habitat quality, that cause decline or impede recovery. Hatchery use should be limited to specific situations where its advantages outweigh its disadvantages. In general, the committee favors the discontinuation of hatchery supplementation for wild salmon when the populations are recovered to a specified degree, as discussed below.

History and Status of Hatcheries for Atlantic Salmon in Maine

Enormous numbers of Atlantic salmon have been produced and stocked in Maine waters for well over a century. In spite of these efforts, salmon runs have continued to decline. Failure to monitor hatchery fish after their release and to compare them with wild populations whose natal streams did not receive hatchery-stocked fish makes it impossible to determine the effect of the stocking program on the continuing decline in salmon abundance. At best, stocking might have retarded the decline; at worst, stocking might have accelerated it.

Overfishing of migratory fish species was recognized as a problem in U.S. waters as early as 1762, concerns being raised about striped bass and sturgeon in the Exeter River of New Hampshire. By 1790, destruction of alewife spawning runs due to dam construction was recognized as a problem (Bowen 1970). By the mid-1800s, some of the southern New England Atlantic salmon runs had been destroyed, and others were declining because of pollution, and commercial fishing (Moring 2000a). The decline stimulated a long period of translocations of Atlantic salmon among widely separated watersheds, and importation of nonnative eggs for hatchery programs began soon thereafter.

By 1870, the Canadian Samuel Wilmot was selling Atlantic salmon eggs, probably from Lake Ontario, to various states in the United States

(Atkins 1874, Milner 1874). Atkins bought 8,000 salmon eggs from the Canadian government in 1871 and stocked the Sheepscot River with about 1,500 fingerlings (Atkins 1874).

The Craig Brook Hatchery in Maine, the first public salmon hatchery in the United States, was established in 1871 (Moring 2000b) to rehabilitate depressed runs of wild Atlantic salmon throughout their range in New England. The U.S. Fish and Fisheries Commission and Maine established another hatchery at Bucksport in 1872. That hatchery produced 876,000 fish derived from Penobscot River stock in its first year of operation for stocking in various states, including Maine (Baird 1876). Additional details associated with early attempts to spawn and stock Atlantic salmon in New England can be found in Baum (1997) and Stickney (1996a,b).

Genetic sources shifted between Maine and Canada from the early history of hatchery use. The Bureau of Fisheries (the successor to the U.S. Fish and Fisheries Commission) obtained salmon eggs from the Miramichi River in New Brunswick and from the Gaspé region of Quebec, between 1920 and 1937, as a result of altercations between the bureau and commercial fishermen who were collecting adults from the Penobscot (Baum 1997). Several reports described key shifts in the life stage stocked and the increasing numbers of hatchery fish stocked throughout the history of Maine's stocking programs (Baum 1997, Smiley 1884). Baum showed the annual take of Atlantic salmon eggs from Maine and Canadian rivers from 1871 through 1995 (see Appendix F).

Both parr and fry were stocked during most years, beginning in 1873, according to Baum (1997), although Moring et al. (1995) indicated that parr stocking began in 1890. Fry and parr stocking continued until the late 1920s, after which annual parr stocking continued until 1958. Fry stocking was conducted only during some years between 1928 and 1941, after which no fry were stocked again until 1972 (Baum 1997). Fry were stocked in alternate years from 1979 to 1986 and annually thereafter (Baum 1997). Modest numbers of smolts were stocked from 1945 to 1947. Smolt stocking began again in 1962 and continued through at least 1995 (Baum 1997).

Current River-Specific Stocking of Fish: A Supportive Breeding Approach

River-specific management was instituted in 1991 for six of the DPS rivers (Sheepscot, Narraguagus, Pleasant, Machias, East Machias, and Dennys) listed under the ESA, a major shift in the strategy of hatchery production in Maine. The other two DPS rivers (Ducktrap and Cove Brook) are not being stocked. Non-DPS rivers still receive hatchery-raised fish from other sources. The Saco River, for example, is being stocked with Penobscot fish (MASC 2001).

The river-specific stocking in the six listed rivers also involves a type of captive brood-stock program, referred to in this chapter as "supportive breeding" (Ryman and Laikre 1991), that intends to increase the population size without introducing exogenous genes into the managed population. Supportive breeding involves bringing a fraction of the wild population into captivity to increase survival of early-life stages in the protective captive environment, followed by release of the offspring into the natural habitat, where they will mix with wild salmon. Starting in 1992, parr were captured in each of the six rivers and maintained at the Craig Brook hatchery until reaching adulthood (Beland et al. 1997; Craig Brook hatchery officials, personal communication, 2001). Initial parr collections for the Pleasant River were held at North Attleboro National Fish Hatchery, but later collections were maintained at Craig Brook Hatchery. The captive adults were used as brood stock, and their offspring were released, primarily as sac fry, back into the streams of parental origin on the premise that if released at about the time fry normally begin searching for food, they would adapt better to their native stream habitat, thus improving survival and future adult return and spawning in the wild. Starting in 2001, managers have aimed at spawning each captive adult only once, preferably at age 4, when the fecundity of females should be high enough to meet production targets for the target DPS river (Buckley 2002a,b). The rationale is to reduce common ancestry of parents and to equalize their reproductive contributions to the next generation, minimizing the loss of genetic variation during the captive breeding phase. Hatchery managers, however, had to include some adults from older age classes in the matings, because fewer females than expected were mature at age 4. Rather than producing mature gametes every year, many of the captive adult brood stock appear to do so only in alternate years, as is the normal pattern in the wild.

Each year since 1995, additional collections of parr have been obtained and reared to adulthood. Sufficient numbers are collected to ensure survival to spawning of at least 50 pairs from each of the six streams (Beland et al. 1997). As soon as two year-classes (cohorts) of adults became available, brood fish from different cohorts were crossed to produce the next generation. The range of crosses was expanded to incorporate additional cohorts in the breeding design for subsequent years. After captive brood stock from one cohort have contributed progeny over a few years, the survivors are released back into their natal streams.

Little is known about survival rates of released swim-up fry to the parr stage, although studies are under way with tagged fry to determine both survival rates and the degree to which hatchery-spawned fish are recaptured as parr to contribute to future cohorts of captive spawners. An experiment conducted in the Connecticut River (no natural population of

Atlantic salmon) was designed to investigate stocked fry survival up through the smolt stage (Orciari et al. 1994). Eggs were obtained from the Penobscot River and fry were stocked in 1982, 1984, and 1985. Stocking Penobscot fry at mean densities of 125/100 square meters (m^2) resulted in mean densities of 34 age-0 parr, 10 fall age-1 parr, and 3.6 smolts/100 m^2 averaged over the 3 years. The study also included stocking fry from an Icelandic strain in 1983, yielding poor survival. The transferability of the results of this study to the restoration program for wild populations in Maine rivers is limited for two reasons. First, both groups of nonnative fry had a relatively low chance of being adequately pre-adapted to environmental conditions in the Connecticut River, whereas river-specific fry produced in the Craig Brook hatchery and released into their own rivers of origin have a higher chance of being adequately adapted. Second, the crucial information needed to evaluate effectiveness of fry stocking is the number of adult returns that contribute to the next generation, but such data are not yet available from this study. Still, for all Maine rivers, 79% of adult returns in 2001 had been stocked as smolts (USASAC 2002). Despite this overall result, there is a need for reliable tests of the relative survival of stocked fry and smolts estimated from matched trials in selected rivers. There also is a need to quantify the fitness of the parr from the hatchery program and compare it with the fitness of wild parr (that is, their presumed fitness had they been left in the river). The hatchery program in Maine rivers has the capability to perform these important comparisons, a capability that emphasizes its value for doing crucial experiments.

Moring et al. (1995) suggested that restoration and rehabilitation of salmon in Maine will not be possible without the use of hatchery-reared fish. However, the establishment of large numbers of stocked fish has not led to the establishment of large runs of fish. In fact, the populations have declined precipitously since the 1970s and have reached historical lows in many streams. At best, stocking may have slowed the decline. More important, nobody can determine whether hatchery stocking has had any effects at all, because controlled, matched trials have never been done.

AQUACULTURE

The potential effects of net-pen salmon aquaculture (salmon farms) on the wild salmon of Maine have been much debated, but little direct evidence from Maine is available. This section evaluates information from Maine and elsewhere, as well as methods to reduce adverse effects of salmon farming. The problem is intensified because one of the DPS rivers, Dennys, empties into Cobscook Bay, which is one of the most concentrated areas for the Maine salmon aquaculture industry. Cobscook Bay also adjoins Passamaquoddy Bay, the focal center of the East Coast Canadian salmon aquaculture industry. Thus, transboundary issues are important as well.

History and Status of Net-Pen Salmon Aquaculture in Maine

Commercial salmon aquaculture is the latest addition to the long list of items that some have pointed to as threats to wild salmon in Maine. Although the industry is being criticized, it has also been experiencing important internal problems, one of which is historically low prices for the product. In 2001, infectious salmon anemia (ISA), which had been present in New Brunswick, Canada, fish farms for several years (Getchell 1997), appeared in Maine (Veneman 2001). By early 2002, all the farms in Cobscook Bay were forced to destroy their fish and begin sanitizing equipment in an attempt to eradicate the disease. The general economic downturn, coupled with ISA, resulted in substantial layoffs of employees and a worsening of the socioeconomic situation in Down East Maine. The federal government has promised $16.4 million over 2 years to help fight the disease and to provide compensation to the industry for some of the financial loss incurred as a result of destruction of the fish. The Canadian federal government also has subsidized the costs of its aquaculture industry. The following information relates to the development of the industry before the appearance of ISA.

According to the Maine Aquaculture Innovation Center (2003), laws governing leasing of public marine waters by the private sector were promulgated in 1973, although the first net-pen operation was established in 1970. That operation, and others, produced steelhead trout (*Oncorhynchus mykiss*) and coho salmon (*O. kisutch*), which were species of choice into the 1980s, a decade that saw rapid expansion of what had been a fledgling industry. The Atlantic salmon aquaculture industry began to develop in the mid-1980s in Cobscook Bay, connected to Passamaquoddy Bay, where the Canadians had established their New Brunswick industry (Conkling 2000). Falling prices for farm salmon led to consolidation of the industry in Maine and to the purchase of many of the farms and hatcheries by feed companies, including multinational corporations.

The Maine industry provides about 800 jobs on the farms, in the hatcheries, and in the processing plants (many of them in Washington County [Alden 1997]). The industry has a production valued at approximately $60 million (Wilson 2000) and produces approximately 13,000 metric tons of Atlantic salmon annually (ICES 2002) and a small amount of steelhead. Seventeen companies hold leases in Maine, 12 of which are in salmon and/or steelhead production, on 42 leased sites covering a total of nearly 300 hectares (ha) (DMR 2001, see Maine Aquaculture Innovation

Center 2003). If production increases in Maine, additional farm sites in the protected waters along Maine's coast will be needed. The current sites apparently do not have room for additional production.

Production is concentrated in Washington and Hancock counties—referred to as Down East Maine—an area that includes five of the eight DPS rivers. ESA provisions could potentially affect aquaculture in the area. Siting criteria for the farms include suitable water temperatures and tidal flushing rates (see Brooks et al. 1998). Regulations allow a maximum of 60 ha (150 acres) of leased water per company, with 40 contiguous ha (100 contiguous acres) maximum per site. Five-year goals (1997–2002) included (1) tripling the contribution of aquaculture to the state's economy to $192 million, (2) doubling employment to 1,620, (3) actively farming 1,000 acres of subtidal habitat, (4) leasing 30 acres for testing the potential for new species, and (5) establishing 10 new aquaculture firms or aquaculture support firms. Table 3-4 lists current leases and sizes of the areas leased.

The original source of fish used to stock commercial net pens in Maine apparently came from Scotland and Ireland, although the ultimate source was probably Norway. According to representatives of the company Atlantic Salmon of Maine, who spoke with members of the committee, brood stock resulting from those European fish (proprietary name Landcatch) were crossed with St. John River, Canada, fish to produce first-generation hybrids. The hybrids were subsequently crossed with Penobscot fish to produce second generation derivatives, which are the source of brood stock currently used to produce fish for stocking the net-pens. Baum (1998) estimated that there is a European genetic influence in 30–50% of the production fish in Maine.

The inclusion of European strains of salmon in the Maine industry has been controversial. These strains have superior characteristics (for example, growth rate) that are desired by the industry, but concern arises from the potential effects that escapes of such genetically foreign fish might have on the wild Atlantic salmon populations of Maine. On May 28, 2003, the U.S. District Court in Maine banned the use of European strains in the decision for *U.S. Public Interest Research Group vs. Atlantic Salmon of Maine and Stolt Sea Farm* (Civil Nos. 00-151-B-C, 00-149-B-C).[2] Moreover, the Agricultural Research Service (USDA) has been directed by Congress to develop a National Cold Water Marine Aquaculture Center in Maine, in part, to address genetic issues related to Atlantic salmon aquaculture.

[2]On August 6, 2003, the 1st U.S. Circuit Court of Appeals upheld the ruling (Docket Nos. 03-1830 and 03-1831).

TABLE 3-4 Atlantic Salmon Net-Pen Lease-Site Locations and Sizes in Maine Waters as of June 2001

Lease Location	Size of Lease (hectares)	Lease Duration	Lease Holder
Hancock County	14.0	03/2000–03/2010	Acadia Aquaculture, Inc.
Washington Country	10.0	09/1997–09/2007	Atlantic Salmon of Maine
Washington County	8.0	01/1997–01/2007	Atlantic Salmon of Maine
Washington County	8.0	04/1995–04/2005	Atlantic Salmon of Maine
Washington County	4.0	09/1994–09/2004	Atlantic Salmon of Maine
Washington County	16.0	04/1992–04/2002	Atlantic Salmon of Maine
Washington County	8.0	11/1993–11/2003	Atlantic Salmon of Maine
Washington County	4.0	12/1996–12/2006	Atlantic Salmon of Maine
Washington County	11.4	04/2000–04/2010	Birch Point Fisheries
Washington County	18.0	12/1996–12/2006	Connor's Aquaculture[a]
Washington County	0.5	09/1997–09/2007	Connor's Aquaculture[f]
Washington County	3.4	07/2000–07/2010	Connor's Aquaculture[f]
Washington County	10.0	12/1996–12/2006	Connor's Aquaculture[f]
Washington County	11.0	03/1998–03/2008	Connor's Aquaculture[b]
Washington County	12.0	05/1997–05/2007	Connor's Aquaculture
Washington County	4.0	06/1998–06/2008	D.E. Salmon[c]
Washington County	4.0	10/1995–10/2005	D.E. Salmon[c]
Washington County	4.0	03/1993–03/2003	D.E. Salmon[a]
Hancock County	6.0	03/1999–03/2009	Island Aquaculture[d]
Hancock County	7.5	06/1994–06/2004	Island Aquaculture[e]
Hancock County	7.2	06/1999–06/2009	Island Aquaculture[f]
Washington County	4.0	07/1996–06/2006	International Aqua Foods[g]
Washington County	4.0	07/1995–07/2005	International Aqua Foods[h]
Washington County	8.8	09/1997–09/2007	International Aqua Foods[i]
Washington County	11.8	03/1992–03/2002	International Aqua Foods
Washington County	10.6	12/1996–12/2006	International Aqua Foods[j]
Washington County	9.9	04/2000–04/2010	L.R. Enterprises
Washington County	9.9	04/2000–04/2010	L.R. Enterprises[c]
Washington County	6.0	12/1996–12/2006	L.R. Enterprises[a]
Washington County	4.0	07/1997–07/2007	Maine Coast Nordic[j]
Washington County	2.5	12/1997–12/2007	Maine Coast Nordic[i]

continues

Genetically engineered salmon are not being produced in Maine. In the United States, the Food and Drug Administration (FDA) has claimed lead regulatory authority over commercial uses of transgenic animals, including fish (OSTP/CEQ 2001) on the basis of its authority to regulate new animal drugs under the Food Drug and Cosmetic Act (FDCA) (21 USC §§ 371-379d and § 321[g]). The Trade Secrets Act (18 USC 1905 and 301[j] of the act) requires FDA to keep secret the investigations, review, and approval of commercial applications and premarket notifications for

TABLE 3-4 Continued

Lease Location	Size of Lease (hectares)	Lease Duration	Lease Holder
Washington County	2.8	09/1993–09/2003	Maine Coast Nordic
Washington County	4.0	03/1992–03/2002	Maine Coast Nordic
Washington County	4.0	05/1992–04/2002	Maine Coast Nordic
Washington County	3.3	01/1999–01/2009	Maine Salmon[k]
Washington County	8.0	07/1994–07/2004	Stolt Sea Farm[j]
Washington County	4.0	06/1997–06/2007	Stolt Sea Farm[a]
Washington County	4.0	06/1998–06/2008	Stolt Sea Farm[a]
Washington County	6.0	12/1997–12/2007	Treats Island Fisheries[l]
Washington County	4.1	09/1997–09/2007	Treats Island Fisheries[m]
Washington County	4.0	05/1997–05/2007	Treats Island Fisheries[n]
Washington County	2.0	01/1998–01/2008	Treats Island Fisheries[a]
Hancock County	10.0	03/1993–03/2003	Trumpet Island Salmon Farm[o]

[a]Lease includes rainbow trout.
[b]Lease includes rainbow trout, Atlantic halibut, soft-shell clams, and scallops.
[c]Lease includes sea urchins and giant sea scallops.
[d]Lease includes Atlantic cod, Atlantic halibut, haddock, and blue mussels.
[e]Lease includes and Donaldson trout (Donaldson strain of rainbow trout).
[f]Lease includes Atlantic cod, Donaldson sea trout, and haddock.
[g]Lease includes rainbow trout, Atlantic halibut, abalone, blue mussels, European oysters, American oysters, bay scallops, hard- and soft-shell clams, seaweed, red algae, and fan worms.
[h]Lease includes Donaldson trout, Atlantic cod, haddock, and Atlantic halibut.
[i]Lease includes rainbow trout, and Atlantic halibut.
[j]Lease includes Atlantic cod, Atlantic halibut, and haddock.
[k]Lease includes rainbow trout, Atlantic halibut, sea scallops, American oysters, and European oysters.
[l]Lease includes rainbow trout, Atlantic halibut, flounder, pollock, sea scallops, and clams.
[m]Lease includes rainbow trout, Atlantic halibut, haddock, sea scallops, and clams.
[n]Lease includes rainbow trout, Atlantic halibut, and red algae-nori.
[o]Lease includes rainbow trout and blue mussels.
SOURCE: DMR 2001.

new animal drugs, including the existence and content of an application unless the applicant chooses to disclose the information. The existence of one first-stage application to FDA for approval of a growth-enhanced genetically engineered salmon line has been publicly disclosed by the applicant (OSTP/CEQ 2001). Under the ESA, 16 USC § 1536(a)(2), any possible approval of genetically engineered salmon for use in commercial aquaculture would be a federal action requiring a determination of whether those salmon jeopardize the continued existence of a threatened or endangered species (OSTP/CEQ 2001). Thus, before FDA approves the pending application for commercial use of genetically engineered Atlantic salmon, the National Marine Fisheries Service (NMFS) and the Fish and Wildlife Service (FWS) would have to determine whether approval would harm federally listed salmon populations, including the listed DPS populations in Maine. However, FDA may require that NMFS and FWS keep secret their written biological opinion due to FDA's lead authority and its need to comply with the Trade Secrets Act.

Possible Threats to Recovery of Wild Salmon

Interactions between farm and wild Atlantic salmon can be classified as ecological and genetic. Ecological interactions can involve the transfer of diseases (including parasites); predation; or competition for space, food, or mates between wild and escaped farm fish. Depending on the direction and strength of these interactions, growth and survival of both wild and farm fish can be affected. Interactions also can involve modification of the timing and pattern of natural migrations and complex interactions during spawning that can affect survival of fish of either origin. Genetic interactions result from exchange of genetic material (hybridization) and the alteration of selection pressures caused by interactions between wild and farm fish (reviewed in Hindar et al. 1991, Verspoor 1998). Ecological and genetic interactions are not mutually exclusive. Rather, ecological interactions can alter selection pressures and the probability of hybridization, and genetic interactions through hybridization can influence the likelihood of ecological interactions in subsequent generations.

Disease Transmission

The transmission of parasites and diseases between farm and wild fish can flow in both directions. In this section, the concern is about the potential impacts on wild populations. The high density of fish in netpens provides the opportunity for rapid spread of diseases within the facility, no matter what the origin, particularly when the fish are stressed. Although documented cases of the spread of diseases and parasites from farm to wild fish are not common, mainly due to a lack of investigation, they are known to occur (Brackett 1991). Most of the cases involve the transmission or introduction of new diseases or parasites. For example, evidence strongly indicates that the planting of infected Atlantic salmon smolts from Sweden by Norwegian hatcheries resulted in the introduction of the freshwater parasite *Gyrodactylus salaris*, and its subsequent spread was facilitated by the movement of smolts among aquaculture sites along the Norwegian coast (Johnsen and Jensen 1986, 1991). The parasite has been responsible for the near extirpation of Atlantic salmon in about 40 Norwegian river systems. Similarly, the importation of Atlantic salmon smolts from Scotland in the mid-1980s to meet aquaculture needs was almost certainly responsible for an introduction of furunculosis (*Aeromonas salmonicida*) to Norway and its subsequent spread. It has been difficult to determine fully the effects of furunculosis on wild fish in Norway, but the effect was clearly negative and significant (Johnsen and Jensen 1994).

A final example relating to the dangers of the importation and movement of fish is the introduction of the salmonid viral pathogen IHN (infectious hematopoietic necrosis) to Japan from a shipment of infected sockeye (*Oncorhynchus nerka*) salmon eggs from a hatchery in Alaska, subsequently causing epizootic mortality in Japanese chum (*O. keta*) salmon and in two species of landlocked salmon that are only in Japan (McDaniel et al. 1994). Accidental disease and parasite introductions are now better controlled, but clearly problems still remain.

Net-pen aquaculture can also biologically increase disease pathogens and parasites. In Europe, farm salmon appear to be increasing the production of sea lice (*Caligus elongates* and *Lepeophtherius salmonis*). The resulting sea-lice epidemics have affected wild salmonid populations (Atlantic salmon and brown trout) in Ireland, Scotland, and Norway (Bjorn and Finstad 1998, 2002; Bjorn et al. 2001; Finstad et al. 2000; Heuch and Mo 2001; Tully et al. 1999). Wild smolts passing lice-infested net pens appear to be highly susceptible, and their mortality can be high.

In Maine, sea lice and bacterial infections, such as furunculosis, have been the source of disease epizootics in salmon farms for many years. Hitra disease, caused by *Vibrio salmonicida*, apparently first became a serious problem in Maine in 1993 (Griffiths 1994). Vaccines have been developed against some diseases and are routinely used by at least some producers. There is concern that ISA may have been transmitted from cage-cultured to wild Atlantic salmon in New Brunswick (Atlantic Salmon Federation 1999). Little research or monitoring has been done on the degree to which diseases and parasites spread from farms to wild salmon in Maine, and consequently little is known.

Behavioral Interactions

Although farm salmon can escape as fry, parr, and smolts into freshwater, most escapes occur in the marine environments as smolts, post-smolts, and adults. Escapees can then move from one habitat to the other and interact directly or indirectly with wild salmon. As farm escapees begin to mature, they tend to migrate into rivers in the vicinity of the site of escape (Hansen and Jonsson 1991, Whoriskey and Carr 2001, Youngson et al. 1997). In Maine, farm escapees have been found in the St. Croix, Penobscot, Dennys, East Machias, and Narraguagus rivers (Baum 1998, USASAC 1996, 1997, 2002). As of 2001, only three of the eight DPS rivers have fish traps that allow direct assessment and exclusion of farm escapees from the runs. The number of farm escapees in the salmon runs of two of these rivers, Dennys and Narraguagus, ranged from 2 to 65 and 0 to 8 fish, representing 44% to 100% and 0 to 22%, respectively, of the runs during 1993–2001. In the St. Croix River, there were 58 farm escapees, constituting 75% of the fish captured during brood stock collections in 2001.

In freshwater, the entry of escaped farm spawners can potentially influence natural migration and spawning, and behavioral interactions can affect mating selectivity and interbreeding, which control genetic interactions and population performance. Escaped farm salmon can spawn successfully in the rivers they enter (Clifford et al. 1998; Lura and Sægrov 1991; Webb et al. 1991, 1993), although their breeding performance at times is inferior to that of wild salmon (Fleming et al. 1996, 2000). There is little evidence to date of farm salmon directly disrupting spawning by wild salmon (Fleming et al. 1996, 2000, but see Garant et al. 2003). Although farm and wild males sometimes compete for spawning females, there is little indication that the competition affects fertilization rates or the performance of females. Occasionally, farm males exhibit inappropriate spawning behaviors that result in reduced fertilization success of a female's eggs when no wild males are involved in the spawning event (Fleming et al. 1996). The most likely negative ecological interaction during the breeding season will be the destruction of early nests by later spawning females (Lura and Sægrov 1991, Webb et al. 1991). In that case, the farm and wild females spawning time, which can vary considerably among populations (Fleming 1996, Fleming et al. 1996, Lura and Sægrov 1993), will be a critical determinant of the impact.

Although interactions between farm and wild fish during breeding may have minimal immediate ecological effects on wild populations (depending on relative spawning times), genetic (gene flow) and subsequent ecological interactions are important to the next generation. Successful breeding and interbreeding by farm salmon will produce the next generation of pure farm and hybrid (farm and wild) offspring that will compete directly with wild offspring. These latter genetic and ecological interactions may profoundly affect the productivity of wild populations.

Interactions among wild, farm, and hybrid juveniles in freshwater are likely to involve one of two main factors: (1) escape from freshwater rearing stations and hatcheries, and (2) successful spawning and production of offspring by farm salmon. The possibility that juvenile farm salmon escape between the fry and smolt stages from hatcheries into rivers has generally been ignored (but see Stokesbury and LaCroix 1997). However, at least two hatcheries supply smolts to the sea cages located within the Maine drainages containing native salmon populations where there is some evidence of juvenile escapees (USASAC 2000, 2001, 2002). Between the fry and smolt stages, competition for food and space can be altered by the introduction of conspecific organisms with a distinct developmental and size advantage. Farm juveniles typically outgrow wild juveniles, even in nature (Einum and Fleming 1997, Fleming et al. 2000; McGinnity et al 1997, 2003), reflecting the directed domestication selection for growth that farm fish have undergone (Gjedrem et al. 1991, Glebe 1998). Although the mortality of farm juveniles may be higher, particularly during the first few months of life in the river, their interactions with wild fish can lead to competitive displacement of the latter (Fleming et al. 2000; McGinnity et al. 1997, 2003). That result probably reflects differences in growth rate and related differences in behavior, such as aggression, dominance, and risk-taking (Einum and Fleming 1997, Fleming and Einum 1997, Fleming et al. 2002, Johnsson et al. 2001). Such interactions can ultimately depress the productivity of the wild population (Fleming et al. 2000).

Genetic Interactions

Farm salmon differ genetically from wild salmon, because the broodstock used to propagate the fish destined for growing cages have origins different from the wild fish (Clifford et al. 1998, Gjedrem et al. 1991, King et al. 1999). The difference is accentuated in Maine because many of the farm strains have incorporated strains of European origin (NRC 2002a). Additional causes of genetic differences are founder effects and genetic drift (Mjølnerød et al. 1997, Norris et al. 1999) and response to the aquaculture environment through intentional and unintentional domestication selection (Fleming and Einum 1997, Fleming et al. 2002, Johnsson et al. 2001). Farming generates rapid genetic change, resulting in large enough differences between farm and wild fish that Atlantic salmon might be considered one species with two biologies (Gross 1998).

When farm salmon interbreed with wild salmon, the resulting offspring (hybrids) can lose fitness, relative to wild fish in the natural environment, due to disruption of local adaptation and of co-adapted gene complexes (outbreeding depression); similar fitness; or even temporarily superior fitness due to hybrid vigor (see hatchery discussion). Information about releases of salmonids shows that the effects are frequently negative (Hindar et al. 1991), and samples taken in 1994–1998 show that genetic infiltration of farm fish into wild Maine populations has been minimal (King et al. 1999). In Europe, where introgression from farm to wild fish has had a longer history than in North America, hybrids are generally behaviorally intermediate between wild and farm juveniles (Einum and Fleming 1997; McGinnity et al. 1997, 2003). Hybrids' growth performance as fry and parr in nature is superior, but their survival in nature is poorer than that of wild juveniles (Einum and Fleming 1997; Fleming et al. 2000; McGinnity et al. 1997, 2003).

Other Potential Interactions

The following environmental effects could potentially develop as a consequence of aquaculture (derived from Stickney 2002, Waknitz et al. 2002), some or all of which might affect wild salmon directly or indirectly. The committee is unaware of information gathered in Maine to determine whether the following factors are important.

- Alteration of predator-prey interactions induced by the presence of large numbers of farm fish, attracting and concentrating predators (Bailey 1998).
- Degradation of water quality through nutrient enrichment in surrounding waters.
- Concentration of cage sites affecting migratory behavior and homing success of wild salmon returning to rivers.
- Benthic pollution (heavy metals) and biological deposits (fish feces and uneaten feed) from farm operations that alter community ecology in the benthos may also affect other trophic levels.
- Effects of therapeutic compounds at net-pen farms on nontarget organisms, including migrating wild salmon.
- Toxic effects of algal blooms, enhanced by the dissolved inorganic wastes in the water column around net-pen farms.

ACIDIFICATION OF STREAMS AND RELATED PROBLEMS

Acidity in streams, mainly due to acid precipitation, has caused concern for the fate of Atlantic salmon in northeastern North America (Cairns 2001). There is widespread acceptance that acid precipitation is acidifying rivers in Scandanavia, Canada, and Maine. Salmon population declines have coincided with pH reductions to 5.0–5.5 (e.g., see Leivestad and Muniz 1976). Some of Nova Scotia's rivers have been seriously affected (Stoddard et al. 1999, Watt and Hinks 1999). Maine's rivers do not appear to show as much acidification, but there is cause for concern there as well. The toxicity of acidity generally manifests itself below a pH of 5.4 (DFO 2000). Fry mortality becomes important at a pH of 5.0; smolts begin to be affected at 5.0, and parr and smolt mortality approaches 100% as pH approaches 4.6 (DFO 2000). Eggs and alevins become affected below a pH of 4.8. Low pH has been blamed for the extirpation of salmon from at least 14 Nova Scotia rivers (DFO 2000). The buffering capacity of rivers varies, and although acid deposition has decreased over the past 20 years, not all rivers have recovered equally well (DFO 2000, Watt and Hinks 1999).

The hazard of acid pH in freshwater systems to Atlantic salmon specifically is well documented. The hazard is exacerbated in the smolt stage because of the challenge they face in transitioning to seawater. The delicacy of the gills of the Atlantic salmon and their important roles in many physiological processes has been noted. The gills are particularly vulnerable to acidic pH and aluminum. Fry-stage (about 1 gram) Atlantic salmon exposed to pH 5.6 in the presence of 107 micrograms per liter (μg/L) of labile monomeric aluminum for 30 days displayed swelling and fusion of the feathery-like lamellae of their gills, whereas in the absence of the aluminum the damage was reduced (Smith and Haines 1995). Smolt-stage Atlantic salmon exposed to pH 5.6 with and without aluminum (158 μg/L) for 16 or 23 days displayed structural and proliferative damage to chloride cells, which are specialized for ion exchange (Jagoe and Haines 1997). Atlantic salmon respond to acidic conditions by feeding and growing less (Farmer et al. 1989, Saunders et al. 1983b); reduced growth may account for lower survival in the marine environment (Friedland et al. 1993). Both endocrine and osmoregulatory physiology are disturbed by acid pH, leading to some mortality (Brown et al. 1990, Haya et al. 1985, Magee et al. 2001, Saunders et al. 1983b, Staurnes et al. 1996).

Haines et al. (1990) measured pH and aluminum in rivers and tributaries in eastern Maine clearly showing acidic conditions concurrent with elevated aluminum that could impair osmoregulation and survival in juveniles, parr, and smolts. A recent study illustrates the detrimental impact of even short episodes of acid pH on smolt physiology and survival. Episodic exposure, concomitant with elevated labile aluminum, is a more realistic event in river systems in Maine. Magee et al. (2003) exposed Atlantic salmon smolts to pulses of a pH reduction from 6.0 to 6.6 (control) to pH 5.2 for 48 hours weekly (episodic) for 4 weeks. They also maintained a group that was exposed chronically to pH 4.4–6.1 (chronic). When smolts were transferred to seawater, even the episodic exposure, with a 30-hour recovery, led to 35% mortality, compared with 0% in control smolts and 100% in chonically exposed smolts. The episodically exposed smolts that survived seawater lost weight in seawater. Magee et al. (2001) had previously observed using ultrasonic telemetry that migratory behavior of acid-exposed smolts could make them more vulnerable to predation than behavior of other smolts, because they wandered in and out of the freshwater and seawater interface, where many predators linger, rather than heading out to sea.

Liming ($CaCO_3$) rivers to neutralize the pH is an immediately available remedy already tested and recommended (e.g., DFO 2000). Liming has the advantage of being amenable to the adaptive management approach. Liming is known to eliminate osmoregulatory disturbances and increase survival of salmon eggs, fry, and smolts (Farmer et al. 1989, Rosseland and Skogheim 1986, Rosseland et al. 1984). Acidification is known to harm salmon populations and is likely a culprit in the poor survival and low returns of salmon in Maine. Liming could be a quick and effective remedy whose efficacy would be clear within years.

RESEARCH AND MONITORING

Research and monitoring are needed to understand the status and trends of populations of wild salmon in Maine and to understand the effects and effectiveness of management and other human actions on salmon. The committee has pointed out knowledge gaps that make managing salmon more difficult. Yet research itself can affect the fish. As the Maine Atlantic Salmon Task Force (1997) pointed out, "Despite careful handling, fish may die from trauma when fisheries biologists capture salmon to collect necessary growth and population data."

In most cases, the number of fish killed by research is so small that it is not a serious consideration, but in several Maine rivers there are so few wild salmon that killing even one parr or smolt could affect the population. In addition, some kinds of handling and sampling seem likely to entail greater risks than others. The committee has concerns in particular about research that requires fish to be anesthetized, samples of blood or scales to be taken from very small fish, and the fish to be caught and held for long periods in strong currents, as might occur in a rotary-screw trap for smolts during high flows. The value of any information obtained needs to be weighed carefully against the possibility of the death of any wild fish subjected to handling, especially where wild populations are very small.

GOVERNANCE

Salmon recovery will depend, to a significant degree, on changing those human activities that are threatening the survival of salmon. The principal human activities that directly or indirectly threaten salmon include dams and hydropower projects, salmon aquaculture operations, fisheries, hatcheries, forestry, roads, land development and use, research and monitoring, among others. Understanding the regulations, incentives, and other forces that shape the nature and extent of these human activities is a prerequisite for designing effective policies that will alleviate the threats they pose to the survival of salmon.

There are three general mechanisms that govern human activities related to the survival of Atlantic salmon: government, markets, and nongovernmental institutions (NGOs) and arrangements, which together constitute a system of governance (Juda 1999).[3] These mechanisms dynamically interact through complex interrelationships, which has been described by Hennessey (1994) as an ecology of governance. Individually and collectively, they influence human interactions with natural environments at various temporal and spatial scales. The governance ecology related to the survival of Atlantic salmon in Maine is comparable in complexity and importance to the natural ecology of Atlantic salmon in Maine.

Government regulations and requirements, at local, regional, national, or international levels, affect the human activities listed above. Governments establish and enforce rules that regulate the use of environmental resources and affect the way goods and services are produced.[4] The government also produces goods and services that cannot be efficiently organized by the market. For example, governments fund and conduct research on fisheries and other environmental and natural resources. These and other government activities may have a profound influence on how environmental and natural resources are used and on the potential for recovery or rehabilitation of degraded environments and endangered species.

Markets generate prices, which structure the incentives faced by business firms and households, and in turn affect humans' choices on how to use environmental and natural resources. Markets for electric power, wood products, food, and land have been major drivers of the nature and extent of the human activities that threaten salmon in Maine. Markets often fail to reflect the full value of nature's services in their prices. For example, wild Atlantic salmon in the water have unpriced values, i.e. values that are not reflected in market prices. Unpriced resource values (e.g., fish in public water bodies and many other ecosystem goods and services) artificially deflate the cost of using such resources. The salmon resource is devalued currently and over time, and markets tend to dis-

[3]Juda (1999) defines governance as "the formal and informal arrangements, institutions, and values that determine how resources are used; how problems and opportunities are evaluated and analyzed; what behavior is deemed acceptable or forbidden; and what rules and sanctions are applied to affect the pattern of resource and environmental use."

[4]Systems of regulation and requirements and the allocation of rights and responsibilities are associated with the development of complex institutions, and this complexity can slow the responsiveness of institutions and may fragment effort, authority, and responsibility leading to a lack of accountability (NRC 1996b).

count the resource benefits to future generations (NRC 1996a). In these circumstances, human users do not face the full social and environmental cost of fishing, habitat destruction, waste disposal, and so forth, which encourages excessive use and results in depleted fish stocks, too little essential habitat, and too much pollution. These and other failures of markets form the basic rationale for government regulation of human activities.

Since government and markets do not always adequately represent individual values, those individuals with sufficient resources to do so often form and participate in NGOs and arrangements.[5] Even people who may be well supported by government may form NGOs to protect that support or to produce more support for their interests. Perhaps the most visible manifestation of these is the voluntary NGOs that are often active in public debates on environmental and natural resource policy. Less visible forms of these values and institutions are the social norms and customs embraced by members of communities, which include such informal rules on the treatment of fish, wildlife, land, and forests as, for example, the local cultures of resource use in northern New England described by Judd (1997). The social forces generated by NGOs and arrangements influence the patterns of use for these resources. They are dependent on the values people attach to their community and neighborhoods, traditions, and long-standing social networks.

The following sections briefly describe the current status of the three governance mechanisms (government, markets, NGOs and arrangements) that influence the human activities related to the survival of Atlantic salmon in Maine. As with other sections in this chapter, this section ideally would begin with a history of governance that explains how government, markets and NGOs and arrangements over time have influenced—both restrained and encouraged—the human activities that have affected the survival of Atlantic salmon in Maine. Unfortunately, the limited time and resources available to the committee has made such historical reconstruction infeasible. Instead, we begin with a description of the existing state of governance as it relates to Atlantic salmon in Maine.

Government Organizations and Programs

There are six levels of government organizations and programs that influence the human activities related to the survival of Atlantic salmon in Maine: local, tribal, state, federal, regional, and international.

[5]NGOs reflect some of the values held by people concerned about Atlantic salmon and the environment (NRC 1996a) that are not necessarily fully accounted for by government and markets.

Local Town Governments

Local town governments have agencies with the authority to regulate, or otherwise influence, aquaculture, fisheries, forestry, roads, agriculture, land use, and boating within their borders.[6] Some town agencies conduct inspections of aquaculture leases and are involved in the leasing process concerning leased areas in the town. Local government agencies also have conservation commissions and fisheries constables who regulate use of local area resources (for example, town agencies such as conservation commissions that regulate forestry). Towns and cities in the state of Maine have municipal agencies, such as the Department of Public Works and others, that maintain, design, and construct roads and bridges within their jurisdictions. Local agencies, such as the board of health, are responsible for monitoring conditions in which crops are maintained and harvested, and charged with monitoring the treatment and care of animal facilities. Local municipalities (e.g., Conservation Commission, Building Department) undertake the role of the Maine Land Use Regulatory Commission (LURC) in organized areas. Some local coastal towns have departments, such as a Harbormaster or Department of Natural Resources, that enforce local, state, and federal laws pertaining to boating.

In the unorganized areas of Maine (for which there is no local government), state agencies (such as LURC and the Maine Department of Environmental Protection) regulate land use, forestry, and several other human activities. Most of these unorganized areas are in the northern inland portion of the state; however, a few unorganized areas exist near the coast in Hancock and Washington counties, the Downeast region of the state.[7]

Native American Tribal Government

Native American Tribes in Maine include the Passamaquoddy Tribe in Washington County, the Penobscot Indian Nation based at Indian Island on the Penobscot River, the Houlton Band of Maliseets, and the Aroostook Band. The Maine Indian Tribal-State Commission (MITSC), an independent commission made up of tribal and state representatives, has exclusive authority to establish regulations that govern fishing within any section of a river for which both sides are within the reservation or trust

[6]An amendment to the Maine state constitution in November 1969 delegated broad "home rule" ordinance powers to cities and towns. Ordinances range from the control of a town's growth, to the review of real estate development projects, to the banning of herbicide spraying, and to the regulation of local timber harvesting (MMA 2002).

[7]For a map that shows the unorganized areas of Maine, see Maine Revenue Services (2003b).

lands (lands owned by the U.S. and held in trust for the tribe). (More information on tribal government in Maine is contained in Appendix B.)

State Government

There are at least 12 state government agencies involved in regulating, or otherwise influencing, the human activities that affect the survival of Atlantic salmon. (These agencies, and the activities that they influence, are listed in Table 3-5; and brief descriptions of each agency's roles and responsibilities related to these activities are given in Appendix B.)

The Maine Atlantic Salmon Commission (MASC) is one of the most important state agencies related to the restoration of Atlantic salmon in

TABLE 3-5 State Agencies Related to Salmon Conservation in Maine

Human Activities That Impact Atlantic Salmon	Dams	Salmon Aquaculture	Fisheries	Forestry
Maine State Departments and Agencies	3	5	5	2
Atlantic Salmon Commission				X
Department of Inland Fisheries & Wildlife				
Bureau of Resource Management		X	X	
Bureau of Fish Warden Service			X	
Department of Marine Resources				
Bureau of Resource Management		X	X	
Bureau of Marine Patrol		X	X	
Department of Environmental Protection				
Bureau of Land & Water Quality (Salmon Rivers)	X	X		
Department of Conservation				
Land Use Regulatory Commission	X			X
Maine Forest Service				X
Public Utilities Commission	X			
Department of Agriculture		X		
Department of Transportation				
State Planning Office				

SOURCE: Compiled from agency information, including Web sites.

Maine. Established in 1999, the MASC is charged with restoration and management of Atlantic salmon throughout its original range in Maine and is involved with all aspects of Atlantic salmon management in coastal and eastern Maine, the MASC has the sole authority to introduce Atlantic salmon to inland waters. Other than commercial aquaculture facilities, the commission has the sole authority to limit or prohibit the taking of Atlantic salmon, may issue licenses for the taking of Atlantic salmon, and may adopt rules establishing the time, place, and manner of Atlantic salmon fishing in all waters of the state.

The MASC manages the Atlantic Salmon Conservation Plan (ASCP) for Seven Maine Rivers. The commission conducts routine monitoring of the abundance and status of salmon in most of Maine's Atlantic salmon

Roads	Agriculture	Land Use	Recreational Boating	Monitoring and Research	Planning and Coordination
2	2	2	4	5	3
				X	X
			X	X	
			X		
			X	X	
			X		
		X		X	
X	X	X			X
	X				
X					
				X	X

watersheds. In addition, the commission supplies brood stock to federal hatcheries, conducts electrofishing surveys to evaluate juvenile fish production in salmon rivers and measures the success of fry stocking programs. The MASC also helps coordinate and support nongovernmental groups of volunteers that have an interest in the restoration and management of Atlantic salmon. For example, in 2001, the MASC provided local watershed councils organizational support and funds to address specific restoration and habitat protection projects.

Ten other state government agencies play prominent roles in regulating fisheries, forestry, agriculture, dams, aquaculture, roads, land use, and recreational boating. (For a full list of state agencies and a description of their responsibilities, see Appendix B.) Prominent among the state agencies are the following:

- The Department of Inland Fisheries and Wildlife (DIFW) establishes and enforces rules and regulations that govern fishing, propagation and stocking of fish, the registration of watercraft and all terrain vehicles, and the issuing of licenses (e.g., hunting, fishing, trapping, guide) and permits. The DIFW also enforces the rules adopted by the MASC.
- The Department of Marine Resources (DMR) regulates marine aquaculture operations, marine fisheries, recreational boating and operates programs for research and monitoring of living marine and resources. For salmon aquaculture, DMR issues permits for aquaculture sites, enforces the Aquaculture Lease Law, administers the Finfish Aquaculture Monitoring Program (FAMP), and monitors for toxic contaminants under and in net-pens. For fisheries, the DMR issues fishing licenses, enforces saltwater fishing laws and regulations, and operates research and habitat conservation programs.
- The Department of Environmental Protection (DEP) governs a wide range of human activities, including hydropower and dams, natural resource protection, shoreline zoning, site development, erosion and sedimentation control, wastewater discharge, and others. With respect to hydropower projects, the DEP, in cooperation with the Land Use Regulatory Commission (LURC), issues permits for the construction, reconstruction, or the structural alteration of a hydropower project; and enforces state laws concerning unapproved hydropower projects. With respect to salmon aquaculture, the DEP tests water for effluent quality from aquaculture sites, and issues permits as part of the Maine Pollution Discharge Elimination System (MPDES). In addition, the DEP issues permits for activities on land adjacent to any freshwater wetland, great pond, river, stream, or brook that could wash harmful material into these resources.
- LURC regulates land use in the state's townships, plantations, and

unorganized areas, and cooperates with the DEP to regulate hydropower projects.

• The Maine Public Utilities Commission (MPUC) enforces all state laws that apply to public utilities, such as hydropower dams. The MPUC shares these responsibilities with the DEP and the LURC, the two agencies that issue permits for the construction, reconstruction, or the structural alteration of a hydropower project; and enforces state laws concerning unapproved hydropower projects.

• The Department of Transportation (DOT) designs, builds, and maintains many of the roads, highways, and bridges in the state and is the main oversight agency for projects involving roads, railroads, and associated facilities. The DOT restores habitat by addressing non-point-source pollution associated with transportation facilities located in salmon watersheds.

In addition, the State Planning Office is charged with coordinating the development of the State's economy and energy resources with the conservation of its natural resources (including Atlantic salmon and its habitat); providing technical assistance to the governor, legislature, and local and regional planning groups.

Federal Government

There are at least 11 federal government agencies that regulate, or otherwise influence, the human activities related to the survival of Atlantic salmon in Maine. (These agencies, and the activities in Maine that they influence, are shown in Table 3-6; and brief descriptions of each agency's roles and responsibilities related to these activities are given in Appendix B.)

Two of the most relevant federal agencies are the U.S. Fish and Wildlife Service (FWS) and the National Marine Fisheries Service (NMFS). As explained above, these two agencies, which share responsibility for administration of the Endangered Species Act (ESA), listed Atlantic salmon as an endangered distinct population segment in November, 2000. The FWS implements ESA programs and regulations for terrestrial and freshwater species, while NMFS implements programs and regulations for marine and anadromous species.

The FWS operates programs to protect and restore fish and wildlife resources and their habitats, including the National Fish Hatchery System, which in Maine consists of two fish hatcheries (Craig Brook and Green Lake).

NMFS also operates programs for the protection, conservation, and

TABLE 3-6 Federal Agencies

Human Activities that Impact Atlantic Salmon	Dams	Salmon Aquaculture	Fisheries	Forestry
Federal Governmental Agencies	4	7	3	2
Fish & Wildlife Service	X	X	X	X
National Marine Fisheries Service		X	X	
Environmental Protection Agency	X	X		
Food & Drug Administration		X		
Department of Agriculture (APHIS, NRCS, USFS)		X		X
Army Corps of Engineers	X	X		
Federal Energy Regulatory Commission	X			
Coast Guard		X	X	
Federal Highway Administration				

SOURCE: Compiled from agency information, including Web sites.

recovery of species protected under the ESA. In addition, NMFS implements the 1988 marine fishery management plan for Atlantic salmon, which applies in federal marine waters. This management plan established explicit U.S. management authority over all Atlantic salmon of U.S. origin to complement state management programs in coastal and inland waters and established federal management authority over salmon of U.S. origin on the high seas. The plan prohibits commercial fishing for Atlantic salmon, directed or incidental, in federal waters (3–200 miles) and prohibits the possession of Atlantic salmon taken from federal waters.

In 2001, the Northeast Fisheries Science Center of the National Marine Fisheries Service opened a field station in Orono, Maine, not far from the University of Maine campus and the Maine Atlantic Salmon Commission. This office serves as home base for several federal researchers and managers who work on anadromous fish in Maine, primarily Atlantic salmon. The move brought researchers closer to their research subjects, but perhaps more important, it brought federal officials closer to local stakeholders, political leaders, agencies, councils, media, and researchers.

Nine other federal government agencies significantly influence fisheries, forestry, agriculture, dams, aquaculture, roads, land use, and recreational boating in the state of Maine. A brief description of some of the other prominent federal agencies follows:

Roads	Agriculture	Land Use	Recreational Boating	Monitoring and Research	Planning and Coordination
3	3	3	1	6	0
X		X		X	
				X	
	X	X		X	
	X			X	
	X			X	
X		X			
				X	
			X		
X					

• The U.S. Environmental Protection Agency (EPA) works with the Maine Department of Environmental Protection, its primary state partner related to Atlantic salmon. EPA has funded a $1.9 million cooperative agreement with the Gulf of Maine Council in its efforts to protect and sustain regionally significant Gulf of Maine coastal and marine habitats. EPA indirectly and directly affects Atlantic salmon farming and agriculture operations by, for example, approving and regulating the use of pesticides around and monitoring the effluent quality from aquaculture facilities.

• The U.S. Army Corps of Engineers (USACE) regulates activities in navigable waterways, including dredging and filling of waterways, and issues permits for dams and dikes placed in interstate waterways. USACE also enforces regulations that require the installation of suitable culverts and bridges, designed to withstand and prevent restriction of high flows and maintain existing low flows, for roads that cross bodies of water.

• The USDA has several programs that affect Atlantic salmon in Maine. Its Animal and Plant Health Inspection Service serves aquaculture, especially those aspects involving disease, pest prevention, and wildlife damage management, and has become involved in facilitating the importation and exportation of aquaculture products. USDA's Natural Resources Conservation Service operates a voluntary program for individuals who want to develop and improve wildlife habitat primarily on private land by providing both technical assistance and up to 75% cost-share assistance to establish and improve fish and wildlife habitat. The Natural Resources Conservation Service also assists local authorities to rehabilitate or remove aging dams by providing 65% of the total cost of a rehabilitation project. Other relevant programs of the USDA include the Small Watershed Program, the Forestry Incentives Program, and the Stewardship Incentive Program. The Forest Service also manages the White Mountain National Forest, which includes part of the drainages of the Androscoggin and Saco rivers. Those rivers have some potential for salmon rehabilitation. (For more information on these and other activities of the USDA, see Appendix B.)

• The Federal Energy Regulatory Commission (FERC) authorizes construction of existing hydropower facilities. FERC issues licenses for a period of up to 50 years and is expected to equally consider developmental and environmental values, including hydroelectric development and fish and wildlife resources (including their spawning grounds and habitat). Small hydro plants that are 5 megawatts or less that use an existing dam, or that utilize a natural water feature for headwater, and existing projects that propose to increase capacity are exempt from FERC licensing.

• The U.S. Coast Guard enforces fisheries laws at sea, such as the Magnuson-Stevens Fisheries Conservation and Management Act, in conjunction with the NMFS. As part of its mission to manage waterways, the Coast Guard participates in aquaculture leasing permit processes and ensures that offshore structures are not hazards to navigation.

Regional Intergovernmental Organizations

The New England Fishery Management Council, with jurisdiction extending from Maine to southern New England, develops management plans that are approved and implemented by the Secretary of Commerce and are implemented by the NMFS. The council developed the Fishery Management Plan for Atlantic salmon, which was implemented by NMFS on March 17, 1988, and explicitly established U.S. management authority over all Atlantic salmon of U.S. origin. The plan prohibits any commercial fishery for Atlantic salmon, directed or incidental, in federal waters (3–200 miles) and prohibits the possession of Atlantic salmon from federal waters.

The Atlantic States Marine Fisheries Commission was formed in 1942 by 15 Atlantic coast states (Maine through Florida, including Pennsylvania) to assist in managing and conserving the states' shared coastal fishery resources. While the Commission's Interstate Fisheries Management

Program aims to promote the cooperative management of marine, estuarine, and anadromous fisheries in state waters of the East Coast through interstate fishery management plans, it currently does not have a fishery management plan for Atlantic salmon.

International

The principal international organization governing Atlantic salmon is the North Atlantic Salmon Conservation Organization (NASCO).[8] NASCO, established in 1984, aims to contribute to the conservation, restoration, enhancement and rational management of salmon stocks. NASCO was organized by the Convention for the Conservation of Salmon in the North Atlantic Ocean. The North American Commission of NASCO requires each of its members, which include Canada and the United States, to implement measures to minimize the bycatch of Atlantic salmon that originate in the rivers of other members. NASCO has developed guidelines on containment of farm salmon, which governs farm site selection, equipment used, and procedures, for each member country to follow.

The St. Croix International Waterway Commission (SCIWC) is an international body established by the Maine and New Brunswick legislatures to manage the St. Croix boundary river corridor (SCIWC 2003). The SCIWC operates the St. Croix's native Atlantic salmon program for research, management, and restoration in this watershed.

The Gulf of Maine Council on the Marine Environment is an international body that promotes and facilitates cross-border cooperation among government, academic, and private groups. The council's action plan for the protection and conservation of coastal and marine habitats in the Gulf of Maine guides state, provincial, and federal policy and budgeting decisions affecting the Gulf's coastal and marine environments.

Nongovernmental Organizations and Institutions

There are several NGOs that are actively engaged in efforts to restore and conserve Atlantic salmon in Maine. These organizations include river and angling conservation groups, Native American, and industry organizations. The Maine Atlantic Salmon Commission lists nearly 50 of these groups and organizations (MASC 2003). (MASC's list of the NGOs is reproduced in Appendix B). These groups and organizations rely heavily

---

[8]The North Atlantic Fisheries Organization governs fisheries in the North Atlantic that exploit species other than Atlantic salmon.

on volunteers and external funding to execute their Atlantic salmon conservation. A few selected examples of these efforts follow.[9]

• Members of the Narraguagus River Watershed Council donated funds and labor to stabilize erosion sites in the Cherryfield reach of the river. Their project was supplemented with funds from the MASC and Maine Department of Environmental Protection. On the Machias River, the River Watershed Council secured landowner permission and coordinated the efforts of volunteers to plant a riparian buffer along a 300-foot section of Dan Hill Brook in Whitneyville.

• In the Ducktrap River Watershed, the Coastal Mountains Land Trust completed three land-conservation projects in 2000. A conservation easement donated by MBNA (a private company) protects 1,467 feet of frontage on the river and 8 acres of steep forested riparian land. A 3.5-acre property with 640 feet of frontage on Black Brook, a primary tributary to the river, was purchased. A second property on Black Brook was placed under a donated conservation easement that protects 66.3 acres and 1,460 feet of frontage. As a result, more than 70% of the riparian buffer of the Ducktrap River is in permanent conservation management and ownership. Funds for accomplishing these permanent conservation protections for Atlantic salmon habitat have been provided by a broad group of local donors, several private foundations, and state and federal agencies.

• Private companies are taking measures to restore and conserve Atlantic salmon. International Paper, a forest products company, provides support to River Watershed Councils and state agencies to identify water quality problems and takes corrective measures when problems are identified. In addition, the company has implemented the Riparian Management Guidelines, originally developed by Champion International, now part of International Paper, for its lands in Down East Maine. According to the company, these measures exceed state regulations. These and many other examples of nongovernmental efforts provide convincing evidence that many people in Maine value the survival of Atlantic salmon. However, the values of those people do not fully match those reflected in actions driven by formal government and market forces.

Markets

Market conditions in general are expected to influence a number of the human activities related to the survival of salmon. For example, if

[9]These examples are drawn from the MASC's 2000 annual progress report on the Atlantic Salmon Conservation Plan for Seven Maine Rivers (MASC 2000).

market prices for electricity rise substantially in response to increased demand, there will be greater pressure to construct new hydropower facilities and re-license existing facilities in Maine. Additionally, world market conditions for seafood, forest products, and blueberries will determine, in part, the level of salmon aquaculture, forestry, and blueberry farming in the state. Declining demand in these markets would likely weaken efforts to expand these sectors, which could benefit salmon conservation. On the other hand, declining demand might also reduce the willingness of these producers to invest in salmon conservation efforts, since soft markets would weaken their financial position.

Maine will likely experience increased demand for land, forest resources, and marine and freshwater areas containing valuable salmon habitat. As in other coastal states, Maine will probably experience increased residential development of land along the coast and rivers that contain valuable salmon habitat. This will increase the pressure to expand Maine's road network, an activity that requires bridge construction or culverts over salmon streams.

The available information on these (and their ancillary) markets, which are powerful drivers of the human activities that affect the survival of salmon, is not sufficient to determine whether the way they are regulated is consistent with salmon recovery. It is unclear at this time whether additional controls on market forces are needed to prevent these threats to salmon from growing stronger over time.

Comanagement

The committee has not been able to document the historical development of this complex ecology of governance or the nature and extent of the relationship between that development and the overall decline of wild Atlantic salmon in Maine. It has been unable to determine if the differential pattern of decline that it identified in the DPS rivers as opposed to the other Maine salmon rivers is related to differences in governance processes between the Down East and other areas. Finally, it has been unable to evaluate the extent to which government agencies and other institutions described in this chapter are capable of learning and adapting to new information and changing circumstances. There is a need for much information to address these matters successfully. However, the committee suggests that experience from elsewhere can usefully be applied in Maine. Much of that experience and the specific kinds of information that would be needed for Maine have been discussed in Burger et al. (2001) and NRC (2002e). Issues related to conflict among interest groups and lack of support for conservation initiatives have also been implicated in resource decline and failed attempts at rehabilitation elsewhere. In order to address this problem, some recommend a shift toward comanagement. *Comanagement* is a generic term used to describe the various ways in which resource users can meaningfully share management-related powers with state agencies. Within comanagement initiatives, government agencies can delegate some or all of their management rights to local authorities, which then comanage with local interested groups (Jentoft and McCay 1995). Both decision-making power and accountability for the consequences of those decisions are shared. Power sharing is often spread among several levels of government as well as nongovernment constituencies.

Comanagement is often recommended for contexts where the ecology of governance is very complex and where the challenges are great and the room for error small, as appears to be the situation with wild Atlantic salmon in Maine. More specifically, comanagement is one strategy for dealing with situations with a heterogeneous group of users with "uneven powers, conflicting interests, unequal bargaining powers and different stakeholder values and rationalities," contexts where deliberation can be cumbersome and where it is difficult to achieve consensus (Hara 2003, Jentoft 2000). Effective comanagement has the potential to develop a heightened sense of acceptance and compliance toward management rules, because rules that reflect the experiences and solutions proposed by users and result from dialogue rather than unilateral imposition by distant agents mean that those affected are less able to rationalize rule violation by treating management regimes as "theirs" versus "ours" (Pinkerton and Weinstein 1995). Compliance also requires, however, that the rules appear to be working. Scientific uncertainty can make it difficult to set and achieve management goals (Holling 1978, Walters 1986), and science is only as good as the data to which it has access. Some evidence suggests that various forms of comanagement can enhance science-based decision making. Thus, scientists are more likely to secure good data and rapid feedback on the ecological effects of management initiatives when resource users are committed to the management process and active participants within it (Felt et al. 1997, Walters et al. 1993). Where comanagement regimes are grounded in local community management traditions and local knowledge, they can benefit from "rules of thumb" developed from past experience and enforced through established social and cultural means (Berkes 1999).

Depending on the context, there can be significant challenges associated with moving toward successful comanagement. For example, it requires a legal framework for both autonomous and shared decision making, as in the case of the "Boldt decision," which required comanagement of salmonid fishes by American Indian treaty tribes and state government agencies (Pinkerton 1994). Like other management regimes, comanagement must include mechanisms for limiting access, resolving conflicting uses, ensuring habitat protection, and ensuring adequate enforcement. It must also promote legitimacy among resource users, as well as compliance and a willingness to exchange information with biologists monitoring the resource (Pinkerton 1994). Where comanagement is deemed to be desirable and needed and where it is possible and feasible to move in this direction, other requirements for successful comanagement also include the presence of appropriate local and government institutions, trust between actors, legal protection of local rights, and economic incentives for local communities to conserve the resource (Berkes 1997). As indicated by the ecology of governance for Atlantic salmon in Maine, there has been a history of delegation of responsibility and resources to lower levels of government and to NGOs related to salmon and their environments. The current management frameworks need to be investigated to see what has worked and what has not worked and whether it would be feasible and appropriate to increase the level of comanagement related to salmon and their habitats.

4

Setting Priorities for Action

RISK ASSESSMENT AND DECISION ANALYSIS BASICS

Risk assessment and decision analysis are tools for organizing and analyzing information in a systematic way and in the face of uncertainty to help identify the best way or ways of tackling a complex problem. By being systematic, these tools also encourage documentation of the methods used. This allows others to apply them in slightly different situations or using different values for a variety of variables. In the logical paradigm developed for evaluating risks to humans from exposure to various contaminants, the National Research Council (NRC 1983) described *risk assessment* as a technique for evaluating the probability and severity of an adverse outcome, while *risk management* is a technique for deciding on the best options for reducing risk.

The information available about the causes of declines of Atlantic salmon populations in Maine is incomplete. Therefore, it is not obvious what actions should be taken and in what order to reverse those declines. The Atlantic Salmon Conservation Plan for Seven Maine Rivers (Maine Atlantic Salmon Task Force 1997) is a thoughtful analysis of the causes of salmon declines and ways to reverse them. This committee agrees in general with that report, with a few exceptions. The value that we have attempted to add here is in (1) our ranking of the factors affecting salmon in terms of their likely severity, (2) our prioritizing the various management options available in terms of their likelihood of being effective and in terms of cost, (3) our suggesting a sequence for undertaking those options, and (4) our suggesting a framework for others to do similar analyses when conditions change, when new information is available, or with different values attributed to various outcomes and costs.

Evaluating and Ranking Threats to Maine Atlantic Salmon

An intractably large number of threats to Atlantic salmon have been identified. For example, a Canadian group of experts recently identified 63 factors threatening the survival of Atlantic salmon in eastern North America (Cairns 2001). No feasible amount of time and resources could be enough to understand and mitigate such a large number of threats. Fortunately, we do know that some threats are more important than others, and furthermore, some threats must be mitigated before others can be addressed. For example, if barriers prevent salmon from ascending a river, then those barriers must be made passable before improving habitat above them could be of any use.

Many documents and presentations read and heard by the committee gave the impression that perhaps the biggest difficulty in knowing how to rehabilitate salmon is seeing the forest for the trees. What are the most important things to do and which of them should be done first?

To approach a solution to this problem, the committee developed a conceptual framework or risk-assessment model for thinking about it that involved identifying and ranking the threats and their contribution to salmon mortality. This framework considers a range of issues that apply across the watersheds in Maine where Atlantic salmon could potentially be restored. However, the committee has not considered in detail mitigation options for the significant issue of at-sea mortality because the committee recognizes the large knowledge gap in being able to ascribe causation. (The hatchery living gene-bank program at Maine's Craig Brook Fish Hatchery is in part an ocean mitigation program. The parr are raised to adulthood in the freshwater of the hatchery, rather than having to become mature in the sea, where survivorship is very low.) The committee acknowledges the importance of at-sea mortality as a threat factor and strongly supports the need for further research to better understand mechanisms and possible remedial measures. The committee similarly has not attempted to evaluate the range of responses to potential threats that could be induced by climate change, because that issue is much larger than conservation planning efforts in Maine can reasonably address.

As noted earlier, the committee's initial work focused on understanding the genetic status of Atlantic salmon in Maine (NRC 2002a) in response to its charge. At the same time, the committee was gathering, organizing, assimilating, and discussing a wide and diverse range of pertinent data and information. Inevitably, we retraced the path of earlier teams of scientists and managers in the United States, Canada, and elsewhere in the world where threatened and endangered anadromous fish (particularly Atlantic and Pacific salmon) have been studied (Cairns 2001; Maine Atlantic Salmon Task Force 1997; NMFS and FWS 1999; NRC 1996a). At the end of each path, we encountered the same obstacles met by earlier efforts—a seemingly intractable number of variables, some of which were quantified with detailed data and others that were clouded by uncertainty. Like our predecessors, we could not rely solely on deductive reasoning and reductionist methods to understand a complex (in both space and time) environmental problem that had taken shape over several centuries. In other words, we faced information overload in some areas and daunting gaps in others with no readily apparent means of reaching sound and timely conclusions. Our individual and collective frustration was increased by the rapid declines in Atlantic salmon populations, despite the best efforts of dedicated scientists and managers, and the corresponding urgency of research and restoration efforts. Once the application of risk assessment and decision analysis methods was proposed by a small subset of our members, the entire committee was unified and energized by the prospect of breaking the impasse and fostering the use of the adaptive management paradigm for conservation of Atlantic salmon in Maine—to help see the forest for the trees.

It has been observed that all models are wrong, but some are useful. In this case, we believe the strengths of risk assessment and decision analysis methods substantially outweigh their shortcomings and weaknesses. Some key strengths include (1) the systematic yet flexible process for diagramming complex systems, (2) the need to consider proportional influences and interaction effects at different levels, and (3) the impetus for improving input data and conducting sensitivity analyses to update and refine estimates. The primary weaknesses are held in common with virtually all modeling methods. First, the process yields an incomplete mathematical abstraction of the environment (natural and anthropogenic). Second, the weakest parameter estimate(s) limits accuracy and utility of the results. Nevertheless, risk assessment and decision analysis methods help to guide and fuel adaptive management efforts.

A few notes on the mechanics of the risk assessment process may be helpful. The bubble diagram of Figure 4-1 illustrates the committee's view of the relationship between humans and a viable wild salmon population, through ecological, direct, and genetic influences. As the group described by Cairns (2001) did, our committee drew on its expert judgment based on personal insights and experience, the information in the literature, the information in the many briefs and presentations received at committee meetings, and so forth, to assign proportional values to the impact factors. For example, the committee estimated that more than half of all the hu-

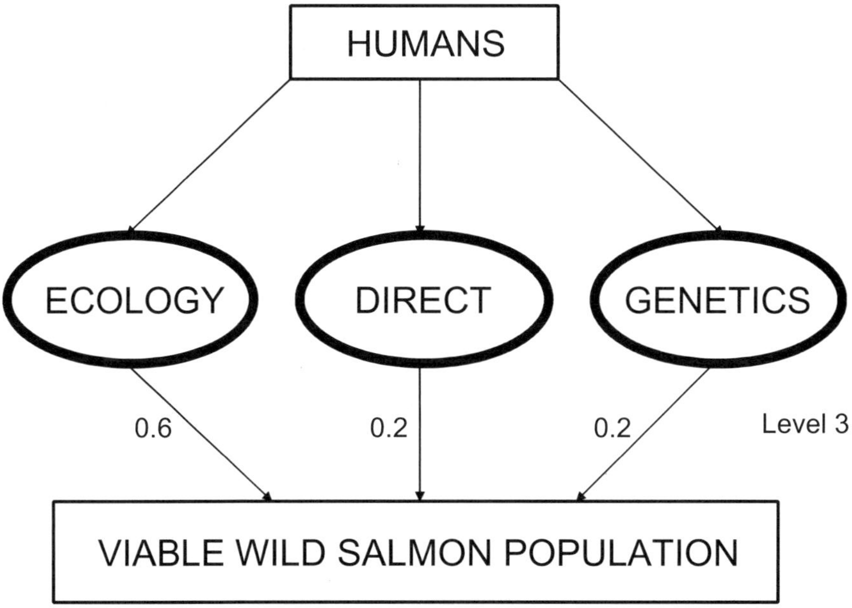

FIGURE 4-1 Categories of human impacts on wild salmon.

man influence on the viability of wild salmon populations is through ecological factors and assigned 0.6 or 60% of the total influence on wild salmon viability to ecological factors. We estimated that direct and genetic influences have roughly equal influences, and therefore each received 0.2 or 20% of the total impact on viability. The fact that the influences sum to 1.0 (0.6 + 0.2 + 0.2) and account for all of the means by which humans affect the viability of a wild salmon population can give the appearance of an artificially high accuracy of assignment (nothing left unexplained; precise allocation to alternatives). But this appearance of precision is not intended by the committee. Instead, our view is that Figure 4-1 provides nothing more than an informed estimate of the relative weighting of impact factors, and that later investigators or new information may well lead to the revision of these estimates. Indeed, this capacity for revision based on improved data is one of the strengths of the risk assessment technique. To fully understand the mechanics of the process, it is useful to see how the analyses are structured and how proportions or probabilities are multiplied and then added to generate the final estimates for the relative importance of impact factors.

Analyzing the Information

The committee has based much of its analysis and many of its conclusions on the bubble diagrams (Figures 4-1, 4-2, 4-3, and 4-4). Before providing additional detail about the process and use of the bubble diagrams, we emphasize again that the numbers and letters in the bubble diagrams were developed for heuristic purposes. They are not random numbers pulled from the air: they are informed estimates. The numbers cannot be considered data but rather help to identify where the greatest impacts to salmon might be and where data are most likely to be useful.

The committee began by asking what the known and potential sources of human-caused salmon mortality are. Using its own experience, general biological judgment, and many publications and other sources of information, the committee listed human-caused threats to salmon. We then categorized them into ecological factors, genetic factors, and direct factors. Ecological factors act by degrading the environment's ability to support salmon productivity (survival and reproductive success, or "fitness"). They include such items as water quality and quantity, obstructions to passage, changes in availability and quality of spawning and rearing habitat, presence of nonnative species that likely compete with or prey on

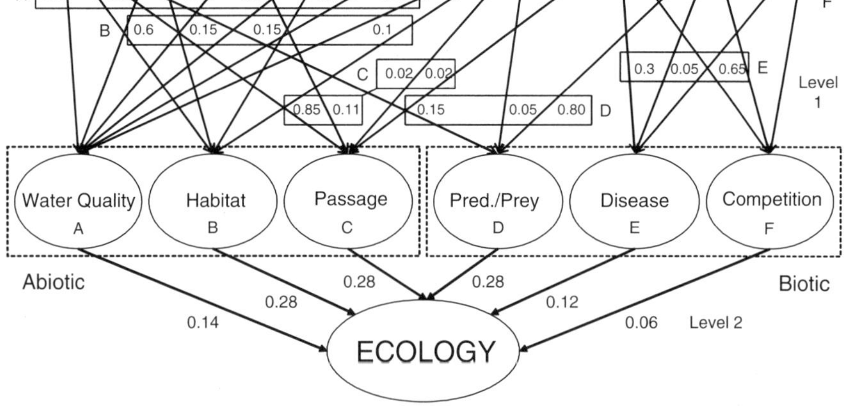

FIGURE 4-2 Subcategories of ecological impacts on wild salmon.

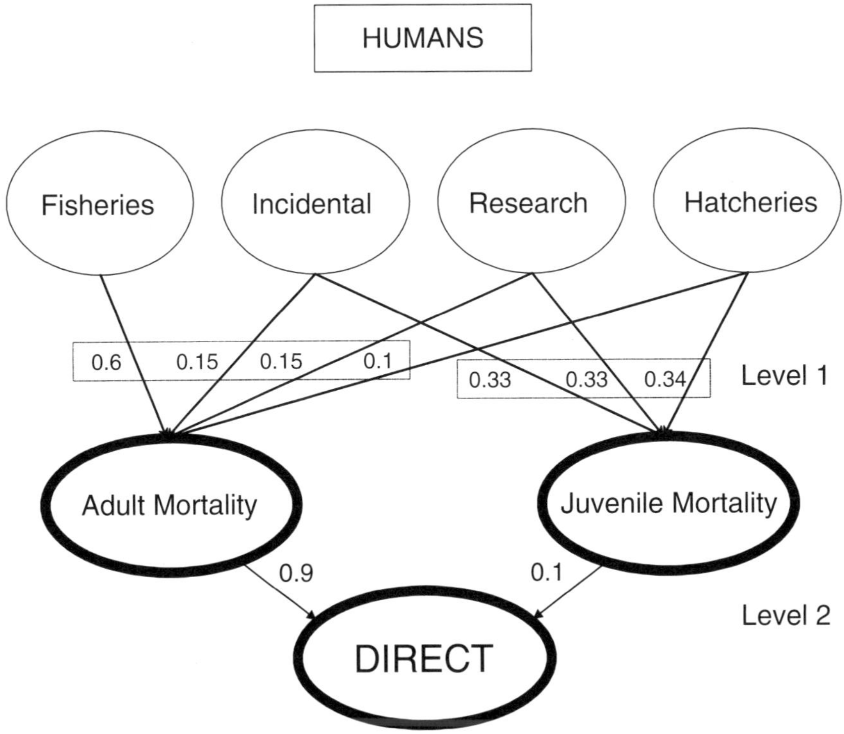

FIGURE 4-3 Subcategories of direct impacts on wild salmon.

salmon, and so on. Genetic factors act to reduce salmon productivity by reducing the quality of their genetic adaptations and thus reducing inherent capacity to respond to their environment within their lifetimes (e.g., appropriate predator avoidance) and, in some cases, the population's ability to respond to environmental change evolutionarily, across generations. They include inbreeding, domestication selection, breakdown of co-adapted gene complexes through lack of mate choice, genetic drift due to small population size, the incorporation of genes into the population from nonnative or nonlocal populations, and so on. Finally, direct factors are human actions that directly kill adult or juvenile fish. They include incidental and targeted fishing, turbines in hydroelectric plants, the killing of fish through research, and so on. These three categories cover all major sources of salmon mortality. Thus, the committee was able to take a rather extensive list of threats to salmon and compile these into three categories, which could then be considered for their proportional impor-

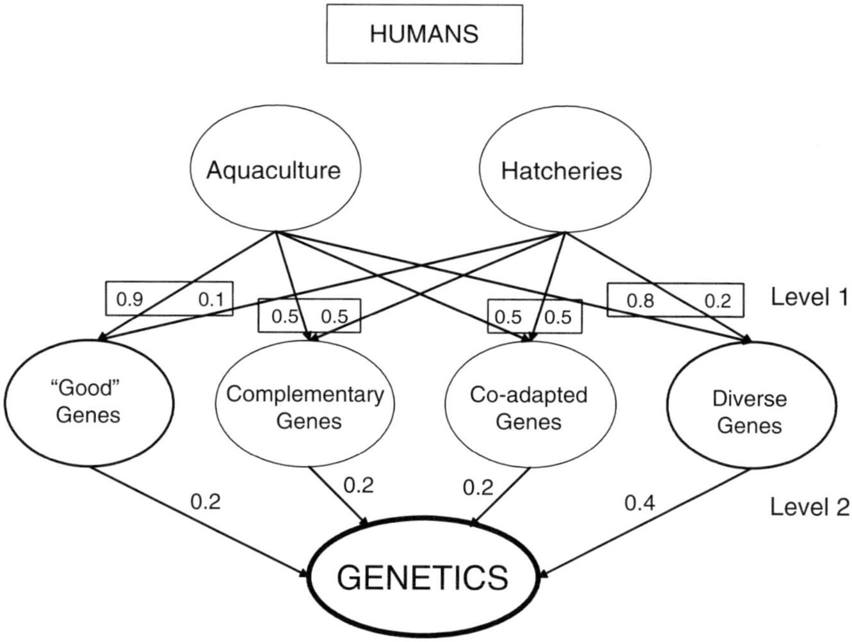

FIGURE 4-4 Subcategories of genetic impacts on wild salmon.

tance as impact factors on salmon viability. The committee considered what reasonable limits could be assigned to the contributions of individual factors in each category. It seemed reasonable that no impact factor should be less than 5% of the total to avoid having to consider too many very small factors.

Given that more than half the original spawning habitat of Atlantic salmon in Maine is no longer available to them because of obstructions to passage, and given the presence of additional ecological factors, it seems clear that ecological factors contribute more than half of all sources of human-caused salmon mortality in Maine. The committee therefore assigned the ecological category a total contribution of 0.6. Given that much direct mortality of salmon has been reduced or even eliminated, especially fishing, it seemed appropriate to allow that factor to be one-third as big as the ecological factor (0.2). Similarly, the genetic factors that have affected salmon are likely to be important but not nearly as important as the ecological factors, so they also were given an overall value of one-third that of the ecological factors (0.2). These three values become level 3 in our analysis—they portion out the relative contribution that all other factors at level 2 and level 1 above them can make to the viability of a wild salmon population. For instance, the myriad of ecological factors will, when considered in sum, account for 0.6 or 60% of the loss of viability of wild salmon.

The committee then addressed the many ecological factors that have been identified in other similar studies (e.g., Cairns 2001, Maine Atlantic Salmon Task Force 1997, NRC 1996a) and tried to position them in a relative sense. It did this by considering the various pathways outlined in the Figure 4-2 bubble diagrams. At level 2, the level that feeds directly into level 3, it was possible to identify three major abiotic factors (water quality, habitat, passage) and three major biotic factors (predator-prey relationships, disease, competition). The committee judged that changes to salmon habitat, physical passage (adults upstream, juveniles downstream), and predator-prey dynamics had large ecological effects, and each received a weighting of 0.28 or 28% of the total contribution to ecological impacts. By contrast, the impact of changes in water quality suggested only half the importance and received a weighting of 0.14 or 14% of the ecological impact on salmon viability. Similar considerations led to assignment of 0.12 to disease and 0.06 to competition. Whenever possible, our estimates were based on the available literature, but in many cases, the values are illustrations of our analysis so that others can provide better numbers if and when better information is available.

Finally, at level one, we identified nine immediate consequence of human activity, including dams, irrigation, roads, agriculture, acidity, logging, hatcheries, introduction of exotics, and aquaculture activities, that feed into level two. Dams, for example, were estimated to account for 0.4 or 40% of the changes to water quality, 0.6 of the changes to habitat other than water quality, 0.85 of the passage problems, and 0.15 of the changes in predator-prey dynamics. By contrast, dams did not appear to contribute to ecological impacts through disease or competition.

Similar methods were used to develop relative values for the other factors at level one. Each factor at level two could thus be attributed to the inputs from level 1. We were therefore able to account for 100% of the ecological impact on viability of wild salmon (0.6 of the total impact on wild salmon population, Figure 4-1) through the action of humans.

Humans also directly affect wild salmon, as illustrated in Figure 4-3. Human activity can directly increase the mortality of adults and that of juveniles, with our rough estimate at level two being 0.9 through adults and 0.1 juveniles. At level one, the impact on adults is through fisheries, including poaching (0.6); incidental activities, such as impacts by boating (0.15); research activities, such as tagging (0.15); and involvement in hatcheries, such as handling (0.1). Juvenile mortality is not affected by directed fisheries, but juveniles can suffer incidental captures (0.33), are frequently in research programs (0.33), and experience direct mortality from hatchery programs in which they are collected from the wild (0.34). For example, some statistics are available on fishing mortality due to catch-and-release fishing, landing statistics are available, rough estimates of removal are available, and so on.

Finally, humans can have strong impacts on the viability of wild salmon through their genetics (Figure 4-4). The genetic quality of wild salmon probably has four components (level 2; see also Appendix D), including good genes (best genes in the population), complementary genes, (ideal matches at a locus within an individual), co-adapted gene complexes (ideal matches among loci within an individual), and diverse genes (heterozygosity across loci). Since diverse genes are so often the target of conservation biology, the committee gave this category twice the weighting (0.4) of the others (0.2 each). In turn, at level one, human activity operates primarily through programs of aquaculture and hatcheries. Aquaculture, for example, has very strong effects on good genes as what is good within the fish farm (e.g., delayed maturity) is often different from what is good in nature. Since hatchery production does not use targeted selection to maximize survival in the artificial environment, its good genes impact (0.1) is less than that of aquaculture (0.9). Aquaculture and hatcheries may have similar impacts on complementary and co-adapted genes, as neither allow mate choice, but aquaculture will have a large impact on diverse genes, as there is usually less attention paid to maintaining heterozygosity and more effort to producing a specific strain of fish. Level 0 in Table 4-3 represents the "penetrance ratio" of hatchery-fish genes as compared with farm-fish genes. For the table, the assumption is that the ratio is 9:1 in favor of hatchery fish, i.e., the value of level 0 for hatcheries is 9 and for aquaculture, or farms, it is 1 (but see Ranking the Threats, below, for a more detailed discussion of this assumption). Level 1 is described above. Level 1b results from calculating level 0 times level 1 using the formula level 1b = (level 0 times level 1) divided by [(level 0 times level 1) aquaculture + (level 0 times level 1) hatcheries].

Throughout its analysis, the committee chose as its target a viable wild populations with a genetically effective population size (N_e) of 1,000 or greater and a probability of surviving for 100 years from now without reliance on a hatchery of 95% or greater. The outcome of the analyses are tabulated in Tables 4-1 for ecology, 4-2 for direct impacts, and 4-3 for genetics. As an example, consider the ecological impacts of dams (Table 4-1). The effects of dams originate at level 1, feeding into level 2 and level 3. Thus, to understand the full impact of dams on viable wild salmon through impacts on water quality, we take the value of 0.4 at level 1 (impact of dams on water quality) and multiply this by 0.14, which is the relative impact of water quality on ecology. We multiply again by 0.6, which is the relative impact of ecology on salmon viability. This suggests

TABLE 4-1 Proportional Impacts of Ecological Components on Viable Wild Atlantic Salmon Populations Based on a Risk Assessment (refer to Figures 4-1 and 4-2)

Component	Level 1	Level 2	Level 3	Product (1x2x3)	Impact (Sum)
Dams					
Water quality	0.4	0.14	0.6	0.0336	
Habitat	0.6	0.28	0.6	0.1008	
Passage	0.85	0.28	0.6	0.1428	
Predator/Prey	0.15	0.12	0.6	0.0108	0.29
Withdrawal					
Water Quality	0.05	0.14	0.6	0.0042	
Habitat	0.15	0.28	0.6	0.0252	0.03
Roads					
Water Quality	0.2	0.14	0.6	0.0168	
Passage	0.11	0.28	0.6	0.0185	0.04
Agriculture					
Water Quality	0.2	0.14	0.6	0.0168	
Habitat	0.15	0.28	0.6	0.0252	0.04
Acidity					
Water Quality	0.1	0.14	0.6	0.0084	0.01
Logging					
Water Quality	0.05	0.14	0.6	0.0042	
Habitat	0.1	0.28	0.6	0.0168	
Passage	0.02	0.28	0.6	0.0034	
Predator/Prey	0.05	0.12	0.6	0.0036	0.03
Hatcheries					
Passage	0.02	0.28	0.6	0.0034	
Disease	0.3	0.12	0.6	0.0216	
Competition	0.3	0.06	0.6	0.0108	0.04
Exotics					
Predator/Prey	0.8	0.12	0.6	0.0576	
Disease	0.05	0.12	0.6	0.0036	
Competition	0.6	0.06	0.6	0.0216	0.08
Aquaculture					
Disease	0.65	0.12	0.6	0.0468	
Competition	0.1	0.06	0.6	0.0036	0.05
Totals				0.6000	0.60

TABLE 4-2 Proportional Impacts of Direct Sources of Mortality on Viable Wild Atlantic Salmon Populations Based on a Risk Assessment (refer to Figures 4-1 and 4-3)

Component	Level 1	Level 2	Level 3	Product (1x2x3)	Impact (Sum)
Fisheries					
Adult Mortality	0.6	0.9	0.2	0.1080	0.11
Incidental					
Adult Mortality	0.15	0.9	0.2	0.0270	
Juvenile Mortality	0.33	0.1	0.2	0.0066	0.03
Research					
Adult Mortality	0.15	0.9	0.2	0.0270	
Juvenile Mortality	0.33	0.1	0.2	0.0066	0.03
Hatcheries					
Adult Mortality	0.1	0.9	0.2	0.0180	
Juvenile Mortality	0.34	0.1	0.2	0.0068	0.02
Totals				0.2000	0.20

that dams, through water quality, reduce the viability of salmon by a relative magnitude of 0.03 or 3%. Similar calculations for the impact of dams through habitat (0.1), passage (0.14), and predator-prey dynamics (0.01) result in a cumulative impact of 0.29. This suggests that 29% of the total impact on viable wild salmon populations by humans is through dams, and we can now map the role that dams have on disrupting the ecology of salmon. The tables summarize similar kinds of analyses for all the possible impact factors that the committee identified.

The analytic procedure described above—multiplying fractions—leads to the appearance of greater precision than is intended. For example, a more appropriate characterization of the importance of dams as a threat is that they are the largest single factor but are responsible for less than half of human effects on salmon.

Finally, the committee performed a very rough sensitivity analysis. How would the results change if a category changed relative size and if factors within a category changed relative sizes? These exercises led the committee to conclude with considerable confidence that the single largest human-caused factor affecting salmon mortality is obstruction to passage. At the same time, the committee concluded that obstruction to passage probably accounts for less than half of all human-caused mortality.

We offer these diagrams and calculations with some trepidation. However, we judge that they provide the best method yet developed for prioritizing actions for rehabilitation in terms of their likely effectiveness (but not their likely costs), and they provide the best method yet developed for prioritizing research. We use them as input to our decision analyses—which include cost considerations—for dam removal and the management of aquaculture. Those decision analyses also are provided for heuristic purposes. No committee constituted as this one can properly identify the societal costs and benefits of various management options. That can be done only by the people who must pay for the options and live with their consequences through the process. This means that all interested stakeholder groups should be represented in the process. A recent report of the National Research Council (NRC 1996b) discusses the issues involved in detail and describes effective methods for addressing them. However, we offer them as an example of how to think through the options.

Model Limitations

The obvious limitations of the model are due to incomplete information. However, if complete information were available, the risk-assessment model would not be necessary. In particular, we point out the following cautions.

TABLE 4-3 Proportional Impacts of Genetics on Wild Atlantic Salmon Populations Based on Risk Assessment (refer to Figures 4-1 and 4-4)

Component	Level 0	Level 1	Level 1b	Level 2	Level 3	Product (1bx2x3)	Impact (Sum)
Aquaculture							
Good Genes	1	0.9	0.5	0.2	0.2	0.02	
Complementary Genes	1	0.5	0.1	0.2	0.2	0.004	
Co-adapted Genes	1	0.5	0.1	0.2	0.2	0.004	
Diverse Genes	1	0.8	0.31	0.4	0.2	0.0248	0.0528
Hatcheries							
Good Genes	9	0.1	0.5	0.2	0.2	0.02	
Complementary Genes	9	0.5	0.9	0.2	0.2	0.036	
Co-adapted Genes	9	0.5	0.9	0.2	0.2	0.036	
Diverse Genes	9	0.2	0.69	0.4	0.2	0.0552	0.1472
Totals						0.2000	0.20

Variation over Time

The model does not take variations over time into account. These include time lags, nonlinear responses, and cumulative effects. For example, there might be a factor that seriously depresses the population at some life stage every few years but does not operate in other years. It could have the same average value as another factor that removes a small portion of the population every year. The first factor will affect the average abundance more than the second, which will affect the year-to-year variation more than the first. The committee judged this to be a second-order problem, i.e., the severity of each factor appears to have a larger impact than its distribution in time. To some degree, the distribution in time of the factor's operation was considered by the committee.

Interactions among Components

The various components in the model interact in nature. Hatcheries affect genetics, which affects survival, which affects the availability of fish to predators, which can affect the number of predators, and so on. To include such interactions would have made the model intractably complex, the more so because there is even less information on most potential interactions than on the primary effects of the factors. The committee had no choice but to ignore many of the interactions, and it judges that it will be a long time before enough information is available for any future analysis to take all of them into account. However, some interactions are obvious. For example, a passage barrier means factors acting above the barrier will have little or no effect on salmon. Similarly, removal of the barrier will affect the action of those factors and will allow human interventions, such as habitat alteration, to be effective.

Density-Dependence of Factors

Perhaps the most difficult problem is knowing which factors are density dependent and to what degree. For example, there appears to be considerable scientific agreement—if not complete consensus—that poor marine survival during their first winter at sea has been a major factor in the recent declines of salmon populations (Maine Atlantic Salmon Task Force 1997). If that factor is operating in a density-independent way—i.e., if it kills a fixed proportion of young fish no matter how many there are—then doubling the number of smolts going to sea will double the number of salmon that survive their first winter at sea. If, on the other hand, the factor is density dependent, it could kill a greater proportion of salmon if there are more of them, or it could increase their growth and survival if there are fewer of them. For example, if the factor is food limitation, then adding to the number of smolts will merely increase the number that starve. But in that case, one would expect better survival and growth if fewer young salmon went to sea. It is critical to know the degree to which various factors are density dependent to develop sensible rehabilitation strategies. This problem is probably most serious in the ocean, where so little detail is known about factors that affect the survival of salmon and how they have changed over time.

Ranking the Threats

Some of the rankings that this model produced are not surprising; others are less intuitive. The largest single factor was ecological, as expected. Dams through their effects on water quality, habitat, passage, and increasing predation contribute to 29% of overall human impacts. The other large ecological factors were exotics (8%) and aquaculture (5%). The committee recognizes that the assignment of apparently precise percentage allocations of impact creates an inappropriate impression of empirical knowledge that goes beyond the available data. The numbers are based on numerical representations of qualitative distinctions based on experience, some data, and expert opinion. When they are multiplied as the model requires, additional apparent precision is generated.

The largest direct impact was fishing (14%); this includes incidental take as well as targeted fishing. Research was estimated to contribute 3% of all mortality. While not large, this is perhaps the most easily controlled of all mortality sources and is large enough to be of concern.

Genetic impacts are due to hatcheries and farming. Table 4-3, which gives a genetic contribution of 15% for hatcheries and 5% for farms, was derived with an assumed penetrance ratio of 9:1 for hatchery and farm fish—nine times as many genes enter the wild population from hatchery fish as from farm fish. The committee made this assumption because hatcheries release many more fish into the environment (deliberately so) than farms do, and because hatchery fish have higher reproductive success than farm fish. However, we do not know how many fish escape from farms, the exact reproductive success of hatchery and farm fish in the rivers, how well their offspring survive, and so on. The committee has therefore recalculated Table 4-3 using penetrance ratios for hatchery to farm fish of 1:1, which yields values of 6% for hatcheries and 14% for farm fish; 1:3, which gives 10% and 10%; and 1:27, which gives 18% and 2%. The different penetrance ratios change the results so that either hatcheries or farms have greater genetic impact. The extreme range of ratios might occur across time, i.e., from one year to another; and across space, i.e., from one river to another. Thus, while it appears from Table 4-3 that hatcheries have a greater genetic effect on wild salmon populations than farms do, the real message is that both are important, and that additional time- and space-specific data would be needed to resolve uncertainties.

Thus, the two largest contributors to human impacts on salmon—together accounting for nearly half (46%) of the impacts—are dams and hatcheries. One of the advantages of this model is that it is easy to change the local estimates of impact and see what the overall outcome is, and this result is quite robust. It is hard to escape the conclusion that the two biggest impacts on wild salmon in Maine are dams and hatcheries.

Despite those rankings, recent information on the effects of acidity in streams indicates that it might also be very important (Chapter 3). That information became available after the rankings were developed, and so it is not incorporated into the diagrams and analyses. It is, however, taken into account in the committee's recommendations. Given the description of the committee's methods, it should be reasonably easy to take advantage of that and other new information to recalculate the rankings.

Using Decision Analysis

Decisions needed to develop and implement restoration strategies for Atlantic salmon will not be straightforward for several reasons discussed below. This ambiguity stems from the need to account for uncertain information and to integrate a variety of complex goals, perceptions, and values, not all of which are scientific. The discipline of decision analysis provides a framework, process, and tools to sort through and analyze these complicating factors to improve the quality of resulting decisions (Clemen 1991).

First, the number of scientific, political, technical, social and economic issues affecting recovery decisions and the interactions among them are complex. It can be difficult to understand the relationships among all these factors and to establish clear, measurable objectives that integrate them into the decision making process. Decision analysis tools originally developed for economic applications that are now being used in the natural resource arena help policy makers to organize these factors and to evaluate their impacts on different alternative strategies.

Second, many of the factors potentially influencing decisions are fraught with uncertainty that is expressed at various levels. What is the likelihood that a particular restoration option is technically feasible? If it depends on an untested method, confidence in the expected outcome is likely to be lower than that stemming from a more reliable technique. Regulatory acceptance and community support for controversial approaches, such as dam removal or restricting water withdrawals, are seldom guaranteed. Stochastic factors, such as short-term weather events (floods, drought, etc.) and longer-term climatic and oceanic circulation trends also complicate predictive ability, especially if their effects on reproduction and juvenile survival are not immediately obvious. Changing land-use patterns within watersheds may be predictable directionally, but their pace, scale, and distribution are often uncertain. Whether the uncertainty is caused by the influence of an irreducible probability factor, such as the chance that a coin will turn up heads or tails for any given toss, or whether it is based on incomplete knowledge of the factors affecting outcomes, decision analysis provides a method to weigh uncertainty elements against preferences for different outcomes (Keeney and Raiffa 1976).

Multiple, potentially conflicting objectives also confound decision makers. For many endangered species, conservation efforts often involve trade-offs between recovery objectives for the protected species and economic or other political interests (Maguire 1986). It can be difficult to sort out priorities among these interests, because the relative values of the different alternatives are often not measured in comparable terms.

Finally, the various possible perceptions that different stakeholders have about the values, priorities, or facts that may be involved are embedded in the diverse problems that affect recovery decisions. For example, which factor—juvenile survival, adult return, or amount and quality of spawning and rearing habitat—might be the most critical issue affecting Atlantic salmon recovery in Maine rivers? Or is there one single factor that warrants priority attention? The risk analysis described above suggests that amount and quality of habitat for spawning and juvenile rearing could be overriding. Even so, other issues are difficult to ignore. For example, do fish farming pens holding nonlocal fish located near the mouths of DPS rivers constitute an unacceptable threat to the recovery of native stocks? The differences in perspective must be taken into account so that the decision is informed by the views of all parties having legitimate interests in the outcome.

The basic rationale for using decision analysis when confronted with these issues is that better decisions generally lead to more favorable outcomes. However, not all good decisions necessarily turn out for the best. Nor do poor decisions always lead to less fortunate results. Many people would opt for having a run of good luck based on intuition rather than living with the results of consistently playing the odds as they enter a Las Vegas casino. However, the advantage of decision analysis is that it provides a systematic and structured approach that explicitly recognizes and accounts for the influence of these complicating factors.

The decision analyses presented here are for illustrative purposes. The results described might be suggestive but cannot be used for prioritizing actions before all the stakeholders have been properly involved in the process.

Analyzing Decisions

The decision analysis process takes place as a series of steps. In general, different authors (Clemen 1991, Keeney and Raiffa 1976, Peterman and Peters 1998) include analogous phases, although they label them somewhat differently. One example is shown in Figure 4-5 (Clemen 1991). The process is iterative because it encourages cycling back to prior steps to redefine objectives or alternatives or to reconstruct models of uncertainty or preference. The process may also result in changes to the decision maker's underlying values and perceptions, which can lead to new insights about the fundamental issues being analyzed.

Examples showing how decision analysis can be used to assist policy makers wrestle with the complexity of natural resource issues are growing (MacGregor et al. 2002, Maguire 1986, NRC 1995, Peterman and Peters 1998). To promote greater understanding of how decision analysis can aid decision makers evaluate possible recovery strategies for Atlantic salmon, we provide two hypothetical examples. The first involves managing the risk posed to wild salmon by aquaculture. The second involves potential measures to increase habitat availability and carrying capacity for adult spawning and juvenile rearing. The examples are simplified to illustrate basic points. We recognize that actual field situations are usually more complex than the depicted case. The first stage of decision analysis is structural. The idea is to frame the problem effectively and to set forth clearly defined management objectives. Explicit objectives are important because they provide benchmarks against which to evaluate the expected outcomes of different management options.

Risks of Farm Fish

An application of decision analysis could be for understanding the importance of various factors that might influence the success of various options for managing the impacts of Atlantic salmon raised in aquaculture pens on wild fish. Rather than proceed through a complete decision-tree analysis, we provide an example that explores how a decision matrix focuses attention on critical factors that affect strategic choices.

The primary concern about raising farm fish in the estuaries of DPS rivers is escapees, because of fears that they will dilute adaptive fitness of native populations by mating with wild fish, and that disease transmission will be increased through exposure of wild fish to escapees and proximity of stocking pens to migration routes. Competitive interactions that displace native spawners from preferred redd sites could also be a problem but one that is more likely to occur when spawning sites are limited. We assume that objectives for alternative management strategies

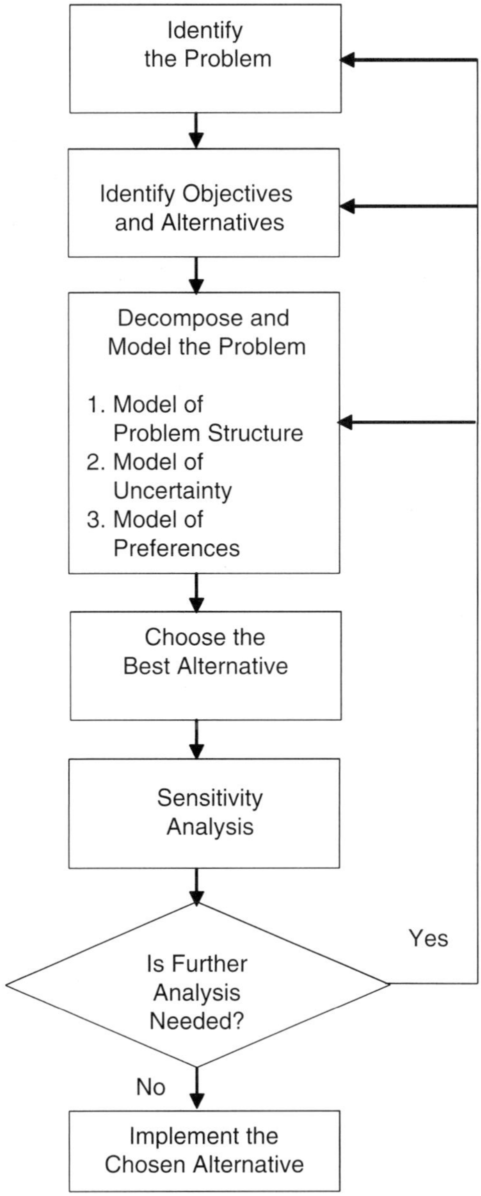

FIGURE 4-5 A decision analysis process flow chart. Source: Clemen 1991. Reprinted with permission; copyright 1991, PWS Kent.

are directed at reducing or eliminating these possible effects. To evaluate the relative merits of different management approaches, it is useful to structure strategic objectives in quantifiable terms.

Objective 1: Reduce ecological and genetic interactions. To address this objective, we would ideally look at the number of matings between farm and wild fish. If genetic tagging needed to identify the progeny of such crosses is not feasible technically or is too costly, then a fallback metric could be the number of escaped farm fish that reach the spawning grounds. For purposes of this example, the committee set an objective that the number of escapees should not exceed 1% of the number of adult fish returning to the spawning grounds. This low threshold takes into account the small population size of wild returnees and the proportion that would be at risk of exposure to reduced fitness matings or competitive interactions.

Objective 2: Reduce disease transmission. Managers setting an objective for disease transmission must recognize the difficulty in using a low rate of infection as a metric. Pathogens are ubiquitous in the environment, and some fish that may be infected will not actually express disease symptoms. Visible evidence of transmission is needed without killing the fish or causing undue stress, so sea lice would be better than pathogenic bacteria as an indicator of transmission. If the major concern is at the zone of closest contact between adult farmed and wild fish, then an ideal metric would compare infection prior to passing pens with that observed upstream of the aquaculture sites. Several possible threshold standards can be envisioned that relate to the following parameters: (1) degree of sea lice infestations on adults before and after passing rearing pens, (2) percentage of outgoing smolts that are infested with sea lice, (3) percentage of smolts infested with some threshold number of sea lice, (4) lice load equal to or greater than an acute level for confined fish, and (5) reduction of the infection level in the pens to some defined threshold. Many of these standards present practical measurement problems, so for purposes of the example and to reflect concern about the harmful effects of disease transmission within a very small endangered population breeding adults, we established a very stringent hypothetical goal of 1/10,000 adults infected with disease or parasites of aquaculture origin. (One could use outmigrating smolts instead of returning adults; doing so would not change the following discussion.) This very low level of infection is impossible to measure because of small population sizes of returning adult salmon, so as a practical matter, it is a surrogate standard for eliminating parasite and disease transfer between farmed and wild fish.

Several alternative approaches to farming discussed in Chapter 5 can be evaluated with respect to the likelihood of success in meeting objectives 1 and 2.

1. *Zoning.* Move aquaculture sites away from the migratory path of adult salmon returning to DPS and other rivers and smolts leaving them.
2. *Bioconfinement.* Sterilize farm fish to minimize mating success between escapees and wild fish.
3. *Tagging/weir.* Mark all farm fish so escapees can be identified and removed at weirs where all upstream migrants are captured.
4. *Land farming.* Move salmon farms onshore.
5. *Solid confinement.* Enclose farm fish pens to prevent escapes.
6. *Status quo.* Current default option.
7. *Remove aquaculture pens.* Discontinue farming of Atlantic salmon in Maine.

Table 4-4 provides the committee's estimates of the likelihood that each strategy would be successful in meeting objectives 1 and 2 and the effects of factors that could influence the ability of each strategy to achieve the desired objective. The estimates of the probability of success were made

TABLE 4-4 Strategic Options and Committee's Estimates of Success Factors for Meeting Aquaculture Escapee Management Objectives[a]

	Strategies						
Success Factors	1	2	3	4	5	6	7
Permitting	X	(–)	(–)	X	(–)	X	0
Political acceptance	X	0	0	X	X	X	XX
Socioeconomic effects	X	0	0	(+)	X	0	XX
Technical feasibility[b]	1.0	1.0	1.0	1.0	0.6	1.0	1.0
Survival impairment	0	0	X	0	0	X	0
Capital costs	X	(–)	X	XX	XX	(–)	0
Management costs	0/X[c]	(–)	X	X	XX	(–)	0
Legal liability[d]	(–)	X	X	(+)	(+)	XX	(–)
Probability of success							
Objective 1 (ecological/genetic)	0.9	0.9	0.99	0.95	0.9	0	1.0
Objective 2 (disease)	0.99	0.5	0.5	0.99	0.9	0.5	1.0

[a]Strategy 1: zoning; strategy 2: bioconfinement; strategy 3: tag/weir; strategy 4: land farming; strategy 5: solid confinement; strategy 6: status quo; strategy 7: remove aquaculture pens.
0 = no real problem or issue.
X = significant problem or issue.
(–) = minor problem or issue.
(+) = improvement from status quo with respective to objectives.
[b]Probability of success.
[c]Depends on relocation site.
[d]Potential for ESA, CWA, or other violation.

as functions of factors such as permitting complexity and success, political acceptability, availability of suitable alternative sites, effects on jobs and local commerce, technical feasibility, impacts on survival of wild fish, capital and management costs including those for monitoring and effectiveness evaluation, and legal liability.

Some conclusions can be drawn from Table 4-4. The first is that bioconfinement, tag/weir, and status quo are not very likely to achieve objective 2. Even if the standard for objective 2 were dropped or significantly lowered, each option carries potential legal liability regarding possible violations of Section 9 of the ESA and perhaps the Clean Water Act. They are therefore unlikely to be sufficient as single approaches for meeting the challenge posed by aquaculture releases.

Among the strategies likely to be successful in meeting both objectives, high capital and management costs and low technical feasibility (0.6) work against solid confinement. Land farming entails higher economic costs than zonal relocation, but the greater availability of suitable sites, lower potential legal liability, and possible socioeconomic benefits argue in its favor. The need to find suitable estuarine or offshore sites for relocating pens where escapees would not threaten wild stocks in DPS and other rivers is a major consideration for relocation in an aquatic setting. Siting factors include an ice-free environment, protection from storms, adequate depth, flushing, ready access, and community acceptance. The impact of displacing an industry and employees from an area of existing operations will also influence its political acceptability. Eliminating aquaculture of Atlantic salmon in Maine altogether would clearly meet both objectives (except of course for any effects of salmon farms in Canada), but would also eliminate employment and economic benefit.

Enhancing Habitat Availability

A more complex example of decision analysis concerns improving access to habitat blocked by dams. Assume that a major goal for recovery efforts is to increase available spawning and juvenile rearing habitat on two rivers that empty into the Gulf of Maine and are separated by about 100 miles. On River A, a large dam located near the mouth blocks upstream passage of returning adults except in occasional years of unusually high stream flow. River A historically supported a substantial run of Atlantic salmon. Although a few fish have ventured upstream to the base of the dam in recent years, spawning is sporadic if it occurs at all because of the lack of suitable habitat below the dam. Salmon found in River A are not protected under the Endangered Species Act (ESA). On River B, where the Atlantic salmon stock is listed under the ESA as a DPS, two moderate sized dams impede access. The dam located further downstream has a marginally effective fish ladder but spawning habitat is poor between the dams because of the quiet water and lack of gravel for redds. The upstream dam is a complete barrier to further migration. On River B a 2-mile reach below the downstream dam is poor habitat for spawning because it is used for gravel mining, but it could be restored, especially if mining were to cease. All three dams provide water for irrigated agriculture and generate electric power during periods of adequate flow. The watershed surrounding River A is mostly forest subject to long-term harvest rotation. Mixed land use consisting of small forest plots, pasture crops, several widely separated small towns, low density rural housing with septic tanks, and cranberry farms occupy the watershed landscape along River B. Riparian corridors that protect the river from adverse impacts of human activity in the watershed only occur along 50% of the length of River B and its tributaries. Consequently the quality of habitat that could be made available above the dam on River A is apt to be greater than that on River B if the dams were breached. Restoration of riparian buffers and control of non-point-source pollution would be needed to maximize the habitat potential on River B.

A well-defined objective for these rivers might be to increase available spawning and rearing habitat by a specified number of habitat units over a certain period of time. A unit of habitat is 100 square meters (see Table 1-1 for habitat units in Maine Rivers). The general term "salmon habitat" refers to riffles and runs. A second success metric could stipulate that some level of spawning by Atlantic salmon should be attained on the newly available habitat. The objective could further distinguish between the relative values of new habitat units on DPS rivers vs. non-DPS rivers. Those expected outcomes that meet the threshold can be further sorted by other measures such as costs, while those that do not reach the standard can be ignored (Peterman and Peters 1998). It is also possible to create an objective that calls for maximizing available habitat independently of time or costs, but this is less realistic in terms of public agency budgetary policy.

The process starts with developing an influence diagram (Figure 4-6) to show important variables and relationships affecting expected outcomes, in this case new habitat units. Other tools, such as a decision hierarchy, ensure that the focus of the decision will be on strategic elements and not directed toward aspects of the problem that are givens or can be resolved later as tactical details (Chevron Strategic Decisions Group, unpublished material, 1991). The main advantage of the influence diagram is to understand the basic structure of a problem (Clemen 1991) and to be able to communicate the essential elements to stakeholders.

The next step is to identify workable alternative strategies that represent choices for action by the decision maker. For this exercise, dam removal should be considered as a possible option in order to attain the

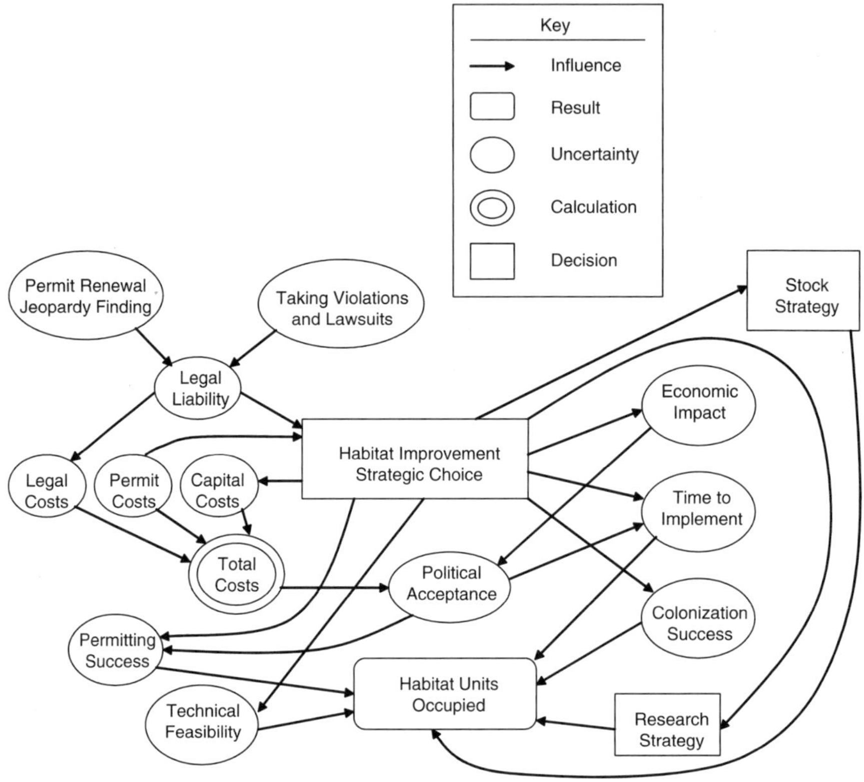

FIGURE 4-6 Influence diagram for factors affecting strategic choices.

biological objective. Although there can be many obstacles for implementing such a strategy (Heinz Center 2002), the recent success on the Kennebec suggests that it could be a viable approach. A strategy diagram shows the range of choices for the series of decisions needed to implement each strategic theme (Figure 4-7). For example, supporting decisions for the habitat augmentation strategies might include whether supplemental stocking is needed, and if so, what life stage should be used. Questions of the preferred sequence for dam removal on River B and whether further research would improve the chances of success might also need to be explored further. The alternative of removing all three dams was not included because we judged the projected costs exceed available funding within the specified time period.

A decision table that represents the various uncertainties shown in the influence diagram can then be used to model the different strategies

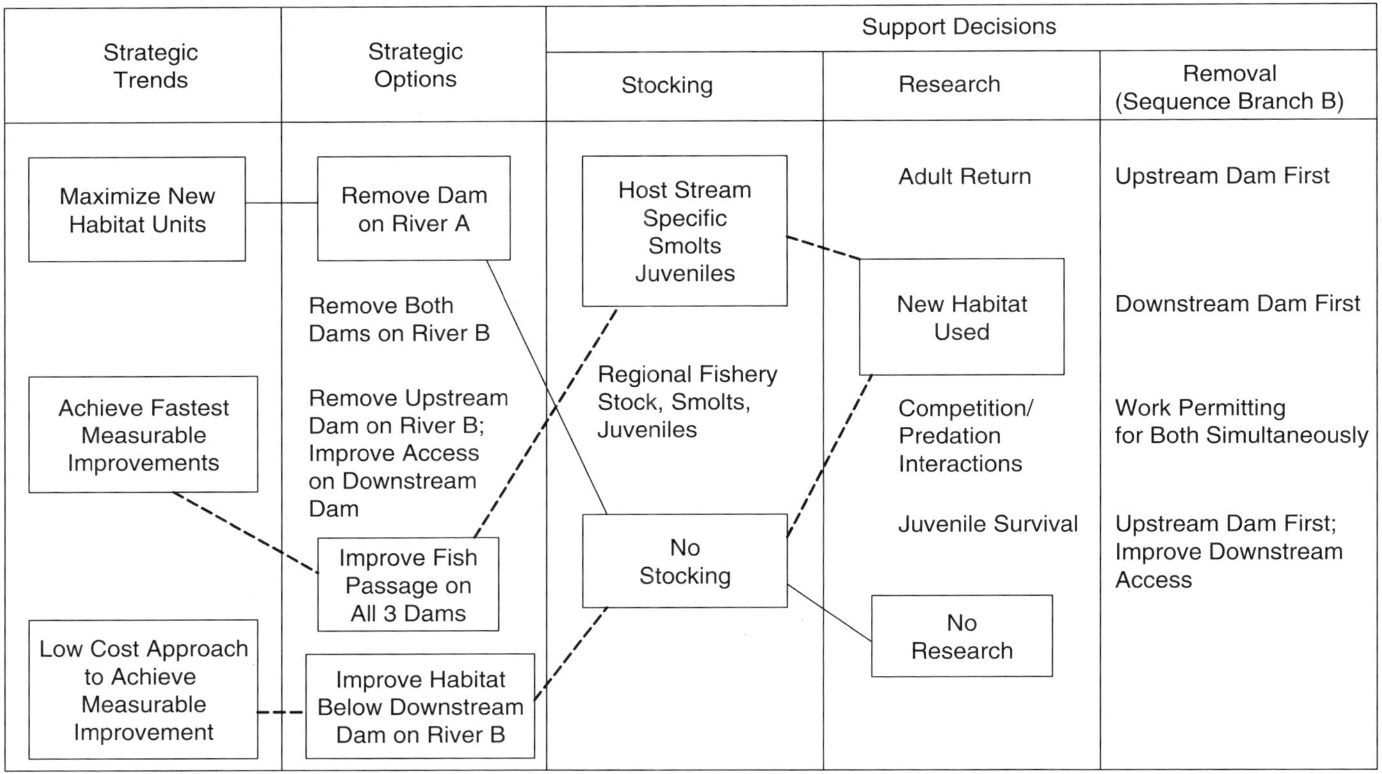

FIGURE 4-7 Strategy table for habitat improvement strategies for Maine salmon.

TABLE 4-5 Decision Table for Habitat Improvement Strategies

Alternative	Technical Feasibility	Regulatory Acceptance	Legal Liability
A. Remove dam on river A	High	Low to Moderate	Low (non-DPS river)
B. Remove both dams on river B	High	Low to Moderate	Moderate (potential for take during demolition)
C. Remove upstream dam on river A; improve access on B	High to Moderate	Moderate	Moderate
D. Improve passage on all three dams	Low to Moderate	High	Low
E. Improve habitat below downstream dam on river B	Low	High	High (potential for jeopardy during permit renewal if access not improved and possible lawsuits over incidental take)

(Table 4-5). Although it is possible to calculate value measures for each strategy if the table is set up deterministically, qualitative estimates provide an informative summary of how key factors are likely to affect outcomes. In this case, the high number of habitat units gained by removing the dam on River A might be tempered by the effect of high costs and high economic impact on the political acceptability of the strategy and the longer time to implement it.

The next step is to understand the critical uncertainties that need to be modeled in the subsequent phases. These uncertain states of nature (Peterman and Peters 1998) are considered by setting a range of values for each uncertainty parameter, whether costs, habitat units, or the likelihood of success (or failure) during different stages of strategy implementation. The influence diagram and the strategy table provide key input for this step of the analysis. In the example, we can identify three main uncertainties that will directly affect success of the various strategies. These are technical feasibility of the alternatives, regulatory acceptance, and the prospect that the new habitat will become occupied for spawning and juvenile rearing by Atlantic salmon. Other factors such as the potential for adverse legal action and political support could also affect outcomes, but

Implementation Cost ($ in millions)	Economic Impact	Time to Implement (Years)	Political Acceptance	Habitat Units Gained (Base Value)
12	High	5–10	Low-Moderate (high cost and high economic impact, but precedent on Kennebec River	High
8	Moderate-High	3–5	Moderate	Moderate to high
4	Moderate	2–4	Moderate	Moderate
4	Low	1–4	High	Low
2	Low	2–3	Moderate (could be pressure from possible legal liability)	Low

their influence may be unpredictable or less direct through their impact on costs, timing or permitting success.

For each uncertainty variable, it is necessary to assign a probabilistic estimate to the different states we choose to analyze. In most cases, sufficient data are not available to develop precise probability estimates with a high degree of confidence. For natural resource problems, the decision analysis process often uses subjective estimates, usually developed by a cross-section of stakeholders. A variety of sources can be used to inform these subjective evaluations—performance history, experimental results, trend analysis, extrapolation, correlations to other variables, scenario modeling, and so forth. As a practical matter, they are often based on personal experience and professional judgment of the team conducting the decision analysis.

Even though hard data are often lacking, decision analysis allows decision makers to consider a range of values to gain a better overall picture of the effect of different uncertainty variables. One of the strengths of decision analysis is that making quantified judgments about uncertainties promotes clear communication (Clemen 1991) and helps to resolve disagreements that can result from differences in belief systems, experience, and biases (Stewart 2000). For example, one of the problems inherent in setting public policy objectives based on imperfect information is to establish whether a decision should favor possible false-positive outcomes or should lean to false negatives. Because these two results are mutually exclusive (Stewart 2000), favoring one or the other usually means having to make trade-offs, e.g., conservation value vs. short-term economic impact, that shift depending on which type of error is more acceptable. Decision analysis can also be used to quantify the value of a trade-off that attends the question of whether to spend more time and money gathering additional information in order to reduce uncertainty about outcomes (Clemen 1991).

After assigning probabilities, a decision tree (Figure 4-8) displays the strategic options, the uncertainty variables, their probability of occurrence, and the outcomes in terms of specified value measures (Peterman and Peters 1998). The estimated probabilities in the example are assigned to illustrate how the process works and are intended to reflect how the hypothetical facts might drive probability estimates. For example, the chance of successful colonization of new habitat on River A would be lower than that on River B because salmon only occasionally occupy the reach below the dam to be removed. However, they consistently appear below the lower dam on River B even though they do not successfully reproduce in the gravel quarry. The decision tree provides a convenient way of ranking the alternatives according to their expected or "net" value. The technique is to weigh the base value of habitat units for each option by the probability factor at each branch of the tree corresponding to the three uncertainty variables. Some management actions lead to more than one outcome that are then summed to give the total net value (NV) for that strategy. In the example, the strategy to improve fish passage on all three dams gives a net value of

$$\begin{aligned} NV &= (0.75 \times 1.0 \times 0.7 \times 1{,}000) + (0.2 \times 1.0 \times 0.5 \times 1{,}000) \\ &= 0.525 \times 1000 + 0.1 \times 1{,}000 \\ &= 525 + 100 \\ &= 625 \text{ HUs.} \end{aligned}$$

According to the decision tree, Strategy 1, removal of the large dam on River A, would create the greatest net value of new habitat units (960 HUs). This is followed by Strategy 2, which calls for removing both dams on River B (800 HUs). This ranking would be reversed, however, if the new habitat units on River A were discounted by 25% because it is not a DPS river, which illustrates the role of perception and relative quality of different options in establishing and choosing among preferences. River B would further benefit if instream habitat improvements were undertaken to maximize the value gained by breaching the dams. This option could

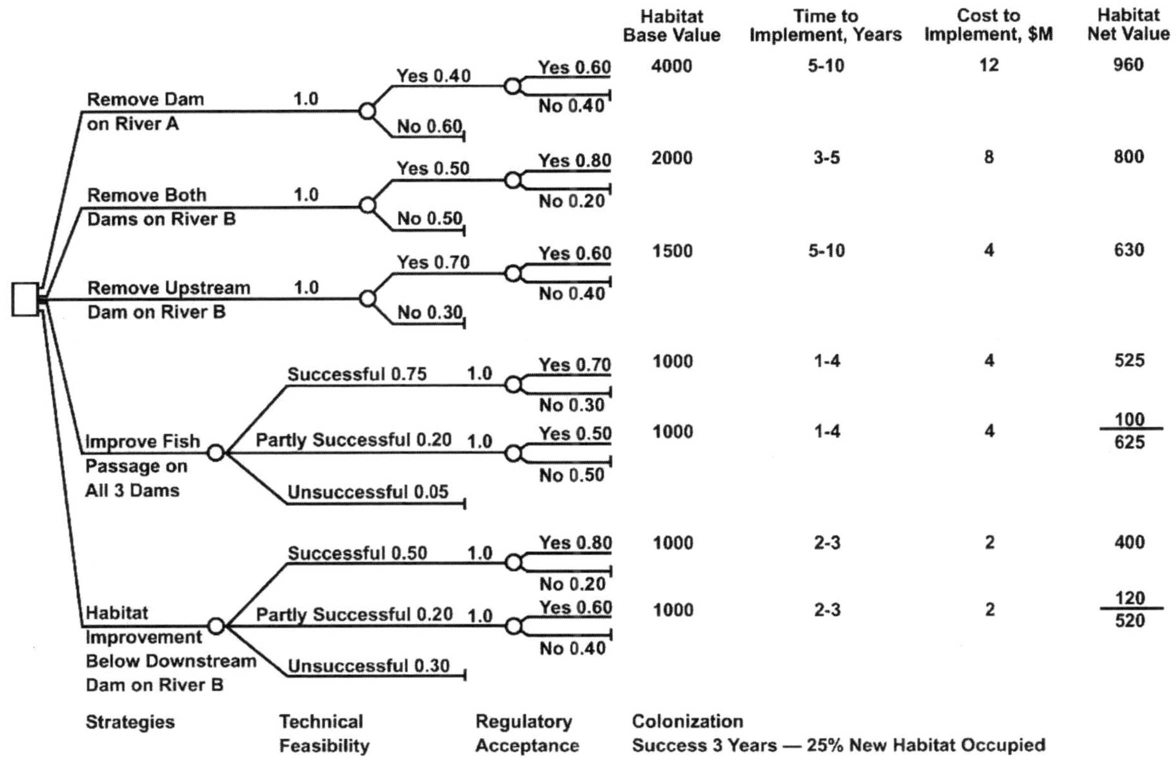

FIGURE 4-8 Decision tree for habitat improvement strategies.

be the subject of a separate decision analysis designed to evaluate its merits in terms of extra costs, time to implement, and likelihood of gaining the landowner support needed for success.

If the objective is to create the most new HUs at the lowest cost, Strategy E, buying out the gravel mining rights and improving habitat below the downstream dam on River B, is most cost effective. Even though the possibility of failure for Strategy E is 30%, it would still represent a preferred approach if cost is a key success driver. If the exact costs to carry out the strategies are not known, they can be incorporated as an uncertainty factor in the analysis. A common approach is to assign probabilities to a high (90%), medium (50%), and low (10%) range of costs for each alternative.

Reviewing the outcomes from another angle, we conclude that Strategies D and E also warrant strong consideration if the objective calls for maximizing occupation of new habitat at the earliest time. Another way to scale the value outcome is to combine the new HUs with the cumulative time that they are available, giving a metric called habitat service years. In this case, a decision maker would include time to implement and time for the new habitat to become occupied as uncertainty variables in the decision tree. The strategy yielding the most habitat service years within a specified period following the decision would rank highest.

The decision tree in Figure 4-8 is based on primary factors influencing all the strategies being ranked. These are determined from an influence diagram like Figure 4-6, and are the variables that must be evaluated to distinguish among the various options. It is possible that factors such as conducting further research and implementing (or disbanding) supplemental stocking programs could improve the success of individual options. But the committee did not include them as part of the primary decision tree, because each would have its own set of variables to evaluate (e.g., life stage, number of individuals, seasonality, and stocking location for supplemental stocking). Thus, they are considered secondary factors in Figure 4-7 to explore maximizing habitat utilization (the primary decision) in Figure 4-8 and Table 4-5. If decisions about whether to implement a stocking plan or to conduct research in conjunction with the strategy to increase habitat availability might differentially affect strategic outcomes, they can be included in the decision tree as decision nodes with yes/no branches.

After the alternative strategies are ranked according to a decision tree, a variety of sensitivity analysis techniques can be used to answer the question, "what matters in this decision?" (Clemen 1991). The primary purpose of this more introspective look is to ensure that the analysis is focusing on the right question to satisfy the original objective. The idea is to avoid making Type III errors, as opposed to the familiar Type I and

Type II errors in statistics (Clemen 1991). Type III errors give rise to the wrong question being asked, given the available information. In the above example, decision makers could be short-sighted if they decided to restrict water withdrawals in an effort to improve water quality above dams that limit access unless plans to improve habitat occupation were also in the works.

These exercises are only illustrative. People with in-depth knowledge of and experience with physical, biological, social, and political environments need to undertake these risk-management decision processes. In addition, people who must live with the consequences of these management decisions should be involved, otherwise the decisions will be difficult to implement.

Conclusions and Recommendations

This discussion has attempted to show how decision analysis could be a helpful tool in sorting through the myriad choices of potential recovery strategies for Atlantic salmon. The approach could be used to understand the value of gaining additional information through baseline assessment, research, and monitoring. The potential value of missing information would become apparent in considering specific decision choices. Another application would be to clarify the role of different stocking strategies. The value of expanding fisheries on non-DPS rivers could be evaluated against the chance of attaining recovery goals on listed rivers. The issue of number, location, and controls on aquaculture facilities also needs to be examined. Habitat restoration measures designed to mitigate the adverse effects of erosion and sedimentation, reduced in-stream flow and elevated temperatures, and pollutant loading should be investigated for their potential contribution to recovery.

The committee recommends that recovery planning efforts for Atlantic salmon in Maine rivers employ structured, systematic, strategically focused decision making processes for developing conservation and recovery objectives and analyzing the optional approaches for achieving them. All stakeholders need to be involved in this process to ensure its validity and acceptability. The committee further recommends that recovery planners engage the services of an expert in the field of strategic decision analysis, especially someone experienced in its application for natural resource problems, to advise them in their endeavors. These activities will need to be repeated when changes in environmental conditions or human interventions change conditions relevant to the analysis. The committee also recommends research on the socioeconomic effects to changes in aquaculture (discussed in Chapter 5).

5

Addressing the Threats to Atlantic Salmon in Maine

A STRATEGY FOR CONSERVATION AND RESTORATION

The complex and dynamic nature of terrestrial, aquatic, and marine ecosystems makes conservation and restoration—especially of threatened and endangered species—a daunting task. Because water connects all three ecosystem types to each other and to Atlantic salmon, to other organisms, and to people, watersheds become the logical unit for an ecosystem approach to conservation and restoration.

The 1997 Conservation Plan (Maine Atlantic Salmon Task Force 1997) provides the foundation for wide range of current efforts in Maine. It describes threats and associated mitigation or management options. Like any plan, it can be improved with the benefit of 5 years of intensive research and operational experience in Maine as well as information from other parts of the world. Principally, it would be improved by more clearly prioritizing, sequencing, and coordinating plans and actions in an adaptive management framework. This means every activity is a field experiment that generates data, information, and experience while sustained progress is made toward conservation and restoration goals. A well-documented cycle of planning, implementation, performance monitoring, and subsequent adjustment or refinement is used to rapidly converge on optimal solutions and methods. Pairing an untreated area, stream reach, or watershed as a reference condition (to account for the complex influences of natural variation) with a similar site where a management action is applied, yields timely information about overall effectiveness (both ecological and economic). Replicated across several sites, the scientific method supplants well-intentioned trial and error as an efficient and systematic way of improving conservation and restoration efforts.

The following sections deal in more detail with specific threats.

DAMS

As described in Chapter 3, dams block passage and later riverine environments both below and above them. Mitigating the threat they pose is usually most completely achieved by removing them, but enhancing passage alone can be at least somewhat effective if they affect only short stretches of river. Mitigating their effects has been discussed in more detail in NRC (1996a) and Heinz Center (2002). The decision analysis example on enhancing habitat in Chapter 4 and the discussion of the costs of dam removal at the end of this chapter provide additional information on addressing the threats to dams, as does the summary of the 1997 Conservation Plan (Maine Atlantic Salmon Task Force 1997) toward the end of this chapter.

HATCHERIES

Possible Goals for Hatcheries

At this stage in the decline of wild populations of Atlantic salmon in the state of Maine, the goals of hatcheries need to be explicit. The recent steep declines in salmon numbers, in spite of increases in hatchery production and the very recent change to river-specific stocking, mean that efforts need to be concentrated on rebuilding wild populations in Maine's rivers. It is helpful to specify immediate goals aimed at dealing with the current extinction crisis as well as ongoing goals that would continue to apply even as signs of rebuilding are seen. It would also be helpful to adapt earlier assumptions and goals to current conditions and scientific knowledge.

Immediate Goals

The goal of hatcheries in response to the extinction crises in Maine should be to conserve genetic quality—a broad term that includes the concepts of genes adapted to local conditions, complementary and coadapted genes, and appropriate genetic diversity—in the remaining wild populations of Atlantic salmon, allowing these survivors to persist. In this respect, the hatcheries might serve as living gene banks. The operation of the Craig Brook National Fish Hatchery is compatible in part with this goal. The large Craig Brook National Fish Hatchery could be altered to fill this role, but it is currently a production hatchery for several stocks separated by natal river. Therefore, changes would be needed in its functioning. Less effort to produce large quantities of releasable fry should make at least some facilities available for careful management of limited brood stock. In addition, some effort could be redirected to working with scientists to address research questions that have already been raised as well as new ones that will emerge as the project proceeds. The most urgent goal is to preserve the genetic structure of the remaining populations, while the longer-term processes of habitat expansion and rehabilitation are pursued. An equally pressing goal should be the acquisition of basic information and research needed to ensure at least two return spawners for each spawning female in the wild.

Ongoing Goals

The ongoing goals of hatcheries should include the preservation of technical knowledge and public education about the biology and ecology of salmon in the wild. The successful production biologists at hatcheries acquire the skill of culturing Atlantic salmon. The skill cannot be fully communicated in technical reports, because it depends on experience and is best taught by practitioners. This skill must be maintained. Many people are fascinated by hatcheries. Hatcheries should be more integrated into public education and designed for site visits. Atlantic salmon have long been an icon for environmental awareness.

Resources should be directed toward adaptive management studies, allowing managers to put research findings into evolving practice in a timely fashion. In the short term, there is a need to better understand how genetic, ecological, and physiological processes affect the ability of hatchery-released fish to survive and successfully reproduce in rivers of Maine, compared with naturally reproduced fish.

Unclear Goals

The goal of providing enough fish to support the commercial or recreational fishery, if such a goal is still imagined, is not clearly articulated. Efforts to subsidize the fishery have been unsuccessful thus far, although fisheries for anadromous salmonids have been subsidized with varying degrees of success through hatchery production elsewhere in North America and other countries. Clearly, current hatchery operations in Maine cannot support recreational or commercial fisheries for anadromous Atlantic salmon. It is possible to establish a small recreational fishery for salmon by rearing fish to adulthood in a hatchery and then releasing them into rivers, but that would not satisfy the Endangered Species Act (ESA) or the stated goals of Maine and federal officials to establish wild salmon populations. If salmon runs in Maine were restored to their pre-dam sizes (before about 1750), they would probably support both recreational and commercial fishing, especially if they were carefully regulated. It is outside the committee's charge to consider other goals than salmon rehabilitation in Maine's rivers, but we have heard comments suggesting that other fish species should be stocked in them if neither recreational nor commercial fishing for salmon can ever be expected.

Reducing Threats Posed by Hatchery Programs

In pursuing the immediate and ongoing goals listed above, it is critically important to consider the growing evidence of genetic and ecological threats posed by hatchery programs. Whenever managers decide to include hatcheries as part of a broader recovery strategy, they need to prevent or reduce those threats through application of practices designed to adhere to "best-practice" genetic, evolutionary, and ecological principles (Miller and Kapuscinski 2002). Although many of the protocols currently used reflect best practices, a more comprehensive vision of how to use hatcheries as part of a program of protection and rehabilitation is needed. That includes recognition of adverse effects that hatcheries can have on the genetic makeup of salmon population, both those than can be reduced by careful practice and those that cannot.

The genetic makeup and phenotypic traits of hatchery-propagated salmonids often differ from those of the wild populations that they are meant to rehabilitate and with which they will interact. Hatchery fish phenotypes commonly differ in ways that will influence ecological interactions between them and wild fish. A meta-analysis of hatchery effects on pre-spawning behavior shows strongly that hatchery rearing results in increased pre-adult aggression and decreased response to predators that may, in part, explain their decreased subsequent survival in the wild (in 15 of 16 case studies) (Einum and Fleming 2001). Somewhat less frequently, hatchery salmonids show changes in growth rates, migration and feeding behaviors, habitat use, and morphology, as reviewed below. Recent evidence of a genetic basis for resistance to pathogens, also as reviewed below, suggests that hatchery programs can inadvertently reduce the genetic quality needed for disease resistance.

Genetic Hazards

Hatcheries used to rehabilitate depressed populations can impose a variety of genetic hazards. Extinction is the extreme hazard from which recovery is impossible. The other hazards are all a form of degradation of what is called *genetic quality*. Genetic quality refers to the overall quality of the genotypes in the population in terms of their effect on the ability of fish to survive, thrive, and respond to changes in their natural environments. (It assumes that the natural environment itself has not been so degraded that it cannot support the populations.) Genetic quality includes individually "good" genes, which confer fitness to individuals that possess them; compatible and co-adapted genes, which provide superior fitness through their complementation of genes at other loci (Andersson 1994, Carrington et al. 1999, Penn and Potts 1999); and appropriate genetic diversity, which confers evolutionary potential by allowing for a variety of genotypes to be produced from various matings but does not counteract other aspects of genetic quality. For example, domestication selection is a well-known hazard of supportive breeding programs (Fleming and Gross 1989; McGinnity et al. 2003; NRC 1996a; Reisenbichler 1997; Waples 1991a, 1999). Domestication selection is a form of degradation of genetic quality by reducing the fitness of hatchery fish in their natural environment.

Many aspects of hatchery programs (supportive breeding) can affect genetic quality. For example, in nature, breeding is not random with respect to genetics (Andersson 1994). By making pair matings or even using other protocols, hatcheries usually limit or work against sexual selection (mate choice) and life-history decisions that help to maintain genetic quality in natural populations (Fleming and Gross 1989, Grahn et al. 1998, Wedekind 2002). Sexual selection can increase fitness by increasing the viability of offspring (Møller and Alatalo 1999). Hatchery protocols typically select against precocious males (e.g., jacks in Pacific salmon and mature parr and grilse in Atlantic salmon), which contribute to genetic quality (Gross 1996; Gross and Repka 1998a,b). Thus, maximizing genetic diversity by preventing mate choice might not be an effective conservation strategy (Wedekind 2002).

Some of the components of genetic quality and the ways that they can be degraded in hatcheries are discussed below. There is increasing documentation of the empirical reality of these genetic hazards (Kapuscinski and Brister 2001, McGinnity et al. 2003, Miller and Kapuscinski 2002, Shaklee and Currens 2002). Hatchery managers can somewhat reduce these risks and can totally avoid certain others by applying appropriate genetic guidelines (Miller and Kapuscinski 2002). Current protocols in place at the Craig Brook hatchery for river-specific supportive breeding of distinct population segment (DPS) brood stocks generally adhere to current guidelines for reducing or avoiding some genetic hazards. Current practices that raise residual concerns are discussed in some detail below. Further information is available elsewhere (Miller and Kapuscinski 2002 and references therein). It is impossible to avoid degrading all aspects of genetic quality at the same time in a hatchery. The committee reiterates that avoiding extinction probably should take priority over all of the other genetic considerations.

Extinction

Demographic processes in the hatchery program can cause extinction under certain conditions. An extreme example would be a hatchery catastrophe in which an entire population of fish brought into captivity is killed. In addition, genetic processes in the hatchery can contribute to extinction risk in subtler ways, as suggested by recent studies attributing increased rates of extinction to reduced levels of genetic variability (Newman and Pilson 1997, Saccheri et al. 1998). The current DPS-river supportive breeding and propagation program at the Craig Brook hatchery reduces the risk of purely demographic extinction by bringing only a portion of a river's parr or returning adults into captivity (Buckley 2002a,b). Additional analyses of extinction risk are being developed with the aim of including them in the Recovery Plan (USASAC 2003).

There is a trade-off between leaving the whole population together and splitting it, however. Splitting an already small population into wild- and captive-reproductive subunits simultaneously increases the risk of losing genetic variability within one or both subpopulations, as discussed in the next section. For example, as run sizes in the Penobscot have declined over the last decade, collections of adults for hatchery breeding have progressively become a greater fraction of the adult returns. Specifically, females spawned in the hatchery rose from 17% of all returning MSW adults in 1986 to 86% in 1998 (K.F. Beland, Maine Atlantic Salmon Commission, unpublished data, 2003), and adults of both sexes collected for hatchery spawning made up over 60% of all returning adults in 2000 and 2001 (Buckley 2002a), up from 17% in 1986. This trend increases the overall exposure of the Penobscot population to loss of genetic quality.

The current supportive breeding program for the six DPS rivers (all except the Ducktrap River and Cove Brook) and the hatchery propagation of Penobscot fish minimizes the extinction risk due to loss of genetic variability by including one-on-one matings and tracking contributions of each family to fry releases and adult returns via genetic markers (Buckley 2002a). However, it does not eliminate loss of genetic quality. By overriding mate selection and perhaps by sampling error, it reduces the likelihood of genetic complementation. For example, the major histocompatibility complex (MHC) is involved in disease resistance (see e.g., Arkush et al. 2002), and one-on-one matings probably reduce genetic complementation at that complex of genes and thus reduce genetic resistance to disease.

Loss of Within-Population Genetic Variability

Loss of within-population genetic variation has several causes, the most important of which is genetic drift due to sampling gametes in finite populations. Loss of genetic variation due to drift occurs at a rate inversely proportional to the genetically effective population size (N_e). The N_e refers to the size of an "ideal" population that has the same rate of loss of heterozygosity (a common measure of genetic variation) as the actual population has, the "ideal" population being defined on the basis of demographic characteristics such as an even sex ratio, stable population size, no immigration, and a Poisson distribution of progeny number. Estimates of N_e for populations of salmonids have typically been smaller than the actual number of reproducing adults, ranging between 4% and 73% of the number of reproductive adults (Ardren and Kapuscinski 2003, Bartley et al. 1992, Heath et al. 2002).

In a process known as the extinction vortex (Gilpin and Soulé 1986), inbreeding and loss of genetic variability due to genetic drift can result in reduced fitness. This loss of fitness may reduce N_e, resulting in greater inbreeding and further loss of variability, which reduces fitness further. The continuing reduction in population size exposes the population to ever-increasing demographic risk of extinction. Considerable interest has been devoted to the threats to wild and captive populations associated with inbreeding and loss of genetic variability, and much of this work refers directly to fishes in general, and salmonids in particular (Allendorf and Phelps 1980; Allendorf and Ryman 1987; Cross and King 1983; Ryman and Ståhl 1980; Ståhl 1983, 1987; Waples 1991a).

Loss of within-population genetic variability is the most common hazard associated with decisions regarding numbers of adults in the hatchery to be mated and how they are to be mated. For instance, the high fecundity of salmon fosters a temptation to produce large numbers of progeny from a few parental fish in each breeding season, artificially creating a "genetic bottleneck" that significantly reduces genetic variability among the progeny. Current protocols at the Craig Brook hatchery for DPS river brood stocks appropriately avoid this obvious pitfall (Buckley 2002a,b). Those protocols include collecting enough parr or adults to ensure reasonable numbers of reproducing adults, one-on-one matings to ensure that each adult contributes, application of genetic profiles to avoid mating close relatives, and use of genetic markers to track families of DPS fish through the hatchery and beyond. Appropriate features of hatchery mating of Penobscot adults include the one-on-one mating design and the collection of genetic data that can be analyzed to avoid matings of close relatives, although the latter is less crucial for this larger population (compared to DPS captive brood stock) and does not appear to have been carried out as of 2002 (Buckley 2002a).

Loss of Genetic Variability from Supportive Breeding

Supportive breeding, as defined above, augments N_e for the hatchery component of the population, but it also entails a potential risk of increasing the loss of within-population genetic variability in the wild. When supportive breeding meets its intended rebuilding goal, it increases the total population size through a higher reproductive output from the captive breeders than from those reproducing in the wild. That increases the reproductive success of the captive (hatchery) segment of the population relative to that of the wild segment of the population. The resulting large increase in the variance of family size within the total population (wild plus captive) is sufficient to reduce the effective population size as a whole (Ryman and Laikre 1991, Ryman 1994, Ryman et al. 1995b, Wang and Ryman 2001). See Appendix C for a detailed discussion of this problem. Often, an overall reduction of effective size cannot be avoided when applying supportive breeding that successfully increases the population census size. However, that problem may not be overly important in the case of declining populations, such as the severely depleted salmon populations in Maine, for which supportive breeding may yield a higher N_e value than would occur in its absence.

Most Atlantic salmon populations in Maine are severely depleted and continue to decline, and for such populations, the positive effects of increasing the actual population size outweighs the potential short-term genetic drawbacks caused by reductions of the N_e. Thus, the need for supportive breeding is urgent. However, no extensive analysis has been done on the genetic impact of supportive breeding on populations that would continue to decline if left on their own (but see Duchesne and Bernatchez [2002] for the special case of binomially distributed family sizes). Clearly, in the extreme situation of a population that would go extinct without supportive breeding, it would be better to maintain a genetically depauperate population than to let it die. It is at least possible that some current populations are small enough for the situation to be considered extreme. Using the general model and variable designation for supportive breeding that is outlined in Appendix C, an example is depicted in Figure 5-1. A declining initial population (N) of 50 is supported with progeny from five captive fish with a much higher average reproductive rate (adult to adult) than the wild fish. The support immediately results in a growing actual population. The N_e stops declining, in-

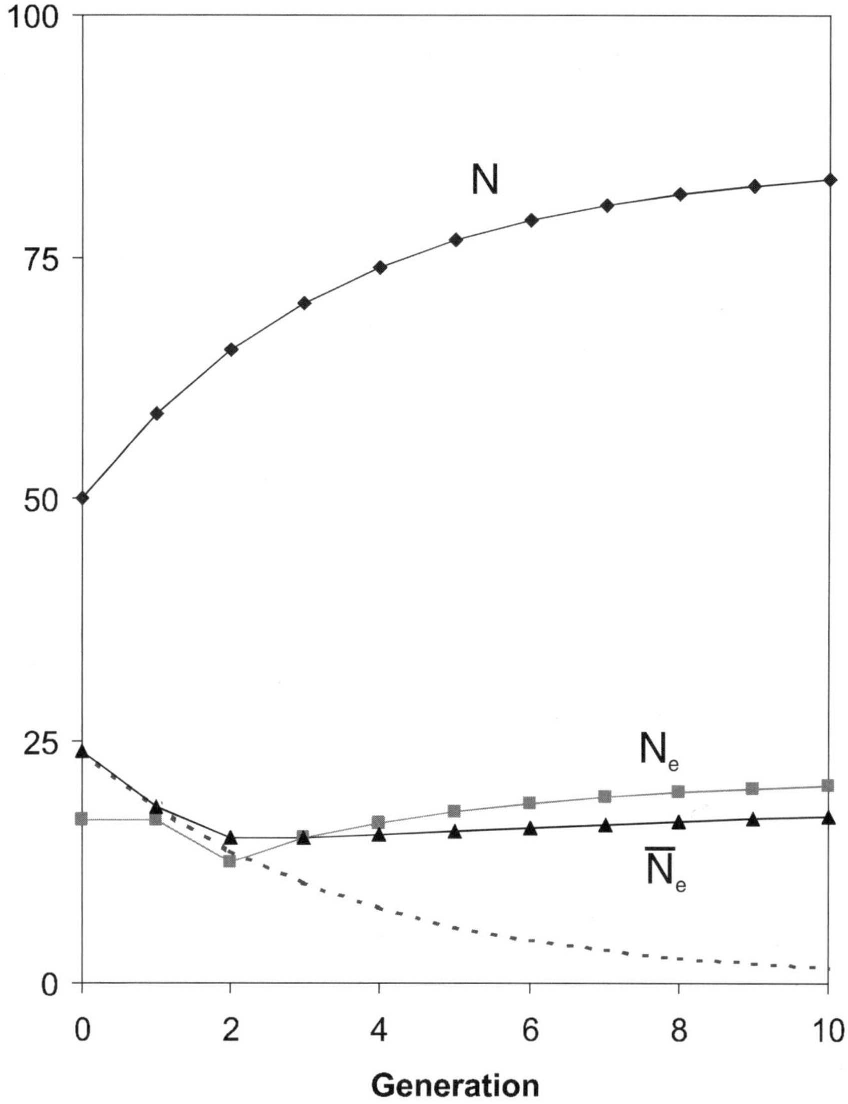

FIGURE 5-1 Census size (N), effective size (N_e), and cumulative harmonic mean of effective size ($\overline{N}_e$) during 10 generations of supportive breeding in a population of 50 individuals that would be declining if left on its own (initial N = 50). In each generation a fixed number of N_c = 5 individuals are caught at random and brought into captivity for reproduction. The mean number of progeny per individual is μ_c = 10 in captivity and μ_w = 1.5 in the wild, with variances σ^2_w = 7.5, and σ^2_c = 50 (five times the corresponding initial μ-value). The dashed line indicates the effective size in the absence of supportive breeding.

creases at a much slower rate than actual population size, and levels out at an N_e that is considerably smaller than N but still much larger than it would have been without support. This example only depicts a particular set of parameter values, and the expected effect of a support program must be evaluated with respect to specific conditions and options. It appears, for example, that the program in Figure 5-1 could be made more "genetically successful" by reducing the variance of family size in captivity or by increasing the number of captive fish in later generations, when the total population size has increased. However, those scenarios have not been evaluated numerically by the committee.

The above considerations lead to advice on how to reduce the adverse genetic effects of supportive breeding by holding the number of progeny to be stocked from each mating to a constant. If the mean and variance of reproductive rate are equal (the usual Poisson assumption), then (ignoring the overlapping generations),

$$N_e = \frac{\bar{k}\left(N\bar{k}-1\right)}{V(k)+\bar{k}\left(\bar{k}-1\right)}$$

where $\bar{k}$ is the mean number of progeny per individual, and $V(k)$ is the variance (Wright 1938). Under Poisson assumptions, $V(k) = \bar{k}$, reducing the equation to

$$N_e = \frac{N\bar{k}-1}{\bar{k}} = N - \frac{1}{\bar{k}},$$

and in a growing ($\bar{k} > 1$) population, N_e is only slightly smaller than N. With much greater than Poisson variance in reproductive output (the usual case, $V(k) >> \bar{k}$), the reduction of N_e is greater, that is, $N_e << N$. However, if $V(k)$ is reduced to 0 by holding the number of progeny per mating to a constant size, $\bar{k}$, then N_e is increased to

$$N - \frac{1}{\bar{k}} < N_e = \frac{N\bar{k}-1}{\bar{k}-1} < N$$

Thus, given that one-on-one matings are being used in the hatchery, maximum N_e is achieved by holding the number of progeny per mating to a constant.

Loss of Genetic Variability among Populations (Population Identity)

Crosses made among fish from multiple populations result in loss of genetic distinctness of each individual population (that is, population identity). One potentially adverse outcome of mixing distinct populations is a reduction in fitness in the admixed population due to disruption of local adaptation or of co-adapted gene complexes (reviewed by Hallerman 2002, Kapuscinski and Brister 2001). Atlantic salmon in Maine, like many fish species, are part of a larger metapopulation, in which relatively isolated subpopulations are connected by low levels of gene flow via straying migrants (NRC 1996a, 2002a). Isolation allows subpopulations to adapt to local environmental conditions. There is almost no hard evidence on the degree to which remnant populations of Atlantic salmon in Maine rivers are locally adapted. The assumption, however, must be that those few fish that return to spawn are at least as well adapted to local conditions as those that fail to return. Sheehan (T. Sheehan, NMFS, personal communication, 2002) conducted a "common-garden" study of three river-specific populations, in which progeny of fish from different rivers are raised in similar environmental conditions. The progeny showed different growth trajectories, a result that is consistent with that expected from locally adapted populations. While Sheehan's study is not definitive, because of design limitations, it is suggestive of the kind of local adaptation that is common in wild populations of salmonids and that forms the basis of the concern for maintaining the remnants of the natural metapopulation structure of wild salmon in Maine.

Low amounts of migration can counter the inevitable loss of genetic variability in isolated populations without overwhelming local forces of adaptation. Massive hatchery mixing of distinct gene pools, however, is likely to overwhelm the local forces of natural selection, because the proportion of breeders coming from another gene pool is typically much larger and the level of genetic differences between the imported and local populations can be much greater (due to ease of transporting salmon from far distant locations). The Craig Brook stocking program avoids this genetic hazard through separate rearing and crossing of river-specific groups for each of the DPS and Penobscot rivers.

Outbreeding between genetically distinct populations can sometimes improve fitness in the wild, but such outbreeding enhancement is most likely when hybridization alleviates pre-existing inbreeding depression within one or both pre-mixed populations (Waples 1995). Although Ferguson et al. (1988) found some evidence of superior fitness of first-generation hybrids between two non-inbred populations of cutthroat trout, superior fitness of hybrids often disappears in subsequent generations when the hybrids backcross to a parental population (Gharrett and Smoker 1991). To date, evidence of inbreeding depression is lacking in Atlantic salmon populations in Maine, despite their depressed status. Natural straying probably occurs often enough to provide gene flow without disrupting local adaptation (NRC 2002a).

Domestication Selection

Domestication selection refers to any change in the selection regime of an artificially propagated population relative to that experienced by the natural population (Waples 1999). Consequently, the genetic composition of a population within a hatchery program is likely to differ from what it would be in the absence of hatchery propagation. The hatchery fish can be expected to adapt genetically to the different selection regime in the hatchery environment, even when hatchery operators do not intentionally practice selective breeding. The basic idea is that significant alterations of the population's genetic composition, due to different selection pressures under husbandry, will reduce a population's subsequent fitness in the wild (e.g., Fleming and Gross 1989, Reisenbichler 1997, Waples 1991a).

Domestication selection can occur in multiple ways (Busack and Currens 1995; Campton 1995; Waples 1991a, 1999). Hatchery practices can involve intentional selection on traits such as size or age at spawning. A recent modification to the DPS brood-stock mating protocol at the Craig Brook hatchery (to mate only 4-year old adults) could increase the risk of this kind of domestication selection, because there probably is a partial genetic basis (heritability) for sexual maturation at a given age, and variability of that trait in Atlantic salmon may have adaptive value in the wild (Hutchings and Jones 1998).

Another potential source of domestication selection is nonrandom collection of hatchery brood stock from a spawning population. That does not appear to be a problem for DPS brood stock, because considerable effort goes into random collection of wild parr to bring into the captive breeding program. Hatchery propagation of Penobscot fish is more vulnerable to this hazard, depending on the extent to which annual collections of returning adults at the Veazie Dam represent all portions of the run. This concern is the basis for a proposal to collect brood stock both in early summer and in the fall (Beland et al. 1997), but no mention of this issue appears in subsequent reports on hatchery brood-stock management (Buckley 2002a,b).

A third form of domestication selection is the unintentional selection that occurs in the hatchery environment. For example, changes in agonistic behavior, probably due to crowding, rearing conditions, or feeding methods, between wild and hatchery fish is observed frequently (reviewed by Einum and Fleming 2001). It may be possible to reduce, though not completely avoid, that source of domestication selection by establishing more natural rearing conditions and applying more natural practices during rearing and at release from the hatchery (Miller and Kapuscinski

2002). Pacific salmon hatcheries are making some efforts in that direction, such as using the Natural Rearing Enhancement System (Maynard et al. 1995, 1996) and various conservation-hatchery strategies (Flagg and Nash 1999). Current protocols at the Craig Brook hatchery do not appear to pay much attention to this form of domestication selection, but the hatchery is still in the first generation of supportive breeding, and there is time to make mid-course corrections (as an adaptive-management adjustment).

A fourth, less-recognized form of domestication selection is the release of juvenile fish from patterns of natural selection that would have been imposed on them had they been in the natural environment (Fleming and Gross 1989, Waples 1991a). Perhaps the greatest concern here is the total removal of sexual selection, through mate choice, which occurs when salmon reproduce naturally in rivers (Fleming and Gross 1989). There is growing evidence of the genetic benefits, including better fitness in the wild, of natural mate choice (reviewed in Appendix D), although the underlying genetic mechanisms are poorly understood. There is also a probable trade-off between increasing the naturalness of sexual selection and decreasing the loss of genetic variability within populations. Reducing domestication selection that is due to loss of mate choice might be achieved by allowing adults to choose their own mates. Reducing loss of variability within populations is best achieved by maximizing N_e by appropriate artificial crosses made in the hatchery. Both genetic hazards cannot be reduced simultaneously. At present, practices at Craig Brook hatchery focus on reducing the loss of genetic variability within populations and ignore the risk of domestication selection that is due to loss of sexual selection.

Domestication and its consequent maladaptation to the wild can happen in a fairly small number of generations of hatchery breeding and has been shown to reduce predator avoidance (Berejikian 1995) and increase aggression or competitive ability of hatchery fish (Holtby and Swain 1992; Johnsson et al. 1996; McGinnity et al. 2003; Mesa 1991; Ruzzante 1991, 1992, 1994; Swain and Riddell 1990, 1991; see also review by Einum and Fleming 2001). The reason for genetically based differences in aggressiveness between hatchery and wild fish might be unintentional artificial selection (imposed when fish are chosen for brood stock) or selection for strong performance under animal husbandry conditions (reviewed by Jonsson 1997). For salmon, increased aggression in wild offspring of matings between hatchery and wild fish would make them more vulnerable to predators (Johnsson and Abrahams 1991). The current strategy at the Craig Brook hatchery of returning adults to the wild after they have been mated once helps to reduce the accumulation of domestication across brood years and generations.

Disease Hazards

One concern about stocking hatchery fish is that they may transmit disease or parasites to wild fish. Disease transmission between cultured salmon (hatchery stocked or commercially farmed) and wild salmon is very likely bidirectional. It has been extremely difficult to determine the incidence of disease transmission from hatchery to wild fish, as well as the impacts such transmission would have on wild stocks (Flagg et al. 2000, Håstein and Lindstad 1991).

Disease can be caused by parasites, bacteria, viruses, or fungi. Many disease-causing organisms tolerate only freshwater or seawater. Epizootics in hatcheries and sea cages are readily observable, whereas epizootics in wild salmon are not. Sick wild salmon quickly disappear. Disease outbreak is considered to occur at the intersection of three components: susceptible host, virulent pathogen, and adverse environment. The ideal method of control is prevention through a clean environment and a healthy, well-fed fish.

Vertebrate Parasites

The sea lamprey (*Petromyzon marinus*) is probably the only vertebrate parasite on Atlantic salmon, but incidence and impacts on Maine salmon are unknown.

Crustacean Parasites

In the marine environment, the most problematic parasites are copepod crustaceans known as sea lice (*Caligus elongates*) and salmon lice (*Lepeophtherius salmonis*). Lice loosen the skin and can expose the flesh. About 30 lice can be enough to kill a smolt (Grimnes and Jakobsen 1996). In British Columbia, salmon lice have been implicated in the decline of pink salmon runs (*Oncorhynchus gorbuscha*). In the Broughton Archipelago, which has many salmon farms, more than 3.6 million adults spawned in 2000, but only 147,000 returned in 2002. (Pink salmon have an obligate 2-year life cycle; thus, roughly the same number of fish that had spawned in 2000 were expected to return in 2002.) Although a cause-effect relationship between lice and salmon numbers was not established, lice were present on Broughton salmon in large numbers. In adjacent areas without farms or lice, the populations did not decline (PFRCC 2003). These crustacean parasites have caused disease outbreaks in Maine salmon net-pens. There is a major effort to control lice on Atlantic salmon in sea cages because, in addition to causing direct harm, they are vectors for the virus causing infectious salmon anemia (ISA) and the bacterium causing furunculosis (*Aeromonas salmonicida*). Sea lice are more common in wild fish in areas with sea cages (PFRCC 2003).

Helminth Parasites

In freshwater, the ectoparasite *Gyrodactylus salaries* is a major disease problem in Norway. *Gyrodactylus* is a monogenean trematode, also known as a flatworm or fluke, that browses on skin mucus. It has almost totally killed off young salmon in some rivers in Norway (Håstein and Linstad 1991). Its distribution in wild salmon is thought to be caused by stocking infected fish (Johnsen and Jensen 1991).

Pathogenic Bacteria

The skin and digestive tract of fishes are colonized by many bacteria, most of which are not pathogenic. The most common bacterial disease affecting Atlantic salmon is furunculosis caused by *Aeromonas salmonicida* (Austin et al. 1989). Furunculosis appears as boils on the sides of salmon in both freshwater and seawater. Johnsen and Jensen (1994) associated the spread of this disease in wild Atlantic salmon in Norway with escapes from fish farms and natural migrations of wild salmon. Enteric redmouth (ERM) is caused by *Yersinia ruckeri*. Epizootics can occur following stress or poor water quality. Vaccines are available against ERM. Coldwater disease is caused by the bacterium *Flavobacterium psychrophilum*) and is a problem in Atlantic salmon in New England. Hitra disease caused by the bacterium *Vibrio salmonicida* became a serious problem in Maine beginning in 1993 (Griffiths 1994).

Pathogenic Viruses

The most alarming viral infection in Maine Atlantic salmon has been infectious salmon anemia (ISA, also known as hemorrhagic kidney syndrome). The ISA virus (ISAV) was first identified in Maine in 2001. The ISAV poses no threat to humans and other mammals. Information about ISA can be obtained from the U.S. Department of Agriculture (USDA 2002) and from Scotland where ISAV was first identified in 1998 (JGIWG 2000). The virus is an influenza-type virus (orthomyxovirus) that mutates rapidly, thus eluding attempts to make a vaccine. It is found in wild and farm Atlantic salmon. Symptoms appear after about 1 year in seawater. Basically, the whole organism is affected. Mortality is estimated by the USDA at 2–50%. Disease has cost Maine salmon growers about $24 million and unknown costs to public agencies for disease control and prevention, and Atlantic salmon are now under careful scrutiny for signs of ISA.

Another virus endemic to Maine is the virus causing infectious pancreatic necrosis (IPN). The IPN virus (IPNV) affects farm salmon but has not caused a serious mortality in salmon in Maine. The final known virus is a lethal retrovirus called salmon swimbladder sarcoma virus (SSSV). This was detected first in 1998 in a hatchery-reared parr captured in the Pleasant River. SSSV causes cancer in salmon.

Pathogenic Fungi

Most fungi encountered by salmon are not pathogenic. Saprolegniasis is a fungal disease caused by *Saprolegnia diclina* type 1. The fungus affects the skin. It is associated with high levels of androgens and therefore has a higher incidence in mature males (Olafsen and Roberts 1993; Gaston 1988).

Genetic Variation in Susceptibility to Disease

Genetic variability in the degree of resistance to disease occurs in salmon as in other vertebrates. Arkush et al. (2002) compared the pathogen resistance of chinook salmon with different genotypes of a gene in the major histocompatibility complex (MHC). In two of five comparisons, "significant genetic effects on disease resistance" were found. The authors concluded that small wild populations and hatchery populations with lowered genetic variability would have increased susceptibility to pathogens. With small wild populations, disease susceptibility is not directly controllable or even necessarily the most important concern, but the protocols described above (also Flagg and Nash [1999], among others) to increase N_e in hatchery populations and their progeny probably will help reduce disease susceptibility as well.

Prevention and Treatment of Disease

Management practices in both freshwater hatcheries and sea cages are designed for disease prevention. The most important elements of prevention are high quality water and a good diet. A major effort is under way in Maine to control sea lice, a known vector of ISA and furunculosis. Salmon are treated for sea lice with hydrogen peroxide, pyrethrin, ivermectin (a neurotoxin), and other pesticides. In British Columbia, the Pacific Fisheries Resources Conservation Council recommended strategic fallowing of net-pens, accelerated marketing of mature fish, and application of chemicals to kill lice as a measure to reduce the incidence of lice infections associated with salmon farms (PFRCC 2003). The Craig Brook facility has protocols in place to prevent disease transmission. The facility brings wild-caught brood stock to the facility and holds them in outdoor tanks. Blood is sampled and tested for ISA virus. There is no on-site expert in fish health; assigning one should be considered.

Behavioral Hazards

Ecological interactions between hatchery and wild fish are problematic. Behaviors of hatchery-propagated fish differ from those of their wild conspecifics because of differences in the genetic or environmental control over expression of behavioral traits. Differences may also be due to different interactions between genetic and environmental controls. Genetically based alteration of behavior in hatchery fish can occur through loss of population identity or domestication selection, two of the genetic hazards discussed previously. Environmental control over behavioral traits occurs because fish phenotypes are strongly shaped by the rearing environment (see Pakkasmaa 2000, Wootton 1995). Hatchery rearing inevitably affects fish development by changing food and feeding regimes, density, substrate, exposure to predators, and interactions with conspecifics.

Numerous studies have found altered behaviors of hatchery fish, compared with their wild counterparts, that are probably both environmental and genetic in origin (reviewed in Einum and Fleming 2001). Hatchery rearing of salmonids frequently results in increased pre-adult aggression and decreased response to predators (reviewed in Einum and Fleming 2001). Differences in aggression have a substantial environmental component, although there are indications of genetic influences as well. The lack of exposure to predators in hatchery populations appears to result in a reduced response to predation risk, both as an environmental effect and as a response to relaxed selection in hatchery populations. Changes in growth rates are common but less consistent. Changes in other fitness-related behavioral traits, such as migration, feeding, habitat use, morphology, and breeding behavior, also occur. Those and other changes are probably responsible for decreased survival of released hatchery fish in the wild.

Altered behaviors of hatchery-reared fish may disrupt or harm the reproductive success or survival of wild fish. Although releases of hatchery fish are often implemented to compensate for reduced production caused by human-induced habitat degradation, a range of potential ecological problems may be associated with this practice. First, stocking of large numbers of fish into a limited habitat will, at least initially, inevitably affect population density. The effects of such stocking can include changes in the frequency of competitive interactions, the amount of available food, or the behavioral response of predators and hence influence growth and survival of the wild fish (Einum and Fleming 2001, reviewed in Flagg et al. 2000). Second, hatchery fish will almost certainly differ phenotypically and genetically from wild fish (see above). Such differences can affect how stocked and wild fish interact beyond those due to pure density dependence (see Nickelson et al. 1986). Third, there may be predatory effects, such as released hatchery fish preying on wild fish and influencing the behavior and dynamics of predator populations, an effect that can indirectly affect wild fish (reviewed in Flagg et al. 2000). Fourth, hatchery fish can transmit disease and parasites to wild fish.

Several other potential behavioral changes in hatchery-reared fish might have detrimental effects on wild fish. For example, released fish might influence the timing of migration of wild fish. Hansen and Jonsson (1985) suggested that wild smolts were attracted to shoals of released smolts and joined them when migrating downstream. Furthermore, releasing fish might increase interspecific (i.e., with brown trout) hybridization rates (Jansson and Öst 1997, Leary et al. 1995). Although little is known about the frequency of early parr maturation among hatchery-reared fish, the high growth rates experienced in the hatchery will probably increase the potential for early maturation following release.

Increasingly, evidence shows that the altered behaviors of hatchery fish are maladaptive, resulting in poor survival and reproductive success in the wild. Hatchery fish experience reduced survival, compared with wild fish (15 of 16 studies reviewed by Einum and Fleming 2001, meta-analysis $p \leq 0.001$). The success of hatchery-produced fish after release is reduced by phenotypic divergence from their wild conspecifics. The reduction occurs because environmental and genetic risks to fish in hatcheries cannot be avoided entirely, and many of the genetically based risks are negatively correlated, so efforts to reduce one risk increase other risks.

Changes in behavioral, life-history, and morphological traits associated with reproduction also occur under hatchery conditions (reviewed in Fleming and Petersson 2001) and may have important implications for the ability of released fish to contribute to natural productivity. A review of 31 studies of introgression of hatchery genetic material into wild populations (Fleming and Petersson 2001) reported that 14 studies showed little or no evidence of incursion of hatchery genotypes into wild populations, despite prolonged hatchery releases. Natural selection may have purged hatchery-origin genotypes from the population due to the maladaptive traits of hatchery fish, although the studies reviewed were not designed to test that possibility in the wild. Many of the studies involved anadromous populations. In contrast, 16 of the 17 studies showing an incursion involved nonanadromous populations, suggesting that anadromous populations are more resistant to introgression (see also Hansen et al. 2000, Utter 2000). That resistance—whatever its underlying cause—will also undermine efforts to rebuild wild populations primarily through release of hatchery fish, although it should also protect them from genetic incursion.

Conclusions—Hatcheries

Hatcheries sometimes give a false sense of comfort about abundance and persistence of natural populations and of positive action toward rebuilding depleted populations. Commitment and allocation of limited resources to other rehabilitation efforts can be sidetracked by this misconception.

- The evidence from over 130 years of stocking is indisputable. Hatchery production has not rescued Atlantic salmon in Maine. The committee judges that hatcheries alone will not be sufficient to prevent extinction, no matter how well they are operated.
- Some of the earlier human adverse effects on the freshwater environment have been ameliorated over the past 20 years, yet runs are still declining, despite continued stocking and improved stocking practices (e.g., using fish from local Maine streams).
- Additionally, hatcheries can have adverse effects on natural populations. We can reason from first principles and numerous case studies, reviewed above, that hatcheries should be used sparingly in rehabilitation of natural populations.
- Due to a lack of appropriate monitoring, there is a dearth of information about the genetic and ecological effects of historical and current stocking of hatchery fish on wild populations in Maine. There has never been an adequate assessment of whether stocked salmon provide a net long-term benefit to natural populations, and that problem is not restricted to Maine. The success of hatchery programs that aim to rebuild depleted populations lies in their ability to allow fish to bypass the high mortality of early life in the wild and then survive, breed, and produce offspring that will contribute to natural reproduction in the wild (Waples et al. in press). In that sense, "contribute" means that the stocked fish should not take away from the production of the wild population but rather add to it.
- Current procedures for management of DPS river and Penobscot brood stock and offspring at the Craig Brook hatchery clearly avoid one genetic hazard posed by hatcheries—loss of population identity.
- The genetic hazards posed by hatcheries other than loss of population identity cannot be completely avoided. Of those, current procedures at Craig Brook hatchery are appropriate for reducing the probability of extinction, loss of genetic variability within populations, and domestication selection. The recent move to mate DPS brood stock only at age 4 may increase the genetic risk of domestication selection.
- As long as the hatchery program relies solely on artificial matings (versus allowing some or all adults to choose their own mates in some sort of spawning channel), domestication selection cannot be avoided. This form of domestication selection can substantially undermine the abil-

ity of hatchery-propagated returning adults to contribute to rebuilding of fish numbers in Maine's rivers. In addition, some degree of domestication selection is inevitable, because the genotypes best adapted to captivity are more likely to survive than others. Inasmuch as the captive environment differs from the natural environment, domestication selection will occur.

Recommendations—
Options for Future Roles of Hatcheries

The committee recommends using hatcheries as only one option in an integrated strategy that includes rehabilitation of habitat, fishery management, and other appropriate strategies. Additionally, any stocking of hatchery fish should include direct monitoring of their performance and their effects on wild fish. Genetic marking based on inherent allelic differences between families (see Eldridge et al. 2002) would be helpful in the current DPS river-specific hatchery program. Some steps in that direction appear to have been taken for fish held and mated at Craig Brook hatchery (Buckley 2002a,b). Making a properly designed monitoring program a central part of hatchery stocking is the only way to determine whether releases of hatchery fish are helping or hurting efforts to rebuild wild salmon in Maine's rivers. The following recommendations address the hatchery component of such an integrated approach. The situation is becoming desperate due to extremely small numbers of returns in 2001–2002 in all rivers except the Penobscot, and numbers of returns in the Penobscot have also been falling fast.

Genetic Management in Hatcheries

Parties responsible for designing and implementing hatchery practices should periodically review existing practices in light of evolving scientific understanding regarding genetic hazards posed by hatcheries. When new insights become available, appropriate mid-course corrections should be designed and implemented. This process could include comparing hatchery practices with genetic guidelines, such as those of Miller and Kapuscinski (2002), specifically designed for hatcheries to rebuild depressed fish populations, particularly migratory salmon species. Those guidelines address four major phases of hatchery operations that can impose genetic hazards on the captive-bred fish or on wild fish with which they interact after release: (1) brood-stock collection, (2) spawning (including mating protocols), (3) rearing, and (4) release into the wild. The discussion of alternative ways to meet a general guideline may be particularly helpful when logistical and unexpected problems prompt hatchery managers to modify practices.

Life Stage of Salmon at Stocking

If decision makers choose to continue the current hatchery-stocking programs, better understanding is urgently needed about the effect that the life stage stocked has on the ability of hatchery-released fish to return as adults and contribute to the next generation of wild fish in the river. Such understanding can only be gained by building into some portion of hatchery-stocking activities an adaptive-management experiment that will allow systematic comparison of results from stocking fry versus smolts. Basically, unambiguous information is needed on whether hatchery-released smolts, after they return to the river as adults, have higher, equal, or lower reproductive success (the average number of in-river parr produced per spawning adult) than hatchery-released fry. The ideal measurement of reproductive success per spawning adult would be the number of offspring that go to sea and return as adults to spawn in the river. The study would greatly advance understanding if it measured reproductive success per spawning adult as the number of parr or outmigrating smolts in the river. It is also important to know what fraction of the released smolts return to spawn, as compared with the average fraction of released fry that return to spawn.

The consideration is based on measuring success as $\lambda \cdot N_0$ (λ is the replacement rate per egg, and N_0 is the number of eggs [say, 7,000] per female). Then, for eggs raised to the smolt stage before release, the question is whether the number of returning adults is less or more than the corresponding value for eggs raised only to the fry stage. There are survival and reproductive trade-offs between these stocking strategies, and the net balance 130 years after stocking began is still not clear.

That question could be examined through the use of DNA-based genetic markers to identify the genotype of all pairs of adults mated in the hatchery (generation 0, G_0), thus providing the information for assigning parentage of offspring that return as adults to the river (G_1) and of their naturally produced offspring that hatch in the river (G_2). Recently, the Craig Brook hatchery appears to have initiated genetic marking, at least for a portion of the matings made in the Dennys River and Penobscot River brood stock, that could be used later to distinguish returns from fry versus smolt releases (Buckley 2002a,b)

Recommendations for Rebuilding Wild Salmon Populations in Maine

The committee recommends two major options for future use of hatcheries as part of a comprehensive effort to rebuild wild salmon populations in Maine.

Gene Banking

The Craig Brook hatchery program for DPS rivers could be revised to provide a gene bank—that is, keeping a representative sample of the remnant populations in captivity as a backup source of germplasm, an insurance policy in case aggressive rehabilitation efforts in other areas, particularly habitat improvements, fail to rebuild numbers of wild fish in the rivers. Thus, the gene bank would propagate and stock hatchery offspring into the river only under the special circumstances discussed below. The committee considered two feasible alternatives for a gene bank.

Single-Generation, Live Gene Banking of Fish in the Hatchery

This alternative is similar to that being done in the DPS rivers, except that no fish would be stocked. A representative sample of fry or parr would be collected from each DPS river to encompass the genetic diversity of the population, as much as possible. Collecting too many juveniles should be avoided. Only enough should be collected to achieve an adequate effective population size (N_e) in the hatchery (see previous description of N_e in this chapter and in NRC 2002a). Determining what is adequate is a judgment based on the number of fish in the river, information about genetic quality, and other considerations described in this report and elsewhere. An adequate number probably would be more than 100.

Having captured the available genetic diversity, the objective would be to avoid spawning the fish in captivity. These fish would be maintained in the hatchery for as long as possible (until they are 6 or 7 years old). Under certain circumstances, for example, if the wild population seems about to disappear or if rehabilitation or other events seem to have substantially improved available habitat in a stream without a surviving run of salmon, the wild fish could be used as brood stock for reintroducing fish into the population. If the wild population maintains itself, however, the fish would not be mated to propagate offspring for release, and the natural process of population adaptation and recovery would not be impeded by any combination of the hatchery-based threats reviewed in this chapter. Rather, after several years, the fish would be sacrificed and a new group of juveniles collected for the living gene bank to begin a second iteration.

The committee assumes that this option would be implemented as insurance, in concert with aggressive pursuit of habitat improvements and other activities (such as dismantling of dams and improvements in fish passage), to give wild fish a better chance of survival. One advantage of this option is that it minimizes impediments to wild-fish adaptation to prevailing local environmental conditions. Another advantage is that it would provide a true indication of the current state of environmental conditions for Atlantic salmon, conditions that hatchery releases might otherwise obscure.

Lacking other approaches to salmon recovery, gene banking alone would ultimately be ineffective. Disadvantages of this option include expense, risk of losing entire banked populations through disease or system failure, the need to periodically tap wild populations for new juveniles, the inevitable loss of genetic quality that would occur, and the difficulty of gaining political support.

Cryopreservation of Sperm

This alternative would involve collecting and freezing milt from adult males to fertilize females at a later date. Because sperm quality (sperm number, ability of each spermatozoon to fertilize eggs, frequency of mutations, and meiotic problems) decreases with the age of the sample, new samples would need to be collected regularly from returning adults. In addition, continuous sampling of sperm would allow the gene bank to represent the ongoing adaptation to natural conditions that is occurring in wild populations. The approach would be much less expensive than live gene banking, because no live fish would have to be maintained in the hatchery. Thus, funds could potentially be redirected to other forms of restoration. However, the rationale and efficacy of the approach would need to be carefully explained. The main disadvantage of this approach is that the female genetic component would be dependent on having a continuous wild population. It would be better to cryopreserve fertilized embryos or both eggs and sperm, but neither alternative is technically feasible at this time. If either one becomes feasible, if should be reevaluated.

Comparison of Stocked and Unstocked Rivers

This option would stress evaluation of the hatchery-stocking program, something that has been lacking. Adult fish from several year classes and from six of the DPS rivers are being maintained and spawned at the Craig Brook National Fish Hatchery, and the offspring are being stocked in a river-specific fashion as swim-up fry. In the current program, even if hatchery fish are shown to contribute genetically to subsequent generations, there is no way of assessing whether they augment natural production within the rivers or displace some wild production. This new option involves maintaining the current stocking program in some streams but not in others. The latter streams would serve as reference sites for more reliable evaluation of the effects of stocking in the maintained streams. The aim would be to assess the contribution of stocking to population persistence, facilitating adaptive management. The committee recommends expanding the program beyond the DPS rivers.

This approach would involve pairing rivers with similar characteristics, one to be stocked and the other not. For the stocked rivers, all released fish should be marked with a physical tag, such as coded-wire tags or adipose fin clips. Marking is possible even shortly after the swim-up fry stage by tagging with half-sized coded wire tags (Kaill et al. 1990, Peltz and Miller 1990). Returning fish could be screened for the presence or absence of the tags, without requiring their sacrifice. Each stream should be monitored annually for returning adults. Some indication of straying rate could be determined if a tagged fish entered a stream that was not stocked. Tissue samples (e.g., fin clips) should also be collected from all adults, both from the brood stock and from returning fish to the river. Given the small population sizes, genetic markers could be used to develop estimates of the genetic contribution of hatchery versus wild adults to subsequent generations. It would be reasonable for a gene bank to contain representative samples of juveniles from all the unstocked rivers, as described in the Gene Banking section. This would provide some insurance against the risk of extinction of fish from these rivers.

Drawbacks of this approach include expense (for rearing and monitoring), potentially harmful effects on certain populations from either stocking or not stocking them, and the diversion of funds for other restoration work. If populations were to disappear in streams where stocking is discontinued, future recovery might depend on introduction of fish from other rivers, natural straying, or both. However, only six of the eight DPS rivers are being stocked (all except Cove Brook and the Ducktrap), and that provides an opportunity to compare stocking and not stocking. However, there could be considerable improvement in understanding the performance of stocking as a restoration tool. Information garnered from this option would significantly enhance the ability of managers to adapt future management plans, determining how best to deploy precious resources and what effort to place on hatcheries, as compared with other intervention actions.

Recommendations for the Penobscot

The Penobscot drainage is the largest in Maine, and it contributes more than half of all the returning Atlantic salmon in most years. The large size and dendritic drainage pattern of the Penobscot watershed provide a diverse array of habitats. As a result, the evidence for genetic differentiation of populations among the various tributaries is compelling (NRC 2002a). The mainstem is much larger than most of the tributaries that have salmon, such as Cove Brook and Kenduskeag Stream. There are various options, but whichever one is adopted, the committee recommends close monitoring of conditions. If sharp declines are seen in unstocked populations, the stocking program can be restarted easily and quickly before the point of no return is reached. If the unstocked populations hold their own or begin to rebound, it might be wise to adjust the stocking strategy for other populations. In any case, an adaptive management strategy should be followed, using the outcomes of the carefully monitored early experiments to guide ongoing management choices.

Recommendations for the Kennebec

NMFS and FWS (1999) characterize the Gulf of Maine DPS as including "all coastal watersheds with native populations of Atlantic salmon north of and including tributaries of the lower Kennebec River (below Edwards Dam) to the mouth of the St. Croix River at the US-Canada border." The agencies later excluded the salmon populations from the lower Kennebec drainage from the DPS. The Kennebec is the second largest watershed in Maine and historically has produced similar numbers of Atlantic salmon (Atkins 1869, Kendall 1935). The largest impact on the survival of Atlantic salmon in Maine will be obtained by conserving and nurturing the Penobscot populations, but the second largest impact can be obtained by restoring Atlantic salmon to the Kennebec.

With the removal of Edwards Dam on the lower Kennebec, the possibility of salmon recovery in the upstream Kennebec main stem has become a matter of considerable interest. Viable populations of Atlantic salmon are in Togus Stream and Bond Brook tributaries, both joining the main stem below Edwards Dam. Strays from other rivers have been documented within the drainage (Beland 1986, Baum 1997). It is not entirely clear whether the current populations represent the remnants of persistent aboriginal populations within the drainage (Baum 1997, Beland 1986, Buckley 1999, Foye et al. 1969, Havey 1968, Vail et al. 1995), but neither Togus Stream nor Bond Brook was incorporated into the DPS (NMFS and FWS 1999).

The report on the genetic status of Maine's salmon (NRC 2002a) included salmon from Togus Stream and Bond Brook (collectively labeled Kennebec) in its comparison of genetic assignment success rates among Maine drainages (King et al. 1999). A close examination of the data (NRC 2002a, Table 3) shows that the salmon populations of the Kennebec drainage are more distinct than are those of the current DPS rivers. The current populations are wild (as defined in Chapter 1), and they should figure prominently in any restoration effort. The committee concludes that there is nothing to lose by not stocking the Kennebec (NRC 2002b). Atlantic salmon seem to be recolonizing the upper Kennebec main stem above the Togus Stream and Bond Brook tributaries. There is preliminary evidence that salmon are already spawning as far upriver as Ticonic Falls, 19 miles above the former dam site (P. Christman, Maine Atlantic Salmon Commission, personal communication, 2002). The opportunity to observe the course of that rebound, in the absence of stocking, should not be missed.

The Kennebec also provides an excellent opportunity for fishery managers and biologists to determine whether dam removal will be sufficient to allow recolonization and expansion of the wild fish populations upstream of previous impediments. A review of accumulated experience in the Bond Brook and the Togus Stream suggests that some recolonization of the upstream Kennebec main stem can be expected. For the short term, salmon should be allowed a chance to rebound naturally in the Kennebec without hatchery augmentation. Conditions should be monitored closely, however. If the population of wild salmon does not rebound naturally in the Kennebec, an enhancement program can be implemented (presumably using Togus Stream and Bond Brook brood stock), but if the main stem population rebounds naturally, subsequent stocking should be avoided. In addition, the Androscoggin—also emptying into Merrymeeting Bay—is blocked by a large dam (although it does have a fishway), thus serving as a control for the Kennebec.

Stocking Related Species

The committee strongly discourages the stocking of landlocked salmon and brown trout into streams containing anadromous Atlantic salmon populations. Problems posed by landlocked salmon include competition for food resources and possibly spawning sites, mistaken retention of anadromous fish by recreational anglers who think or claim that they are landlocked salmon, bycatch of anadromous fish, and potential hybridization with anadromous fish. Stocking of other nonnative fishes, such as large- and smallmouth bass, should also be avoided.

AQUACULTURE

Options for Aquaculture

The committee performed a decision analysis of the options given below as an illustrative example; that analysis appears in Chapter 4. The purpose of the example is to illustrate how to think systematically about the options while including technical, societal, and economic factors. Because the appropriate weightings for those factors can be determined only by the people who have an interest in the outcomes, we have not based recommendations on the analysis.

- **On-land and other physical containment for salmon farms.** This option allows for full separation and nearly complete containment. Like land-based production facilities, closed or contained floating facilities, water recirculation or controlled inflow and outflow of water, and other containment technologies can reduce disease and parasite transmission and escapes. The option allows for the protection of wild populations alongside the development of aquaculture. However, although closed systems are more secure than net-pen, no system is escape-proof, and land-based recirculating systems can be uneconomical. Current prices for salmon might be too low to support this option (and some others).
- **Zoning.** This option allows for the relocation of cage sites away from important Atlantic salmon populations. The magnitude of most environmental impacts on wild salmon diminishes as distance is increased between the cage site and the natal rivers and migratory routes. This option is being considered by Norway. The establishment of protection areas where salmon aquaculture is restricted or prohibited may protect wild populations of salmon. Such protection areas may minimize genetic, behavioral-ecological, disease, parasite and environmental impacts. Offshore cage aquaculture, which is now being considered, is another possibility. If and when that becomes a practical option, the committee recommends careful risk and benefit assessment. Brooks et al. (1998), however, suggested that the net-pens in Maine are in the best available locations for dispersal of nutrients and solids released from the pens. Thus, moving the pens could produce adverse effects on water quality elsewhere, even if it solved other problems.
- **Biological containment.** Making farm fish sterile is a biological containment strategy for reducing the likelihood of their interbreeding with wild salmon. The present approach to sterility, called induced triploidy, involves tricking newly fertilized eggs to retain an extra pair of chromosomes by applying a mild temperature or pressure shock at the right moment. Methods to induce triploidy are easy to learn and require relatively inexpensive, simple equipment. Protocols for large-scale induction of triploidy have been worked out for Atlantic salmon. Although the effectiveness of triploidy induction varies greatly (e.g., 10–95% success rates [MacLean and Laight 2000]), success can be determined through relatively inexpensive and nonlethal screening of treated fish before transfer to net-pens (Kapuscinski 2001). In one of few field tests of this approach, triploid adult salmon migrated back to natal freshwaters at a much lower rate than control salmon, thus reducing the numbers that could compete or try to mate with wild fish (Cotter et al. 2000). Triploids

may have enough sex hormones in their bloodstream to enter into normal courtship and spawning behavior, interfering with the reproduction of wild relatives. This concern appears to be mostly with triploid males (Inada and Taniguchi 1991, Kitamura et al. 1991, Cotter et al. 2000), and making the farm fish all female in addition to making them sterile may reduce the concern.

Induced sterility, however, addresses only some concerns (genetic and behavioral-ecological) and not others (such as disease). Moreover, it does not fully eliminate potential behavioral-ecological interactions, because farm salmon will enter the environment on a recurring basis where competition with wild relatives and predation on other species may occur (Kitchell and Hewitt 1987). Disadvantages may also exist in terms of yield, fish health, and other marketing factors.

- **Tagging (physical and genetic) all farm fish.** Physical tagging or marking could be used to identify farm salmon in the wild and facilitate their separation from wild fish. This option can be used to determine the source of escapes and to assess the interactions of escaped farm salmon with wild populations. Genetic tagging would allow for the tracking of genetic introgression and potential removal of farm and hybrid offspring.
- **Weirs.** Weirs are used to separate wild and farm fish during upstream migration and thus reduce impacts of escapes. Currently, weirs are on the Dennys, Pleasant, and Narraguagus rivers, with plans for collection facilities on the East Machias and Machias rivers. Ideally, they would be used in conjunction with tagging of farm fish. Public funds would probably be used to construct and maintain these and additional structures. In addition, this option would entail increased handling of wild fish and migratory delays, both of which might affect survival and reproductive performance. Those effects might be reduced if it were possible to identify tagged farm fish by video stationed at the weir (e.g., Lamberg et al. 2001) and have an electronic gating system to separate them. However, weir systems, and particularly those involving video, become nonfunctional in high-water conditions, which often coincide with peaks in salmon migration. In addition, ice formation in the fall requires dismantling parts of the weir to prevent damage before the upstream salmon runs are complete.
- **Genetic makeup of farm fish.** This option would require the use of local North American genetic material. There is a deep phylogeographic discontinuity in genetic structure (based on allozymes and mitochondrial and nuclear microsatellite DNA) between North American and European Atlantic salmon (reviewed in NRC 2002a). All things being equal, reducing the genetic distance between the farm and wild fish would likely reduce potential genetic impacts. However, all things may not be equal, and local North American strains may be more successful at interbreed-

ing with wild Maine salmon, resulting in a more rapid introgression of nonadaptive domestic traits into wild populations. Any reduction in reproductive performance in the wild from using nonlocal (European) strain fish would have to be great enough to compensate for the additional genetic risks imposed by using such strains (Fleming 1996). However, if there were successful interbreeding, the offspring of farm and wild fish would be easier to detect genetically if the genetic makeup of farm fish were very different from that of local wild fish.

The committee sees a need for additional research and analysis on the effects of escapes of farm fish of differing genetic origins. Until that research and analysis are complete, the committee judges it safer for farms to use local North American fish. Neither tactic would eliminate the effects resulting from the introgression of domesticated (farm) traits into wild populations. Moreover, potential ecological impacts remain.

- **Disease management.** Disease could be reduced by better management of stocking density in pens and by area-management strategies. Aquaculture production should be conducted in accordance with appropriate fish-health protection and veterinary controls, including the application of appropriate husbandry techniques to minimize risk of diseases (vaccination, use of optimal stocking densities, careful handling, frequent inspection of fish, proper diet and feeding regimens, detailed health inspections, and strict controls over transport of fish). There should be incentives or regulations to promote disease and parasite treatment beyond a cost-benefit perspective to maximize production while minimizing expenses. Current practices need to more fully integrate the costs of the impact of disease transference and magnification from farm to wild fish. Conditions would improve, but the dangers of disease outbreaks would not be eliminated, and other ecological and genetic concerns would not be addressed.
- **Effluent guidelines.** These guidelines would cover biological pollutants, as well as nutrients, organic matter, and chemicals, and provide incentives to prevent water pollution by establishing settling ponds, recirculation systems, floating bags and tanks, polyculture systems, and other cost-intensive measures. This option would not address concerns associated with escapes.
- **International agreements.** Cooperative agreements with Canada should be implemented to reduce the impacts of salmon farming on wild salmon, especially in the Cobscook and Passamaquoddy Bay areas.

Some of the measures that provide opportunities for coexistence between cultured and wild fish are initially costly to the industry. But maintenance of genetic diversity in wild populations may be crucial in the long run both for wild populations and for cultured strains. Thus, it remains to be seen what the final costs will be if effective measures to protect native populations are not taken immediately.

Research on the Socioeconomic Effects of Changes in Aquaculture

In Chapter 4, the committee considers several options for reducing the risk to wild Atlantic salmon of salmon farms. It also describes a decision analysis based on those options. As the discussion of the decision analysis points out, the people who will have to live with the consequences of the decisions and who might have to pay for them should be involved in the analysis. But that analysis will be difficult even for people with local knowledge and a stake in the outcomes because much is unknown about the consequences of those decisions. For example, nobody knows whether more or fewer Maine residents would be employed in salmon farming if it moved inland than are employed now. Nobody knows what the mixture of employment would be among those currently working on the farms and new employees—how far would they have to move, if at all; what would be the socioeconomic consequences to individuals of such moves; and how difficult would it be to find and train new workers if they were needed? Similar questions could be asked about most of the options described in Chapter 4.

However, changes to the aquaculture industry are inevitable, even if it does no more to reduce risk to wild salmon than is being done now. Technology and economic factors change, as do political and environmental ones. To the degree that socioeconomic factors associated with the industry are understood, it will be less difficult to adapt the industry to reduce risks to wild salmon. Even if it does not change, many socioeconomic factors related to aquaculture have not been quantified, and better knowledge of them could be used to the benefit of Maine's residents and the industry itself. Therefore, the committee recommends research into the socioeconomic factors associated with the aquaculture industry.

FISHING

Fishing conducted in Maine and elsewhere was and has the potential to be a source of direct mortality for anadromous Maine Atlantic salmon, as described in Chapter 3. Directed fishing for anadromous Atlantic salmon in Maine and its adjacent marine waters has been prohibited since 2000, although some directed fishing continues in Greenland and St. Pierre and Miquelon (Chapter 3). Also, bycatch and poaching continue to cause the deaths of an unknown number of anadromous Atlantic salmon in Maine and at sea. Any fishing mortality is serious for populations of salmon as depleted as Maine's are, and any reduction in that mortality would help.

Adult anadromous Atlantic salmon can be confounded with landlocked Atlantic salmon and brown trout, which they strongly resemble. Anglers can believe or pretend that they have caught a landlocked Atlantic salmon or a brown trout—fish that can be legally retained subject to regulations in Maine—when they in fact have caught an andromous Atlantic salmon, which cannot legally be retained. Juvenile Atlantic salmon, especially as they approach the smolt stage, can be mistaken for small landlocked salmon or brown, rainbow, and brook trout. The committee has seen no data on the frequency of such mistaken retention, but it has heard anecdotes. The mistakes are likely to occur at least occasionally. Even if the accidentally caught salmon are not retained, hooking them can cause some deaths even if the fish are released, especially at high temperatures. At sea, various fishing methods have the potential to capture Atlantic salmon. Again, little information is available on the frequency of such captures.

Prohibiting all fishing for all species in waters inhabited by anadromous Atlantic salmon is not acceptable currently, and is unlikely to produce large benefits for Atlantic salmon. However, several approaches short of total prohibition could be helpful.

Stocking gamefish that resemble anadromous Atlantic salmon or compete with or prey on them in streams with imperiled anadromous Atlantic salmon populations is probably detrimental to Atlantic salmon and should be carefully evaluated wherever it occurs. Seasonal closures, at least for other salmonids at times when anadromous Atlantic salmon are most likely to be accidentally taken, also could reduce bycatch mortality in such waters. Size limits also can be protective, as described in Chapter 6.

The committee has heard the view that it would be better to use Maine's streams as habitat for gamefish that are easy to establish than to attempt to restore salmon runs in them. This consideration is not within the committee's statement of task. It and related considerations are more appropriately within the purview of local and national decision makers. However, the listing of Atlantic salmon as endangered under the ESA in the eight DPS rivers and the task of this committee both derived from the view that the conservation of biological diversity, including genetic diversity, is an important societal goal. The stocking of gamefish that would adversely affect the survival and restoration of wild Atlantic salmon in those rivers is clearly contrary to that goal.

MORTALITY OF SALMON IN ESTUARIES AND THE OCEAN

As described in Chapter 3, declining rates of returns of adult anadromous Atlantic salmon to Maine's rivers indicate increased mortality after the young salmon leave freshwater. While it is not possible to determine from return rates alone how much of the increased mortality occurs as smolts transition from freshwater to saltwater in the estuaries and how much occurs at sea, there are reasons to be concerned about both environments.

Changes in ocean conditions could affect salmon in many ways. They could affect the migration routes salmon take, their physiology, the amount and kinds of food available to them, and the degree to which they are preyed on. While most of those factors are not easily dealt with by human intervention, knowledge of how they affect salmon would still help to focus efforts on appropriate restorative actions in other environments used by salmon, and they could help to understand the likely effects and urgency of such interventions.

If the increased mortality is associated with the interaction of contaminants in freshwater with the physiological stress of the transition from freshwater to saltwater, it is probably amenable to human intervention. It is of great importance to establish first whether there is such an interaction, and second what the main contaminants are. Contaminants can interact with salmon transitioning from freshwater to saltwater through changes in pH or temperature or through direct toxic effects. Knowing whether they are present and how they are acting on salmon is critical to a successful effort to rehabilitate salmon populations in Maine.

RESEARCH AND MONITORING

Research and monitoring are needed to understand the status and trends of populations of wild salmon in Maine and to understand the effects and effectiveness of management and other human actions on salmon. The committee has pointed out knowledge gaps that make managing salmon more difficult. Yet research can affect the fish. At the Maine Atlantic Salmon Task Force (1997) pointed out, "Despite careful handling, fish may die from trauma when fisheries biologists capture salmon to collect necessary growth and population data."

In most cases, the number of fish killed by research is so small that it is not a serious consideration, but in several Maine rivers there are so few wild salmon that killing even one parr or smolt could affect the population. In addition, some kinds of handling and sampling seem likely to entail greater risks than others. The committee has concerns in particular about research that requires fish to be anesthetized, samples of blood or scales to be taken from very small fish, and the fish to be caught and held for long periods in strong currents, as might occur in a rotary-screw trap for smolts during high flows. The value of any information obtained needs to be weighed carefully against the possibility of the death of any wild fish subjected to handling, especially where wild populations are very small.

Noninvasive Methods of Estimating Numbers of Wild, Hatchery, and Farm Salmon in Streams

Accurate estimates of the annual abundance of various life stages of wild, hatchery, and farm salmon in Maine rivers and knowledge of other aspects of their genetic makeup are important for adaptive management. However, obtaining such information entails varying degrees of risks to the fish, especially when the fish are small. Therefore, the benefit of obtaining such data via electroshocking and rotary screw traps must be balanced against the risks of increased physiological stress and decreased survival posed by these collection methods and the subsequent handling of collected fish. It seems undesirable to add such stressors to wild salmon at a time when their numbers are as desperately low as they are at present. Therefore, the committee recommends the development of noninvasive fish counting (using visible external marks of hatchery and farm fish) to be used on a carefully selected representative sample of stream sites. For instance, in Norway underwater video systems for monitoring anadromous salmonids migrating up rivers are effective in registering fish and providing data on species and fish size (Lamberg et al. 2001). Underwater video recording is best developed for adults migrating up river but warrants consideration for adapting to counting a sample of parr in streams or outmigrating smolts.

If noninvasive sampling is infeasible or too costly at present, the committee suggests that until wild fish numbers rebuild substantially, invasive sampling be limited to counting smolts migrating down river with minimal holding time when rotary screw traps are used and that the collection of blood and other tissues be discontinued. That would reduce further stress to wild fish. Although genotyping of sampled fish would also be precluded, the committee judges that increasing the survival rates of wild salmon is more important in most cases than gaining additional data because of the low population sizes.

GOVERNANCE

As explained previously, the committee has not been able to assess the overall effectiveness and efficiency with which government agencies are contributing to the restoration and conservation of Atlantic salmon in Maine. Nor has the committee been able to evaluate the extent to which government agencies and other governance institutions and arrangements are capable of learning and adapting to new information and changing conditions in the natural and human environments. Barriers to learning from policy and other initiatives within and across institutions may have constrained the effectiveness of previous efforts to reverse the decline of wild Atlantic salmon in Maine as elsewhere (NRC 1996a). Such barriers need to be documented and addressed. Examples of such analyses are given in Burger et al. (2001) and NRC (2002e).

One strategy for dealing with this problem is to design policies based on the principles of adaptive management. From this perspective, policy initiatives need to be designed as experiments so that their impacts can be monitored, and lessons learned from these experiments can be used to inform future policy initiatives. Adaptive management could be particularly valuable in the design of initiatives, such as dam removal, that are unlikely to have negative impacts on remaining salmon. It is more appropriate, however, for situations where resource decline or extinction are not yet major issues. Adaptive management is not a no-cost or no-risk strategy because experiments can have unanticipated negative impacts and because there are costs associated with monitoring the effects of policy initiatives.

Additionally, the committee sees a need for the State Planning Office, or other legitimate authority, to conduct a systematic assessment of governance to determine whether there are gaps in authority; overlapping authority; conflicts of goals, interests, and values among agencies and groups; and adequate cooperation among government agencies as well as between these agencies and NGOs. Among other things, the study should determine whether the current ecology of governance contains disincentives or incentives for experimentation or other forms of learning; it should also determine the extent to which the public processes used to date have contributed to the development of effective strategies for conservation and rehabilitation of salmon habitat and salmon populations that are perceived as legitimate and credible by the different interest groups affected by these strategies. This is especially important since governance will play a major role in determining the success of efforts to restore and conserve Atlantic salmon in Maine.

To help guide this investigation, the committee notes that research done elsewhere on the rehabilitation of badly depleted salmon stocks has found that governance can pose a threat to salmon (and to other species) when governance institutions and their jurisdictional boundaries do not match the spatial, temporal, and functional scales of the salmon problem. One consequence of this mismatch is poor coordination of local, regional, national, and international rehabilitation efforts. One potential solution to this problem is to reshape governance structures to be consistent with salmon biology. This could involve developing multistakeholder governance institutions for each drainage basin, each nested within larger-scale governance bodies to address larger effects such as climate change and aquaculture (NRC 1996a).

The complexity of the natural history of Atlantic salmon, the extremely small remaining populations, and the broad range of threats to their survival identified in this report point to the challenging nature of the risk situation confronting any program for recovery of Atlantic salmon in Maine. This report contains a preliminary risk assessment for Atlantic salmon in Maine carried out by the committee that identifies many risk factors and ranks those factors. It also contains some partial decision analysis trees related to two key threats to Atlantic salmon: dams and aquaculture.

The development of a successful recovery program for Atlantic salmon in Maine will require a deeper and more sustained process of risk characterization and risk assessment than it was possible or feasible for this committee to undertake. Contrary to general practice, risk characterization involves much more than the translation of results of technical analyses into accessible language for decision makers. To date, this appears to have been the central component of efforts to diagnose the problem of Atlantic salmon in Maine. To be effective, risk characterization requires diverse and sustained participation by the full range of interested and affected parties throughout the process of diagnosing the situation, characterizing risks, risk assessment, decision analysis, and implementation of the recovery program (NRC 1996b).

A broad range of participants needs to be involved in the risk-characterization, risk-assessment, and decision-analysis process to design and implement an effective recovery program. Risk characterization is the outcome of an "analytic-deliberative process," with analytic referring to the collection of reliable, replicable information on hazards and exposures and deliberative referring to informal and formal processes for communication and collective consideration of issues (NRC 1996b, pp. 3–5). Those participating in the risk-characterization, risk-assessment, and decision-analysis process need to consider the magnitude of uncertainty and its sources and character. They need to get the right science, the right participation, and the participation right; and they need to develop an accurate, balanced, informative synthesis characterizing risk.

ADDITIONAL CONSERVATION OPTIONS WITH MULTIPLE ENVIRONMENTAL BENEFITS

The committee carefully considered a number of well established options that are not specifically targeted to help recovery of Atlantic salmon, but have been used to restore or enhance habitat for aquatic biota. The idea of adopting strategies that are likely to benefit wild Atlantic salmon and that are even more likely to improve the condition of other aquatic resources is particularly appealing. As is true for the other recommended options, all interested stakeholders should be involved in these decisions. Most of the options have been used effectively in other environmental and natural resource management programs, often in combination. The committee offers these recommendations not to compete with or displace the central tasks described above but to complement and reinforce them. The need for action on the ground expressed in other parts of this report clearly extends to these recommendations.

Among the 14 goals of the 1997 Conservation Plan for seven Maine Rivers (Maine Atlantic Salmon Task Force 1997), several focus on salmon habitat. These include (1) habitat protection, (2) water quality monitoring and management, (3) regulation of water withdrawals, (4) removal or mitigation of barriers to fish passage, and (5) protection or restoration of wetlands—an interrelated set of ecosystem attributes linked by water. This section provides detailed examples of strategies that relate to spatial data and management information systems, roads, irrigation withdrawals and return flow, agricultural chemicals, riparian forest buffers, forest management planning, forestry best management practices (BMPs), and recreational use and that reduce the adverse impacts of residential, commercial, and industrial development.

Although implicit in the 1997 conservation plan and other documents published by state and federal agencies and NGOs, it is not clear whether management objectives are being inventoried and analyzed in a way that systematically compares the merits, costs, benefits, and likelihood of success between watersheds. If it does not already exist, developing and maintaining a spatially referenced database of Maine rivers (and the principal tributaries of major rivers, such as the Penobscot and Kennebec) that includes the following attributes would be useful for strategic planning and comparison of watersheds.

- Area
- Mean daily stream flow
- Minimum and maximum flow of record (or estimates)
- Stream flow normalized by watershed area ($[m^3/sec]/km^2$)
- Number of National Pollutant Discharge Elimination System permits

- Total wastewater discharge
- Generalized land cover and land use (percent forest, percent agriculture, percent urban)
- Water withdrawal permits
- Number of dams
- Total height of dams
- Total area of Atlantic salmon habitat
- Maximum number of salmon returning, 1960–present, etc.

After completing the first iteration of the watershed assessment, a more detailed functional inventory of dams and other obstructions (culverts, bridges, channelized reaches, waterfalls, any hydraulic conditions or structures that inhibit fish passage) to fish passage could be developed to evaluate cumulative effects and design optimal conservation strategies. (The development of a detailed database for dams could occur in parallel with the watershed database to reduce delays in planning and implementation and could build on existing inventories, such as that by Elder (1987b).) The functional inventory of dams could include a number of key attributes.

- Location (global positioning system [GPS] coordinate)
- Proportion of watershed area above dam
- Height
- Condition (breached, leaking, intact)
- Fish passage structure (Yes/No? type, condition, effectiveness, etc.)
- Total habitat units above dam
- Habitat units between dam and next upstream obstacle
- Current use
- Historical use
- Potential for contaminated sediments
- Any other useful metrics

Both databases could be queried, sorted, and routinely updated to provide an objective foundation for project planning, sequencing, and implementation.

Roads

With the exception of large dams on the lower reaches of rivers, no human alteration of the landscape has a greater, more ubiquitous impact on aquatic habitat than roads. Every road-stream crossing has the potential to be a barrier to fish passage and a major source of sediment. A well-designed road, either paved or unpaved, has a slight crown along the centerline to direct rain or snowmelt off to the sides. In some cases, stormwater flows harmlessly off into the adjacent forest or fields and is termed "country drainage" by engineers. More often it is collected in ditches or swales that parallel the road, sometimes for long distances. As the volume and velocity of flow increases, so does the quantity of sediment that can be transported. Clay, silt, and fine sand that accumulate in road ditches arethe first to be transported to streams during rain and snowmelt events. Soil particles also carry nutrients, metals, and other potential non-point-source (NPS) pollutants on their charged surfaces. In addition, fine sediment increases turbidity in streams. Unless deliberate efforts are made to divert or store water and sediment along the way, they flow unimpeded into streams at every road crossing.

Even in large forested areas with low road densities, the alteration of natural pathways of flow can be significant. Removing forest cover increases the amount of precipitation reaching the surface. The earthwork, compaction, and surfacing (e.g, crushed stone, clay caps, bank-run gravel) needed to construct roads greatly limits the rate at which water can enter the soil. As a result, larger quantities of lower quality water are generated, concentrated, and directed downstream. These pulses of stormwater and sediment can destabilize stream channels, fill or cover redds, and contribute to eutrophication and acidification of streams.

A wide range of BMPs can be used to prevent and minimize the adverse impacts of roads on aquatic habitat. They include, but are not limited to, (1) careful route planning to keep roads on resistant terrain and minimize the number of road-stream crossings, (2) bridge and culvert designs with hydraulic characteristics that permit fish passage in both directions for different life stages, (3) bioengineering techniques to stabilize embankments (either cut or fill slopes) associated with road construction, (4) stormwater management practices to eliminate or reduce the hydraulic connections between roads and streams, (5) aggressive soil erosion control on new construction or unstable areas, and (6) regular preventive maintenance to prevent debris dams or beaver from blocking culverts. Although unglamorous, the last item is especially important to maintaining aquatic habitat quality. When a culvert is blocked, the road embankment becomes an earthen dam at least until the water flows over the road or pressure causes the saturated fill to give way. When the embankment fails, it sends a torrent of water, sediment, and debris downstream. In areas with multiple road-stream crossings, this can lead to a domino effect involving downstream structures. When true-cost accounting of long-term forest management is used, due diligence with BMPs and preventive maintenance are a bargain compared with replacing culverts, bridges, and road fills; dealing with enforcement orders and lawsuits for environmental and property damage; and the increased risk of motor vehicle accidents.

Irrigation Withdrawals and Return Flow

To remain competitive in international markets glutted with cultivated blueberries from more temperate areas, some farmers and most large commercial operations in Maine have begun to irrigate wild blueberry heaths throughout the growing season. This practice typically produces a threefold increase in crop yield and greatly reduces the fluctuations usually associated with the vagaries of New England weather. In fallow fields (berries are produced every other year), irrigation leads to more vigorous growth, an increase in root reserves for the following year, and a subsequent increase in flowering and fruit production. The season of peak blueberry irrigation usually corresponds with the annual minimum flows in Maine's streams and rivers (July and August). Direct withdrawal from streams causes unavoidable increases in water temperature, associated decreases in dissolved oxygen concentration, and as a result, increased stress for Atlantic salmon and other aquatic organisms. Some large operations (e.g., Cherryfield Foods, Inc.) have installed deep wells to supply irrigation water. Other growers (e.g., Lincoln Sennett) have constructed ponds to store snowmelt and spring rain for growing season application. As long as wells and ponds do not intercept appreciable quantities of water that would have entered streams and rivers, these forms of supply are clearly preferable to direct pumping.

As with any crop, when irrigation water is applied in excess of the plants' physiological requirements, the surplus water percolates through the root zone carrying whatever chemical constituents it has mobilized. If, for example, the fields are located on deep glacial outwash deposits, water from the root zone flows vertically (10 to 30 meters) until it reaches deep groundwater systems. By contrast, in areas of shallow (e.g., 1 or 2 meters) glacial till, water flows laterally over impermeable bedrock. This "return flow" to streams can be rapid and problematic if it carries nutrients, pesticides, or other agricultural chemicals. Because the blueberry farms in the Down East rivers are located in large blocks along the lower reaches, their influence is concentrated in the area traversed by all adult fish on their way upstream to spawn and all smolts on their way to the sea.

Agricultural Chemicals

Low-bush wild blueberry (*Vaccinium angustifolium*) is a small woody shrub that once grew in the understory of sparse forests, openings created by wildfires, or larger patches when soil and site conditions were too poor to support trees. It now grows in expansive fields (totaling about 40,000 acres across the Down East watersheds) that are intensively managed to maximize yields. Wild blueberries exhibit substantial clonal variation, which helps to limit the severity of insect and disease impacts in a monoculture. While not strictly an organic crop, blueberry growers are eager to promote the health benefits (high antioxidant content) and "wild mystique" of their product especially in the bakery trade and European and specialty markets (WBANA 2001). Therefore, most growers, especially large commercial operations, strive to minimize the use of agricultural chemicals.

Blueberry growers have supported the University of Maine's research and extension efforts since 1945. As a result, traditional practices and trial and error approaches have been supplanted by integrated pest management (IPM), integrated crop management (ICM), and other methods and approaches aimed at increasing efficiency and reducing cumulative environmental impact. Current research on water use efficiency holds promise for the improvement of irrigation practices, particularly the reduction or elimination of return flow. The establishment or enhancement of riparian buffers and windbreaks also shows an increasing awareness of potential off-site impacts. Prescribed fire is used to limit weed competition and prevent natural regeneration of trees and other forest vegetation in the blueberry fields. Although its effects should be quantified, it is likely that burning is more desirable than the use of herbicides especially in the Down East watersheds. Water quality data are so limited in the Down East region that it is not possible to quantify the effect, if any, of agricultural chemicals on Atlantic salmon and other parts of the aquatic ecosystem. A multiyear program of soil solution, groundwater, and stream chemistry, in an "above and below" or paired watershed (reference and treatment) design that includes flow proportional sampling is needed. Biomonitoring methods using aquatic macroinvertebrates also may help to assess mechanisms, patterns, and trends.

Riparian Forest Buffers

The riparian area is the transition between terrestrial and aquatic ecosystems (NRC 2002d). Vegetation in the riparian zone is critically important to the biotic integrity of aquatic ecosystems. Trees and other forest vegetation provide a suite of ecological services:

- Shade that helps to regulate water temperature
- Root support to stabilize banks and floodplains
- Inputs of organic carbon that comprise the base of the food web
- Leaf litter to protect soil from erosion and maintain high surface permeability
- Large woody debris to form pool habitat
- Hydraulic roughness to dissipate the energy of flood flows

- Nutrient uptake and assimilation
- Travel corridors for terrestrial wildlife and amphibians

To maintain these ecological services, the width of riparian forest buffers should be modified in relation to landform (both the floodplain and adjacent uplands) and the character and condition of the forest (Verry et al. 2000). While fixed-width buffer strips (usually 100 feet) are certainly preferable to gaps, one-size-fits-all does not fit most situations. Contemporary methods use the height of mature trees, slope, and landform to devise an appropriate and conservative (in both senses of the word) riparian forest buffer. The largest landowner in the Down East region, International Paper Company (formerly lands of Champion International), maintains a 1,000-foot buffer along the main stem of rivers that traverse its forestland. In an area where trees rarely exceed 100 feet, this represents corporate decision making in the face of ecological, regulatory, and political uncertainty. Notably, International Paper's mapping and harvest planning also includes riparian forest buffers on headwater tributaries. This avoids the common approach of designating large buffers on large rivers while neglecting small headwater streams that constitute the majority of the system. As a result, NPS pollution that enters in upstream areas flows right past large downstream buffers.

Project SHARE is undertaking a regionwide assessment of riparian forest buffers (RFBs). Using aerial photography, satellite imagery, geographic information systems (GIS), and field inspections, they will identify stream reaches that lack RFBs and devise site-specific restoration plans. They also have established a native plant nursery to produce growing stock (both trees and shrubs) that is appropriate for local conditions. The USDA Forest Service Northeastern Area is providing funding and technical assistance for this project.

Forest Management Planning

A brief summary is needed to explore the potential interaction of forestry and Atlantic salmon in Maine. Contemporary forest management involves the harvesting of trees to generate a sustainable supply of wood fiber for paper, lumber, and other forest products while avoiding or mitigating adverse impacts on other resources—water, fisheries, wildlife, recreation, aesthetics, and spiritual values. Long-term forest management on large public and industrial landholdings typically uses a 20-year strategic planning horizon (with detailed forest growth and yield projections that extend 100 to 200 years into the future) to systematically organize operations at the landscape scale. A 5-year business plan is used to optimize interrelated components and to determine sequencing of key components, such as (1) harvest areas and silvicultural prescriptions, (2) road construction, reactivation, or reclamation, (3) harvest schedules and expected volumes, and (4) plans and practices to protect other forest resources. Annual operating plans contain detailed schedules, contracts, budgets, health and safety, and staffing requirements, and contingency plans for unseasonable weather, natural disturbances (e.g., wildfires, floods), and short-term fluctuations in mill production schedules. Five-year plans are updated annually to reflect changes in the forest, including natural disturbance events. The 20-year plan serves as the benchmark as the 5-year plan is implemented.

Recent advances in computing and mapping technology have enhanced the detail and accuracy of forest management plans in several important ways. GIS have largely replaced conventional maps and aerial photographs that were the foundation of management planning from the 1930s through the late-1980s. GIS databases allow planners and managers to intersect, combine, or overlay themes or digital maps that represent multiple attributes of forest ecosystems. Digital imagery from satellites (10 to 30 meter resolution) or conventional aircraft (0.5 to 1 meter resolution) provides accurate depictions of forest vegetation types, wetlands, streams, rivers, and lakes. When coupled with field surveys using sample plots located with GPS and/or low-altitude flyovers with helicopters or light planes, the species composition, biomass, character, and condition of forest stands can be accurately mapped over large areas. This includes tree, shrub, and herbaceous cover in recently harvested areas. Sample plot and aerial survey data are extended over the remainder of the forest using the GIS and a wide range of statistical methods. Other ecosystem measurements are used to quantify the influence, positive or negative, of forest management and compliance with environmental laws and regulations. These efforts may include road stability surveys, stream reach assessments, water quality measurements, biomonitoring with aquatic macroinvertebrates, wildlife and recreational user surveys. How these data are used in planning and operations varies widely in the public and private sector. Whether environmental monitoring is proactive or reactive is largely a function of the corporate philosophy of the firm or agency.

There are several ways that state-of-the-art forest management planning could help to conserve Atlantic salmon populations in Maine. The first is simply by using terrain (digital elevation model), soils, land-cover data, and the GIS to map areas with management restrictions. These include, but are not limited to, (1) the designation of conservative riparian buffers along streams, lakes, and rivers, (2) contract restrictions on equipment and operating conditions (e.g., frozen or dry season only, slopes less than 15%), or (3) acceptable silvicultural systems (e.g., small group selection, patch cuts, patch retention). The second is to distribute the spatial pattern and temporal sequence of harvesting in a way that anticipates and avoids adverse cumulative effects on aquatic ecosystems.

A recent review and synthesis of long-term paired watershed studies by Hornbeck and colleagues (1993, 1997) suggest that reductions of forest biomass *or* forest area of 20% to 30% are needed to generate significant changes in water yield (stream flow volume and timing). Without significant increase in soil moisture and stream flow, nutrients mobilized by decomposition of organic matter are used by the trees and other forest vegetation adjacent to the openings or patches left by harvested trees. Even if the volume or area harvested exceeds 20% to 30% of any given watershed, the hydrologic influence of timber harvesting is short-lived in temperate climates. As the total leaf area of the regenerating stand approaches the mature trees that were cut, water yield returns to pre-harvest levels, usually in 3 to 5 years. The 1997 conservation plan notes that harvest areas for the period 1990 to 1994 ranged from 2% to 10% of the Down East watersheds. Depending on the spatial distribution, regeneration success, and growth rates, this may be far below the threshold identified by Hornbeck and colleagues (1993, 1997) or exceed thresholds at the subwatershed scale. In the latter case, the influence of timber harvesting near smaller tributaries with unobstructed, high-quality salmon habitat could be substantial even though they are protected with riparian forest buffers.

After delineating watersheds across a range of spatial scales—from first-order streams, to second- and third-order tributaries, up to the entire watershed for each river—analysts could use the GIS to test the spatial arrangement and temporal sequence of harvesting operations in proposed annual, 5-year, and 20-year plans. Using a spatially distributed model such as SNAP (Scheduling and Network Analysis Program, Sessions and Sessions 1997), a decision rule of, for example, 30% forest biomass removal would restrict subsequent harvests for a 5-year period in that headwater area. By summing all the harvested areas at intermediate and landscape scales, the same space and time thresholds could be evaluated. Of course, this requires landholdings of sufficient size to balance constraints on harvested area and time between entries, losses of fiber to natural disturbance, forest growth and yield, and the volume and grade requirements of the mills. It also adds additional complexity to road network design, use, and maintenance. In other words, since roads are clearly a more significant cause of adverse impacts than harvesting, a spatial and temporal harvesting pattern that requires a greater net road mileage would be counterproductive. In fact, minimizing the length of the active road network and the number of road-stream crossings could be used as additional objective functions in the model. Iterative or Monte Carlo simulation methods can be used to enumerate a broad range of possible management scenarios.

Forestry Best Management Practices

The profession of forestry was established in North America in response to the waste and destruction caused with industrial logging, floods, and catastrophic fires in late-1800s. While many associate best management practices (BMPs; more appropriately named conservation-management practices [CMPs] in Canada) with the Clean Water Act and other 1970s-vintage environmental laws and regulations, they have always been a central part of a professional forester's work. The work of Civilian Conservation Corps (CCC) in the 1930s could be largely characterized as the landscape- or even national-scale application of BMPs. For example, the reforestation of eroding farm fields, pastures, cutover and burned areas; stabilization and improvement of roads; construction of bridges over perennial streams (to replace fords and undersized box culverts); and a wide range of other activities transformed millions of acres in a decade of unprecedented effort and commitment. Unfortunately, World War II, the post-war building boom in the 1950s and 1960s, rapid mechanization of logging and road construction, coupled with the erosion of management standards and a strong conservation ethic, led to a general relapse to 1890s standards of practice. Progressive companies and diligent government agencies now require a suite of BMPs to protect the functions and values of forest ecosystems.

A comprehensive system of BMPs is needed to reinforce the effectiveness of individual practices and ensure that overall efforts are cost-effective and durable. Key principles for the adaptation or development of BMPs for regional and site-specific conditions include the following:

1. BMPs should be integrated with routine planning and operations; they should not be an after-the-fact addition or reaction to undesirable conditions.
2. Leaf litter and soil surface should be protected because it helps to retain the favorable hydraulic properties of forest soils (e.g., permeability and infiltration rate) and to avoid overland flow, soil erosion, nutrient mobilization, and sediment transport.
3. Whenever overland flow occurs, it should be deliberately dissipated or dispersed before it increases in volume and momentum.
4. Hydrologic connections between roads and harvest units and streams, lakes, and wetlands should be avoided.
5. Timber harvesting, road construction, road reclamation, and post-harvest site stabilization efforts should be adjusted to terrain and weather conditions.
6. Biological and physical control measures should be combined to enhance their effectiveness.

Forestry BMPs have been developed, tested, and refined for decades and number in the hundreds. Some examples of BMPs derived from the principles enumerated above, in addition to those already discussed for roads and riparian areas, include the following:

- Contract specifications, terms, and conditions that clearly state acceptable start and end dates, provisions for delays and extensions based on field conditions, performance standards for all aspects of the operation, performance bonds held in escrow accounts to motivate such factors as compliance and equipment type, size, and weight limits.
- Temporary bridges or brush mats to cross ephemeral streams or wetlands.
- Seeding of exposed soil with annual winter rye to ensure rapid revegetation while limiting the permanent introduction of exotic grasses and herbaceous plants (the rye dies and adds organic matter to soil as native species recolonize the site).
- Limiting the size of log landings by matching the log haul to harvest production rates (maximizing throughput to minimize the size of the disturbed area).
- Strict hazardous materials handling procedures in relation to heavy equipment maintenance and refueling operations.
- Gates on temporary logging roads to limit access by all-terrain and four-wheel-drive vehicles . . . and associated damage.
- Supervision by professional foresters on an as-needed basis (e.g., daily, weekly, random unannounced visits) to ensure compliance with contract specifications.

Recreational Use

Many forms of outdoor recreation (snowshoeing, cross-skiing, snowmobiling, canoeing, kayaking, hiking on well-designed trails, hunting, etc.) generate little or no impact on soils, water, and aquatic ecosystems. All-terrain and off-road vehicles (ATVs and ORVs) are a recent and notable exception. ATVs ("quads" or "four-wheelers") and ORVs (four-wheel-drive trucks and sport-utility vehicles) can cause substantial damage to soils, water, and aquatic ecosystems unless their use is carefully planned and managed. Whenever people reenact television commercials by fording streams, climbing steep banks or hills, and mixing, rutting, and compacting soil, they cause a host of environmental impacts. This damage may be inadvertent or intentional, but in either case, their actions can negate months or years of work to control NPS pollution in one Saturday afternoon.

COSTS OF OPTIONS

Estimating the costs of the options the committee has recommended for improving the survival prospects of Atlantic salmon in Maine is complex. The least difficult aspect of them—and the only one the committee addresses below—is the direct monetary costs of executing the options. Even those costs are accompanied by uncertainty, but a rough idea of their order of magnitude is provided below for some of the options, along with a discussion of the uncertainties associated with the estimates. The committee cannot provide any estimates of indirect costs and benefits, but they are important when considering the costs of various actions, and so they are discussed briefly here.

Many costs and benefits are not directly associated monetarily with a particular option. For example, time often is spent in lobbying for various outcomes, negotiating, legal activity, reviewing permit applications, consulting with colleagues and experts, and so on. These are real costs but only rarely are they directly accounted for. Other costs accrue over time, for example, as an accumulation of adverse effects of pollution or dams, adverse financial effects on businesses that are required to contribute to costs of executing options, or the accumulated effects on planning of uncertainty over what measures will be taken and when.

Different groups, organizations, and individuals have various interests. They can be affected differently by factors related to these options, some benefiting more than others from the status quo, others benefiting more than others from the proposed options. Most of the human activities that affect the survival of Atlantic salmon in Maine generate benefits to at least some people. To the extent that those activities are constrained for any reason, including protecting salmon, some costs will occur in the form of foregone benefits. In a few cases, such as a dam in disrepair that generates no power and provides no flood protection or recreational benefits, an action to protect salmon will probably have only direct costs and benefits, but such cases will be in the minority. Similarly, liming a small acidified stream probably has few hidden costs. But for the others, the hidden or indirect costs and benefits can be substantial.

For example, if a dam that blocks fish passage is retained, the dam's owners benefit from any net revenues generated by the dam and property owners adjacent to the pool behind the dam benefit from owning waterfront property. On the other hand, other groups and individuals suffer from the absence of migratory fish above the dam and from the loss of a free-flowing river there.

Different groups bear costs and enjoy benefits differently. For example, if a dam is removed, any loss of revenue associated with that removal directly affects the dam's owners, and any loss of tax revenue affects the relevant taxing jurisdiction. Property owners adjacent to the pool behind the dam lose the benefit of owning waterfront property. Other groups, however, benefit from the presence of migratory fish in new stretches of the river and from the existence of a free-flowing river. If an option affects the profitability of a salmon farm, its owners bear the loss. In addition, there are broader societal effects of options. In the case of the salmon farm, jobs could be lost if it loses profitability, and shareholders could be affected economically. In addition, jobs likely would be lost by those who provide products such as feed to the industry, and its demise could also affect retail and real-estate sales. But salmon anglers, commercial fishers, and the tourist industry could perhaps benefit from increased populations of wild salmon.

An additional complication is the uncertainty surrounding the effect of an option on salmon and its effect on other species of interest. There is no guarantee that implementing any of the options the committee recommends, or even all of them together, will lead to a recovery of wild salmon populations in Maine. That uncertainty is at least partially offset by the high probability that other species as well as a variety of ecosystem goods and services, such as provision of clean air and water, will benefit from the options. Other complications include the difficulty of taking into account the costs and benefits that might accrue to future generations; the costs and benefits of secondary effects, such as coming into compliance with environmental laws and regulations or the consequences of altering commercial operations; and other societal consequences. Many of the above issues are discussed in greater detail in Heinz Center (2002), especially with respect to dam removal.

The above and other factors should be considered for a full evaluation of the costs and benefits of the options and decisions about what actions to take. Even though the committee cannot provide quantitative estimates of those factors, they are important when considering the costs of various actions, and they should be taken into account.

Dam Removal

The cost of removing a dam depends on many factors, including the dam's size; how it was constructed; the need for compensation to its owners or users or beneficiaries; the amount of administrative, political, and legal work that is done. The societal costs and benefits of removing dams are also difficult to quantify (American Rivers et al. 1999, Heinz Center 2002). Below we provide some examples.

Edwards Dam

This privately owned dam, 917 feet long and 24 feet high, on the Kennebec River was removed in 1999. The Federal Energy Regulatory Commission (FERC) denied the request for relicensing. Following an appeal, a settlement was reached whereby the owners avoided building a $9 million fish ladder that would have been required by agreeing to the dam's removal. They paid the city of Augusta, a co-licensee, $100,000 to make up for lost revenue. Bath Iron Works, a shipbuilder, agreed to contribute $2.5 million in exchange for favorable consideration of its request to expand its shipyard on the river, and the Kennebec Hydro Developers Group of upstream dam operators contributed $4.75 million in return for extra time allowed for the installation of fish passage devices at their dams (Associated Press 1998). The money was used to remove the dam and to restore fish habitat. American Rivers et al. (1999) reported that it cost $2.9 million to remove the dam, including $800,000 for engineering and permitting, and that $4.85 million was provided for associated fish restoration efforts in the basin.

The costs listed above total more than $7 million. However, that is not the total cost of removing the dam. The Kennebec Hydro Developers Group has saved money by being allowed to postpone the installation of fish-passage devices, and the Bath Iron Works had the opportunity to increase revenue by expanding its shipyard. The time spent by all the people involved in reviewing license applications, filing appeals, lobbying, and other related activities is not included in the total. Societal benefits and costs are not included.

The Edwards Dam was one of the larger Maine dams obstructing the passage of salmon and adversely affecting their habitat. It took approximately 6 years to remove the dam: the license expired in 1993, the relicensing application was first denied in 1997, the agreement was signed in 1998, and the dam came down in 1999. Smaller dams, especially those that do not generate any power, would cost less and probably take less time to remove than Edwards, although there often are objections to the removal of dams that have large pools behind them. The objections often focus on loss of recreational opportunities and loss of water-front by property owners.

Grist Mill Dam

The Grist Mill Dam (GMD) on Souadabscook Stream is at the head-of-tide on this tributary to the Penobscot River and is the first obstacle anadromous fish encounter on returning to freshwater in this drainage. The dam was 14 feet high, 75 feet wide, and its removal in October 1998 cost $56,000 (American Rivers et al. 1999). Additional upstream dams were breached as well. Four salmon-spawning sites were discovered upstream of GMD in December 1998 (American Rivers et al. 1999). The process that led to the dam's removal took approximately 3 years.

Other Dams

Estimated costs of dam removal have exceeded $100 million for the Glines Canyon and Elwha dams on the Elwha River in northwest Washington (NRC 1996a). The NRC report (1996a) indicated that the large main-stem Columbia and Snake river dams would be much more expensive to remove; perhaps that cost could exceed $1 billion for each of those larger dams. The costs can be as low as thousands of dollars for removing small brush or even earth dams (e.g., $1,500 for the removal of the 3-foot-high Amish dam on Muddy Creek, PA, reported by American Rivers et al. [1999]). Several dams removed in Wisconsin, at least one of which was 13 feet high, cost a few hundred thousand dollars each, including restoring adjacent lands (American Rivers et al. 1999, Wisconsin River Alliance 2001). The recent agreement to remove two Penobscot River dams has an agreed-on initial cost of $25 million to be raised over 5 years (Richardson 2003).

Estimated Cost of Removing Maine Dams

Dams blocking Maine's rivers and streams range widely in size and construction materials. Most are smaller than the Edwards Dam. Assuming a cost of from $100,000 to $3 million per dam and the removal of three to five dams per year, the cost of this option would be between $300,000 and $15 million per year. The bearers of the cost would have to be determined by negotiation, legal action, or other processes. More information on estimating costs of dam removal is provided by the Heinz Center (2002).

Liming (Deacidifying) Streams

Liming is a method of reducing the acidity of streams by adding limestone, primarily calcium carbonate ($CaCO_3$). It often is regarded as one of the lower-cost methods of rehabilitating acid streams (Helfrich et al. 2000, Weigmann et al. 1993). However, costs vary according to the size of the stream and the equipment used. The cost of the limestone is the smallest expense, about $25–$100 per ton in 1993, including transportation. A rotary-drum limestone dispenser capable of dispensing 500 tons of limestone per year would have cost about $132,000 plus $16,500 for maintenance and perhaps $25,000 per year for the limestone, or an annual cost of a little more than $40,000. For 2,200 tons of limestone per year, the estimated costs in 1993 were $55,000 for an electric doser plus $12,100 per year to maintain and $110,000 for the limestone for an annual cost of $122,100. It would thus appear that this option, which would probably not incur significant ancillary political and societal costs, would be on the order of $100,000 initial cost plus $50,000–$100,000 per year for each stream treated.

Hatcheries

The committee's recommendations for improving hatchery operations would not require major additional expenditures in addition to what is currently being spent on federal hatchery operations for Atlantic salmon in Maine. However, there would be some additional costs. Tagging fry would cost some money and determining whether they are tagged and reading the tags would cost as well. A properly conducted research program involving paired streams might require additional employees and support and equipment.

Salmon Farms

The cost of many of the committee's suggested modifications of salmon farming cannot be reliably estimated because the costs of salmon farming operations are proprietary and because many factors—for example, the willingness of employees to move to work at a new site, the costs of various permitting and other legal and political requirements—are unknown. Nonetheless, it is clear that most of the modifications would likely cost enough to eliminate the profitability of salmon farms. Tagging all the fish reared on farms could be done most economically with an otolith tag, such as Terramycin, but even so, it would add significant additional expense to the operations. In addition, it would not provide a way to determine the source of any captured escapees. Coded-wire tags would allow identification of the origin of a particular fish but would be more expensive than otolith tags. This means that requiring most of the suggested modifications to salmon farms would result in the elimination of the salmon-farming industry in Maine, with the attendant costs of unemployment and other societal costs or it would require public or private subsidies.

6

Findings and Recommendations

FINDINGS

The decline of Atlantic salmon populations in Maine has been pervasive and substantial over the past 150 years, despite some periods in which they increased in numbers. The decline has brought them close to extinction in recent years. The combination and interaction of factors influencing salmon populations have been changing as well. Although salmon have declined over much of their natural range in Europe and North America in recent decades, suggesting that some factors affecting them operate over large areas, the severity of the declines in Maine warrants special attention. Maine's rivers and streams once had the capacity to support much larger salmon populations than they do now, so the potential exists to substantially increase the populations of wild salmon in Maine. In other words, rehabilitating salmon populations in Maine is challenging but appears possible.

The evidence suggests that regional climate change in Maine—mainly winter warming—has increased the difficulties encountered by salmon populations. Climate change, along with probably associated changes in oceanic conditions, appears to be an important factor affecting salmon, and it cannot be directly influenced by human intervention over the short and medium terms. The question arises as to whether the climate changes are so great that attempts to restore salmon are futile. The committee cannot answer that question, but there is no doubt that the changes make it more urgent to improve other aspects of their environments if wild salmon populations are to persist in Maine. In the absence of additional warming or other adverse climate changes, comprehensive efforts to rehabilitate salmon populations probably could be successful. Most of the measures designed to restore salmon populations would also benefit other native aquatic resources that depend on ecosystem services in these same watersheds.

Although genetic problems are important for Atlantic salmon in Maine, they appear to be less urgent than demographic problems. Given the choice of reducing an adverse genetic effect or reducing an adverse population effect, initial priority should be given to the population effect.

Dams appear to be the single most important class of impediments to salmon recovery that can be influenced by human actions in the short and medium terms. Although they are perhaps of smaller importance on the eight DPS streams than elsewhere in Maine, they are very important throughout the state.

Local populations of Atlantic salmon inhabiting different rivers and tributaries are demographically and genetically connected into so-called metapopulation systems through exchange of individuals. To the degree that this structure can be retained by maintaining or reestablishing salmon runs in many of Maine's rivers, the evolutionary future of salmon in Maine will be enhanced.

Aquaculture also appears to have an important and generally adverse effect on wild salmon populations, although reliable data are not available for Maine. Elsewhere, aquaculture has been shown to affect native salmon populations through ecological competition from escaped farm fish and through a large increase in the population density of parasitic copepods (sea lice). Other diseases can become concentrated in net-pens and affect wild fish as well. Even if the diseases are originally transferred to farms from wild fish, the concentration in the net-pens aggravates the problem. Although reliable data for such effects are lacking in Maine, similar effects are likely to occur there.

The evidence from over 130 years of stocking leads to the conclusion that hatchery production has not rescued Atlantic salmon in Maine. The evidence does not allow an objective assessment of whether, or to what degree, hatcheries have slowed the decline of Atlantic salmon in Maine. There has never been an adequate assessment of whether stocked salmon, when they return to spawn in Maine's rivers, successfully contribute offspring to the next generation. Reliance on hatcheries as the sole or primary intervention will not be sufficient to prevent extinction for very long.

Additionally, large releases of hatchery fish can have adverse effects on natural populations, as reviewed in Chapter 3. Current procedures for management of DPS river and Penobscot brood stock and offspring at the

Craig Brook National Fish Hatchery commendably avoid some of these hazards and could reduce additional hazards with feasible modifications. Some of the known hazards, however, are inherent to hatchery operations and cannot be fully avoided or substantially reduced. The committee concludes that hatcheries should be used sparingly in rehabilitation of natural populations and that published guidelines for reducing the adverse effects should be followed.

Survival of salmon at sea appears to be significantly depressed below that required to maintain robust populations. Other than the possible adverse effects of salmon farms and fishing, the factors involved, such as predation, competition, and adverse water temperatures, are not well understood and do not appear to be accessible to human control, at least for the short or medium terms.

The use of deep groundwater wells and storage ponds to irrigate agricultural crops (principally blueberries) does not appear to adversely affect stream flow and Atlantic salmon. By contrast, direct water withdrawals from streams, interacting with climate-induced changes in stream flow, could substantially degrade salmon habitat.

Timber harvesting does not currently appear to be a substantial problem for salmon. However, some forest practices (inappropriate road construction and deferred maintenance) have the potential to adversely affect salmon habitat quality and availability.

Some research that entails the collecting or trapping of fish appears to increase the risk of salmon mortality in streams with very small populations. Given the urgency of demographic problems, the committee questions the value of obtaining detailed genetic and physiological data on wild fry, parr, and smolts from such depleted populations.

Fishing has historically been a major source of mortality of Atlantic salmon. Currently, directed fishing for Atlantic salmon is prohibited in Maine and in most of the ocean that Maine salmon use. The Greenland salmon fishery is currently operated at a low level, but if it increased, it probably would affect Maine Atlantic salmon adversely. Recreational angling for brown and rainbow trout and landlocked salmon in waters that harbor wild anadromous Atlantic salmon is likely to add to the mortality of Atlantic salmon through bycatch. The amount of bycatch of Atlantic salmon in ocean fisheries is not known.

Water-quality degradation caused by atmospheric deposition (and subsequent acidification and metals mobilization) and pesticides (irrigation return or aerial drift) may threaten Atlantic salmon in subtle and pervasive ways. Historical and current monitoring programs are not sufficient to detect and evaluate these threats.

RECOMMENDATIONS

Many recommendations have been made for the rehabilitation of Atlantic salmon populations in Maine. Most of them are sound, but there are too many recommended actions to take at once. Moreover, not all of them are equally urgent. Most of the actions have been recommended by others, such as the Maine Atlantic Salmon Task Force, but here an attempt is made to set priorities for them and to recommend those actions most likely to be effective.

Urgently Needed Actions

There is an urgent need to reverse the decline of salmon populations in Maine if they are to be saved. Other than the salmon that returned to the Penobscot River, only 80 adult salmon were recorded to have returned to Maine's rivers in 2002. The serious depletion of salmon populations in Maine underscores the need to expand rehabilitation efforts to as many of Maine's rivers as possible. Since most Maine salmon are now in the Penobscot River, that population should be a primary focus for rehabilitating the species in Maine. The committee recommends the following urgent actions:

- A program of dam removal should be started. Priority should be given to dams whose removal would make the greatest amount of spawning and rearing habitat available, meaning that downstream dams should be considered for removal before dams upstream of them. In some cases, habitat restoration will likely be required to reverse or mitigate some habitat changes caused by a dam, especially if the dam is many decades old. The recent agreement to remove two Penobscot River dams (Richardson 2003) is encouraging.
- The problem of early mortality as smolts transition from freshwater to the ocean and take up residence as post-smolts needs to be solved. If, as seems likely, that the difficulty of the transition is due in part to water chemistry, particularly acidification, the only methods of solving the problem are changing the water chemistry and finding a way for the smolts to bypass the dangerous water. Liming has had considerable success in counteracting acidification in many streams, and the techniques are well known. Examples of its application are in nearby Nova Scotia. Liming should be tried experimentally on some Maine streams as soon as possible. Bypassing the dangerous water is best achieved by rearing smolts and acclimating them to seawater in controlled conditions. This approach is not appealing because of the degree of human intervention required and because of the adverse selection that must result from it.

Given the extreme depletion of salmon populations, however, desperate measures are called for.

• Hatcheries need to continue to be used, at least in the short term, to supplement wild populations and to serve as a storehouse of fish from the various rivers. There is an urgent need to understand the relative efficiency of stocking of different life stages in the rivers in terms of adult returns per brood-stock fish and their reproductive success. Additional research on hatcheries and scientific guidance for their use is needed, because hatchery-based restoration of wild salmon populations remains an unproven technology. Indeed, hatcheries themselves should be used adaptively as scientific tools for obtaining additional information.

The approximate costs of these options are discussed in Chapter 5.

Actions Important over the Longer Term

• Over the longer term, the committee recommends a comprehensive decision analysis approach to the rehabilitation of Atlantic salmon populations in Maine. The analysis should be conducted along the lines of the examples in Chapter 5 of this report but in more detail and with all major groups of stakeholders involved. Taking a Maine-wide view is more likely to be successful than focusing only on some rivers.

• No anadromous Atlantic salmon of any life stage should be stocked in rivers that have populations of wild Atlantic salmon unless those rivers are specifically identified as part of a hatchery-recovery program that uses river-specific stocks. Stocking of nonnative fish species and landlocked salmon also should be avoided in those rivers. Other rivers that once supported wild Atlantic salmon runs, but which lack them now, will probably become repopulated by strays from nearby streams if populations in those nearby streams recover. The advantages over stocking of such natural repopulation, which would be more likely to lead to local genetic adaptation, should be given serious attention before any decision is made to stock streams that currently lack wild Atlantic salmon, runs.

• The current prohibition of commercial and recreational fishing, including catch-and-release fishing, for salmon in Maine should be continued. Any further reduction in the take of Maine salmon at sea would be helpful. Maximum and minimum size limits for trout and landlocked salmon should be established in rivers that have anadromous Atlantic salmon. The minimum size for retention should be large enough to protect Atlantic salmon smolts, and the maximum size should be small enough to protect adult Atlantic salmon. Any fishing that might take a wild Atlantic salmon constitutes an additional risk to the species. This risk should be carefully evaluated for all Maine rivers with Atlantic salmon and additional measures should be taken if the risk is judged to be important. Habitat zones most heavily used by Atlantic salmon young and adults should be closed to fishing for all species until salmon populations have recovered.

- Research that increases the risk of death to wild fish should be curtailed. The value of any information obtained needs to be weighed against the likelihood of increased death of wild fish subjected to handling.
- Every effort should be made to further curtail the escape of salmon from farms. If accumulation of parasitic copepods (sea lice) or other pathogens is found to be a problem for wild salmon, the aquaculture facilities should be moved to a place where they will not adversely affect wild salmon.
- Hatchery practices should be evaluated in an adaptive-management context to further reduce adverse genetic and ecological effects, and modified as needed.
- The monitoring of water quality and gauging of streams should be augmented. A network of metereological-monitoring, stream-gauging, water-quality-monitoring, and biological-monitoring sites should be linked to a geographic information system and an online database within 2 years.
- Government, industry, and private organizations and landowners should cooperate to evaluate forestry best-management practices and forest-road networks. Mitigation and pollution prevention should be organized on a priority basis to maximize the effectiveness of stormwater management and sediment control and the removal of barriers to fish passage.
- The State Planning Office should conduct a systematic governance assessment to see whether there are gaps in authority, overlapping authority, conflicts of goals and interests among agencies, and adequate cooperation among agencies.
- The State Planning Office, in cooperation with all other agencies, should implement adaptive management to monitor performance of governance activities related to Atlantic salmon, to experiment with alternative institutions for salmon recovery, and to systematically learn and adapt to the results of new information.
- The Maine Atlantic Salmon Commission should consider shaping governance structures so that they are consistent with salmon biology. That could involve developing multistakeholder governance institutions for each drainage basin, each nested within larger-scale governance bodies to address effects, such as climate change and aquaculture, that are larger than individual basins.
- The suite of conservation options with multiple environmental ben-

efits outlined in Chapter 5 should be adopted. Those strategies are likely to help Atlantic salmon in Maine, and they will have other environmental benefits even if they do not help salmon. The energy and commitment of the members of many local watershed and river-specific groups focused on restoring salmon and their habitats is an important asset and should be included in any overall approach to rehabilitating Atlantic salmon and their habitats in Maine.

References

Alden, R. 1997. Maine's Aquaculture Strategy. Maine Department of Marine Resources.

Allendorf, F.W., and S.R. Phelps. 1980. Loss of genetic variation in a hatchery stock of cutthroat trout. Trans. Am. Fish. Soc. 109:537-543.

Allendorf, F.W., and N. Ryman. 1987. Genetic management of hatchery stocks. Pp. 141–159 in Population Genetics and Fishery Management, N. Ryman, and F. Utter, eds. Seattle: University of Washington Press.

American Rivers, Friends of the Earth, and Trout Unlimited. 1999. Dam Removal Success Stories: Restoring Rivers Through Selective Removal of Dams That Don't Make Sense. American Rivers, Washington DC. [Online]. Available: http://www.foe.org/camps/reg/nw/river/dam.html [Nov. 12, 2002].

AMS (Agricultural Marketing Service). 2003. AMS Mission Statement. Agricultural Marketing Service, U.S. Department of Agriculture. [Online]. Available: http://www.ams.usda.gov/admin/mission.htm [accessed May 19, 2003].

Andersson, M.B. 1994. Sexual Selection. Princeton: Princeton University Press.

Antonsson, T., G. Gudbergsson, and S. Gudjonsson. 1996. Environmental continuity in fluctuation of fish stocks in the North Atlantic Ocean, with particular reference to Atlantic salmon. N. Am. J. Fish. Manage. 16(3):540–547.

APHIS (Animal and Plant Health Inspection Service). 2002. APHIS Strategic Plan, Mission. APHIS Strategic Plan, 2000–2005. Animal and Plant Health Inspection Service, U.S. Department of Agriculture. [Online]. Available: http://www.aphis.usda.gov/lpa/about/strategic_plan/archive/strategic_plan_00-05/strategic_mission.html [accessed May 19, 2003].

APHIS (Animal and Plant Health Inspection Service). 2003. National Aquaculture Program, APHIS Services for the Aquaculture Industry. Veterinary Services, Animal and Plant Health Inspection Service, U.S. Department of Agriculture. [Online]. Available: http://www.aphis.usda.gov/vs/aqua/aquaphis.html [accessed May 19, 2003].

Ardren, W.R., and A.R. Kapuscinski. 2003. Demographic and genetic estimates of effective population size (Ne) reveals genetic compensation in steelhead trout. Mol. Ecol. 12(1):35–49.

Arkush, K.D, A.R. Giese, H.L. Mendonca, A.M. McBride, G.D. Marty, and P.W. Hedrick. 2002. Resistance to three pathogens in the endangered winter-run chinook salmon (*Oncorhynchus tshawytscha*): Effects of inbreeding and major histocompatibility complex genotypes. Can. J. Fish. Aquat. Sci. 59(6):966–975.

ASMFC (Atlantic State Marine Fisheries Commission). 2003. Interstate Fisheries Management Program, ASMFC Programs. Atlantic State Marine Fisheries Commission, Washington, DC. [Online]. Available: http://www.asmfc.org/serv02.htm [accessed May 8, 2003].

Associated Press. 1998. In Maine, Old Timber Dam to Come Down so Fish Can Flourish. Associated Press, Section-Domestic News. May 27, 1998.

Atkins, C.G. 1869. Reports of the Commissioners of Fisheries of the State of Maine for the Years 1867 and 1868. State Printing Office, Augusta, ME.

Atkins, C.G. 1874. On the salmon of eastern North America, and its artificial culture. Pp. 226–337 in Report of the Commissioner for 1872–1873, S.L. Baird, ed. Washington, DC: U.S. Govt. Print. Off.

Atlantic Salmon Federation. 1999. ASF Finds Infectious Salmon Anemia (ISA) in Aquaculture Escapees and Wild Atlantic Salmon. Atlantic Salmon Federation Research and Environment Department, Research Update. October 1999. [Online]. Available: http://www.asf.ca/Research/ISAinwild.html [accessed Dec.19, 2002].

ATSDR (Agency for Toxic Substances and Disease Registry). 2003. Public Health Assessment: Region 1. Maine, Eastern Surplus Company, Meddybemps, ME. [Online]. Available: http://www.atsdr.cdc.gov/HAC/PHA/easternsurplus/ [accessed Oct. 20, 2003].

Austin, D.A., D. McIntosh, and B. Austin. 1989. Taxonomy of fish associated *Aeromonas* spp., with the description of *Aeromonas salmonicida* subsp. *smithia* subsp. nov. Syst. Appl. Microbiol. 11:277–290.

Baggerman, B. 1960. Salinity preference, thyroid activity and the seaward migration of four species of Pacific salmon (*Oncorhynchus*). J. Fish. Res. Board Can. 17(3):295–322.

Bailey, R.G. 1995. Descriptions of the Ecoregions of the United States, 2nd Rev. Ed. Misc. Pub. No.1391. Washington, DC: U.S. Dept. of Agriculture, Forest Service.

Bailey, J. 1998. Options for Containment of Farmed Atlantic Salmon. DFO Canadian Stock Assessment Secretariat Research Document 98/165. Division of Fisheries and Oceans, Canada.

Baird, S.F. 1876. Pp. VII–XLVI in Report of the Commissioner for 1873–1874 and 1874–1875, Part 3. U.S. Commission of Fish and Fisheries. Washington, DC: U.S. Govt. Print. Off.

Ballou, J.D., and T.J. Foose. 1996. Demographic and genetic management of captive populations. Pp. 263–283 in Wild Mammals in Captivity: Principles and Techniques, D.G. Kleiman, M.E. Allen, K.V. Thompson, and S. Lumpkin, eds. Chicago: University of Chicago Press.

Bardonnet, A., and J.-L. Baglinière. 2000. Freshwater habitat of Atlantic salmon (*Salmo salar*). Can. J. Fish. Aquat. Sci. 57(2):497–506.

Barten, P.K., T. Kyker-Snowman, P.J. Lyons, T. Mahlstedt, R. O'Connor, and B.A. Spencer. 1998. Massachusetts: Managing a watershed protection forest. J. For. 96(8):10–15.

Bartley, D., M. Bagley, G. Gall, and B. Bentley. 1992. Use of linkage disequilibrium data to estimate effective size of hatchery and natural fish populations. Conserv. Biol. 6(3):365–375.

Baum, E., M. Colligan, and J. McKeon. 2002. Development and implementation of habitat protection and restoration plans according to the NASCO plan of action in the USA. Pp. 53–74 in Habitat Protection and Restoration, Report of Special Session of NASCO, Torshavn, Faroe Islands, June 2002. [Online]. Available: http://www.nasco.int/pdf/nasco_hpr_report2002.pdf [accessed May 21, 2004].

Baum, E.T. 1997. Maine Atlantic Salmon: A National Treasure, 1st Ed. Harmon, ME: Atlantic Salmon Unlimited.

Baum, E.T. 1998. A Description of the Salmon Aquaculture Industry in Maine. DFO Canadian Stock Assessment Secretariat Research Document 98/152. Division of Fisheries and Oceans, Canada.

Baum, E.T., and R.M. Jordan. 1982. The Narraguagus River. Atlantic Salmon River Management Report No. 7. Bangor, ME: State of Maine Atlantic Sea Run Salmon Commission.

Bean, M.L. 1983. The Evolution of National Wildlife Law, 2nd Ed. New York: Praeger.

Beland, K.F. 1984. Strategic Plan for Management of Atlantic Salmon in the State of Maine. Bangor, ME: Atlantic Sea Run Salmon Commission.

Beland, K.F. 1986. Atlantic Salmon Management in the Kennebec River: A Status Report and Interim Management Plan. Maine Sea-Run Atlantic Salmon Commission, Bangor, ME.

Beland, K.F., P. Gaston, T. King, J. Kocik, and I. Kornfield. 1997. Management and Spawning Protocols for Atlantic Salmon Broodstocks at Craig Brook National Fish Hatchery. Report to Maine Atlantic Salmon Technical Advisory Committee. Atlantic Salmon Commission, Bangor, ME.

Berejikian, B.A. 1995. The effects of hatchery and wild ancestry and experience on the relative ability of steelhead trout fry (*Oncorhynchus mykiss*) to avoid a benthic predator. Can. J. Fish. Aquat. Sci. 52(11):2476–2482.

Berkes, F. 1997. New and not-so-new directions in the use of the commons: Co-management. Common Property Resource Digest 42:5–7. (Reproduced in Collaborative Management News (IUCN), No. 2, January 1998).

Berkes, F. 1999. Sacred Ecology: Traditional Ecological Knowledge and Resource Management. Philadelphia, PA: Taylor& Francis.

Bigg, G.R. 2000. Historical and potential long-term climatic change in the north Atlantic. Pp. 137–152 in the Ocean Life of Atlantic Salmon: Environmental and Biological Factors Influencing Survival, D.H. Mills, ed. Malden, MA: Fishing News Books.

Biological Review Team for Anadromous Atlantic Salmon. 1999. Review of the Status of Anadromous Atlantic Salmon (*Salmo salar*) Under the U.S. Endangered Species Act. National Oceanic and Atmospheric Administration and U.S. Fish and Wildlife Service, Washington, DC. [Online]. Available: http://news.fws.gov/salmon/asalmon.html [accessed Nov. 12, 2001].

Bisulca, P. 1996. Penobscot: A People and Their River. Conservation Matters. [Online]. Available: http://www.penobscotnation.org/articles/bisulca.htm. [accessed May 8, 2003].

Bjørn, P.A., and B. Finstad. 1998. The development of salmon lice (*Lepeophtheirus salmonis*) on artificially infected post smolts of sea trout (*Salmo trutta*). Can. J. Zool. 76(5):970–977.

Bjørn, P.A., and B. Finstad. 2002. Salmon lice, *Lepeophtheirus salmonis* (Krøyer), infestation in sympatric populations of Arctic char, *Salvelinus alpinus* (L.), and sea trout, *Salmo trutta* (L.), in areas near and distant from salmon farms. ICES J. Mar. Sci. 59(1):131–139.

Bjørn, P.A., B. Finstad, and R. Kristoffersen. 2001. Salmon lice infection of wild sea trout and Arctic char in marine and freshwaters: The effects of salmon farms. Aquacult. Res. 32(12):947–962.

Björnsson, B.T. 1997. The biology of salmon growth hormone: From daylight to dominance. Fish Physiol. Biochem. 17:9–24.

Blackwell, B.F. 1996. Ecology of Double-Crested Cormorants Using the Penobscot River and Bay, Maine. Ph.D. Thesis, University of Maine, Orono, ME.

Bley, P.W., and J.R. Moring. 1988. Freshwater and Ocean Survival of Atlantic Salmon and Steelhead: A Synopsis. Biological Report 88(9). Washington, DC: U.S. Fish and Wildlife Service, U.S. Dept. of the Interior.

Boeuf, G. 1993. Salmonid smolting: A pre-adaptation to the oceanic environment. Pp. 105–135 in Fish Ecophysiology, J.C. Rankin, and F.B. Jensen, eds. London: Chapman and Hall.

Bosch, J.M., and J.D. Hewlett. 1982. A review of catchment experiments to determine the effect of vegetation changes on water yield and evapotranspiration. J. Hydrol. 55(1/4):3–23.

Bourque, B.J. 1995. Diversity and Complexity in Prehistoric Maritime Societies: A Gulf of Maine Perspective. New York: Plenum Press.

Bowen, J.T. 1970. A history of fish culture as related to the development of fishery programs. Pp. 71–93 in a Century of Fisheries in North America, N.G. Benson, ed. Washington, DC: American Fisheries Society.

Brackett, J. 1991. Potential disease interaction of wild and farmed fish. Bull. Aquacult. Assoc. Can. 91-3:79–80.

Brennan, W.J. 1999. Aquaculture in the Gulf of Maine: A Compendium of Federal, Provincial and State Regulatory Controls, Policies and Issues. Prepared for The Gulf of Maine Council on the Marine Environment. [Online]. Available: http://www.gulfofmaine.org/library/aquaculture_compendium.htm [accessed May 16, 2003].

Brooks, D.A., M.W. Baca, and Y-T. Lo. 1998. Sustainable salmon aquaculture and tidal flushing in a macro tidal ecosystem: Cobscook Bay, Maine. Pp. 977–982 in Oceans '98: Conference Proceedings, September 28 – October 1, 1998, Nice, France, Vol. 2. Piscataway, NJ: Oceans '98 IEEE/OES Conference Organizing Committee.

Brown, S.B., R.E. Evans, and H.S. Majewski. 1990. Responses of plasma electrolytes, thyroid, hormones, and gill histology in Atlantic salmon (*Salmo salar*) to acid and limed river waters. Can. J. Fish. Aquat. Sci. 47(12):2431–2440.

B.S.H. 1876. Sea and river fishing. Fish in season in June. Forest and Stream 6:319.

Buckley, D.B. 1999. Summary of Maine Atlantic Salmon Collections for Broodstock and Genetic Analyses 1990–1998. U.S. Fish and Wildlife Service, East Orland, ME.

Buckley, D. 2002a. Summary of Activities in 2000 and 2001 Related to Maine Atlantic Salmon Broodstock Management and Genetic Sample Collections, April 2002. U.S. Fish and Wildlife Service, Maine Fishery Resources Office, East Orland, ME.

Buckley, D. 2002b. Summary of Spawning Procedures at Craig Brook National Fish Hatchery 1999–2001, June 2002. U.S. Fish and Wildlife Service, Maine Fishery Resources Office, East Orland, ME.

Burger, J., E. Ostrom, R.B. Norgaard, D. Policansky, and B.D. Goldstein. 2001. Protecting the Commons: A Framework for Resource Management in the Americas. Washington, DC.

Busack, C.A., and K.P. Currens. 1995. Genetic risks and hazards in hatchery operations: Fundamental concepts and issues. Pp. 71–80 in Uses and Effects of Cultured Fishes in Aquatic Ecosystems, H.L. Schramm, and R.G. Piper, eds. American Fisheries Society Symposium 15. Bethesda, MD: American Fisheries Society.

Cairns, D.K. 2001. An Evaluation of Possible Causes of the Decline in Pre-fishery Abundance of North American Atlantic Salmon. Canadian Technical Report of Fisheries and Aquatic Sciences No. 2358. Charlottetown: Dept. of Fisheries and Oceans, Science Branch.

Campton, D.E. 1995. Genetic effects of hatchery fish on wild populations of Pacific salmon and steelhead: What do we really know? Pp. 337–353 in Uses and Effects of Cultured Fishes in Aquatic Ecosystems, H.L. Schramm, and R.G. Piper, eds. American Fisheries Society Symposium 15. Bethesda, MD: American Fisheries Society.

Carlson, C. 1988. Where's the salmon? A reevaluation of the role of anadromous fisheries to aboriginal New England. Pp. 47–80 in Human Holocene Ecology in Northeastern North America, G.P. Nicholas, ed. New York: Plenum Press.

Carlson, C.C. 1993. The (In)significance of Atlantic Salmon in New England. Presented at New England's Creatures: 1400–1900, August 6–8, 1993, Historic Deerfield, MA.

Carrington, M., G.W. Nelson, M.P. Martin, T. Kissner, D. Vlashov, J.J. Goedert, R. Kaslow, S. Buchbinder, K. Hoots, and S.J. O'Brien. 1999. HLA and HIV-1: Heterozygote advantage and B*35-Cw*04 disadvantage. Science 283(5408): 1748–1752.

Chadwick, E.M.P. 1987. Causes of variable recruitment in a small Atlantic salmon stock. Pp. 390–401 in Common Strategies of Anadromous and Catadromous Fishes, M.J. Dadswell, R.J. Klauda, C.M. Moffitt, R.L. Saunders, R.A. Rulifson, and J.E. Cooper, eds. American Fisheries Society Symposium 1. Bethesda, MD: American Fisheries Society.

Chase, S. No date. Closing the North American Mixed Stock Commercial Fishery for Wild Atlantic Salmon. Atlantic Salmon Federation, St. Andrews, New Brunswick and Calais, Maine.

Clemen, R.T. 1991. Making Hard Decisions: An Introduction to Decision Analysis. Boston: PWS Kent.

Clifford, S.L., P. McGinnity, and A. Ferguson. 1998. Genetic changes in Atlantic salmon (*Salmo salar*) populations of Northwest Irish rivers resulting from escapes of adult farm salmon. Can. J. Fish. Aquat. Sci. 55(2):358–363.

Colebrook, J.M. 1991. Continuous plankton records: From seasons to decades in the plankton of the north-east Atlantic. Pp. 29–45 in Long Term Variability of Pelagic Fish Populations and Their Environments, T. Kawasaki, S. Tanaka, Y. Toba, and A. Taniguchi, eds. Oxford: Pergamon Press.

Collette, B.B., and G. Klein-McPhee, eds. 2002. Bigelow and Schroeder's Fishes of the Gulf of Maine. 3rd Ed. Washington, DC: Smithsonian Institution Press.

Conkling, P.W. 2000. Fish or foul? Will aquaculture carve out a niche in the Gulf of Maine? Maine Policy Review 9(2):12–19. [Online]. Available: http://www.umaine.edu/mcsc/MPR/Vol9No2/Conkling.htm [accessed Nov. 12, 2002].

Cotter, D., V. O'Donovan, N. O'Maoileidigh, G. Rogan, N. Roche, and N.P. Wilkins. 2000. An evaluation of the use of triploid Atlantic salmon (*Salmo salar* L.) in minimizing the impact of escaped farmed salmon on wild populations. Aquaculture 186(1/2):61–75.

Cronon, W. 1983. Changes in the Land: Indians, Colonists, and the Ecology of New England, 1st Ed. New York: Hill and Wang.

Cross, T.S., and J. King. 1983. Genetic effects of hatchery rearing in Atlantic salmon. Aquaculture 33:33–40.

Cunjak, R.A. 1988. Physiological consequences of overwintering in streams: The cost of acclimatization? Can. J. Fish. Aquat. Sci. 45:443–452.

Cunjak, R.A., T.D. Prowse, and D.L. Parrish. 1998. Atlantic salmon (*Salmo salar*) in winter: "The season of parr discontent"? Can. J. Fish. Aquat. Sci. 55(suppl.):161–180.

CVM (Center of Veterinary Medicine). 1999. What does FDA do with aquaculture? In Transgenic Fish Developments: Are Transgenic Fish & Shellfish in Our Future? Presented at Genetically Engineered Advances in Aquaculture: Reshaping America's Seafood Supply? San Francisco Seafood Show, November 4, 1999, Center of Veterinary Medicine, U.S. Food and Drug Administration. [Online]. Available: http://www.fda.gov/cvm/biotechnology/shellfish/sld013.htm [accessed May 19, 2003].

CVM (Center of Veterinary Medicine). 2002. Information about Center of Veterinary Medicine. Center of Veterinary Medicine, U.S. Food and Drug Administration [Online]. Available: http://www.fda.gov/cvm/aboutcvm/aboutcvm.html [accessed May 19, 2003].

Dempson, J.B., and D.G. Reddin. 2000. Newfoundland and Labrador Atlantic Salmon (*Salmo salar*) Stock Status for 1999, and Methods to Address Possible Causes for Continuing Low Abundance and Survival. Canadian Stock Assess. Secretariat Res. Doc. 2000/112. [Online]. Available: http://www.dfo mpo.gc.ca/csas/Csas/Francais/Recherche_Annee/2000/2000_112f.htm [accessed May 9, 2003].

DEP (Maine Department of Environmental Protection). 1996a. Planning Projects to Meet Permit-by-Rule Standards. Fact Sheet, June 1996. Maine Department of Environmental Protection. [Online]. Available: http://www.state.me.us/dep/blwq/docstand/fs plan.htm [accessed May 14, 2003].

DEP (Maine Department of Environmental Protection). 1996b. DEP Issue Profile: Regulation of Dams. Maine Department of Environmental Protection, Augusta, ME. July 1996. [Online]. Available: http://www.state.me.us/dep/blwq/docstand/ip-dams.htm [accessed Dec. 18, 2002].

DEP (Maine Department of Environmental Protection). 1996c. Wetland Protection: A Federal, State and Local Partnership. Issue Profile. Maine Department of Environmental Protection, Augusta, ME. September 1996. [Online]. Available: http://www.state.me.us/dep/blwq/docstand/ipwetfsl.htm [accessed May 14, 2003].

DEP (Maine Department of Environmental Protection). 2000. Hydropower Projects in Maine. DEPLW-82-C00. Maine Department of Environmental Protection. June 1, 2000. [Online]. Available: http://www.state.me.us/dep/blwq/docstand/hydropage.htm [accessed May 14, 2003].

DEP (Maine Department of Environmental Protection). 2003. Atlantic Salmon Rivers. Maine Department of Environmental Protection. June 1, 2000. [Online]. Available: http://www.state.me.us/dep/blwq/docmonitoring/salmon/index.htm [accessed May 14, 2003].

Department of the Secretary of State. 2002. 94-409—Maine Indian Tribal-State Commission, 2001–2002 Regulatory Agenda. Department of the Secretary of State Maine. [Online]. Available: http://www.state.me.us/sos/cec/rcn/apa/agendas/agenda02/94-409 .htm [accessed May 7, 2003].

DFO (Department of Fisheries and Oceans-Canada). 2000. The Effects of Acid Rain on the Atlantic Salmon of the Southern Upland of Nova Scotia. DFO Maritimes Regional Habitat Status Report 2000/2E. [Online]. Available: http://www.mar.dfo-mpo.gc.ca/science/rap/internet/hsr2000-2e.pdf [accessed May 27, 2003].

Dickson, R. 1997. From Labrador Sea to global change. Nature 386:649–650.

Dickson, R.R., and W.R. Turrell. 2000. The NAO: The dominant atmospheric process affecting oceanic variability in home, middle and distant waters of European Atlantic salmon. Pp. 92–115 in the Ocean Life of Atlantic Salmon: Environmental and Biological Factors Influencing Survival, D.H. Mills, ed. Malden, MA: Fishing News Books.

Dickson, R.R., J. Lazier, J. Meincke, P. Rhines, and J. Swift. 1996. Long-term coordinated changes in the convective activity of the North Atlantic. Progr. Oceanogr. 38(3):241–295.

Dickson, R.R., T.J. Osborn, J.W. Hurrell, J. Meincke, J. Blindheim, B. Adlandsvik, T. Vinje, G. Alekseev, and W. Maslowski. 2000. The Atlantic Ocean response to the North Atlantic oscillation. J. Clim. 13(15):2671–2696.

DIFW (Maine Department of Inland Fisheries and Wildlife). 2002a. Fishing on Open Waters in Indian Territory. Maine Department of Inland Fisheries and Wildlife. [Online]. Available: http://www.state.me.us/ifw/fishing/f-indian.htm [accessed May 8, 2003].

DIFW (Maine Department of Inland Fisheries and Wildlife). 2002b. Fish Health Laboratory. Maine Department of Inland Fisheries and Wildlife. [Online]. Available: http://www.state.me.us/ifw/fishing/fishlab/intro.htm [accessed July 13, 2002].

DIFW (Maine Department of Inland Fisheries and Wildlife). 2002c. Saltwater Fishing Information. Maine Department of Inland Fisheries and Wildlife. [Online]. Available: http://www.state.me.us/ifw/fishing/f-salt.htm [accessed June 5, 2002].

DIFW (Maine Department of Inland Fisheries and Wildlife). 2002d. Maine Department of Inland Fisheries and Wildlife. [Online]. Available: http://www.state.me.us/ifw/index. [accessed June 23, 2002].

DIFW (Maine Department of Inland Fisheries and Wildlife). 2002e. Fishing. Maine Department of Inland Fisheries and Wildlife. [Online]. Available: http://www.state.me.us/ifw/fishing/fishing.htm [accessed June 23, 2002].

DIFW (Maine Department of Inland Fisheries and Wildlife). 2003a. About Us. Maine Department of Inland Fisheries and Wildlife. [Online]. Available: http://www.state.me.us/ifw/aboutus/aboutus.htm [accessed Oct. 22, 2003].

DIFW (Maine Department of Inland Fisheries and Wildlife). 2003b. Maine Online Sportsman's Electronic System (MOSES), a project of the Maine Department of Inland Fisheries and Wildlife. [Online]. Available: http://www.state.me.us/ifw/mainemoses/ [accessed May 14, 2003].

DiGiulio, R.T., and D. Tillitt. 1999. Reproductive and Developmental Effects of Contaminants in Oviparous Vertebrates: Proceedings from the Pellston Workshop on Reproductive and Developmental Effects of Contaminants in Oviparous Vertebrates, 13–18 July 1997, Fairmont Hot Springs Resort, Anaconda, Montana. Pensacola, FL: Society of Environmental Toxicology and Chemistry.

DMR (Maine Department of Marine Resources). 2001. Aquaculture Lease Inventory. Maine Department of Marine Resources, West Boothbay Harbor, ME. [Online]. Available: http://www.state.me.us/dmr/aquaculture/aqualeaseinventory.htm [accessed Nov. 13, 2002].

DMR (Maine Department of Marine Resources). 2002. Marine Aquaculture. Maine Department of Marine Resources. [Online]. Available: http://www.maine.gov/dmr/aquaculture/index.htm [accessed June 6, 2002].

Doremus, H. 2000. Delisting endangered species: An aspirational goal, not a realistic expectation. Environ. Law Report. News Anal. 30(6):10434–10454.

Drinkwater, K.F. 2000. Changes in ocean climate and its general effect on fisheries: Examples from the north-west Atlantic. Pp. 116–136 in the Ocean Life of Atlantic Salmon: Environmental and Biological Factors Influencing Survival, D.H. Mills, ed. Malden, MA: Fishing News Books.

Duchesne, P., and L. Bernatchez. 2002. An analytical investigation of dynamics of inbreeding in multi-generation supportive breeding. Conserv. Gen. 3(1):45–58.

Dudley, R.W., and G.A. Hodgkins. 2002. Trends in Streamflow, River Ice, and Snowpack for Coastal River Basins in Maine during the 20th Century. WRI 02-4245. MAINE. Open-Files, U. S. Geological Survey, Augusta, ME.

Duston, J., and R.L. Saunders. 1990. The entrainment role of photoperiod on hypoosmoregulatory and growth-related aspects of smolting in Atlantic salmon (*Salmo salar*). Can. J. Zool. 68(4):707–715.

Duston, J., and R.L. Saunders. 1999. Effect of winter food deprivation on growth and sexual maturity of Atlantic salmon (*Salmo salar*) in seawater. Can. J. Fish. Aquat. Sci. 56(2): 201–207.

Duston, J., R.L. Saunders, and D.E. Knox. 1991. Effects of increases in freshwater temperature on loss of smolt characteristics in Atlantic salmon (*Salmo salar*). Can. J. Fish. Aquat. Sci 48(2):164–169.

Dutil, J.D., and J.M. Coutu. 1988. Early marine life of Atlantic salmon, *Salmo salar*, post-smolts in the northern Gulf of St Lawrence. Fish Bull. 86(2):197–212.

e-Hormone. 2003. e-Hormone Web site. Center for Bioenvironmental Research at Tulane and Xavier University. [Online]. Available: http://e.hormone.tulane.edu/ [accessed Oct. 18, 2003].

Einum, S., and I.A. Fleming. 1997. Genetic divergence and interactions in the wild among native, farmed and hybrid Atlantic salmon. J. Fish Biol. 50(3):634–651.

Einum, S., and I.A. Fleming. 2001. Implications of stocking: Ecological interactions between wild and released salmonids. Nord. J. Freshwat. Res. 75:56–70.

Elder, B. 1987a. Existing and Former Dams. (Map). State of Maine Office of Energy Resources, Augusta, ME.

Elder, B. 1987b. Inventory of Existing and Former Dams in Maine. State of Maine Office of Energy Resources, Augusta, ME.

Eldridge, W. H., M.D. Bacigalupi, I.R. Adelman, L.M. Miller, and A. R. Kapuscinski. 2002. Determination of relative survival of two stocked walleye populations and resident natural-origin fish by microsatellite DNA parentage assignment. Can. J. Fish. Aquat. Sci. 59(2):282–290.

EPA (U.S. Environmental Protection Agency). 2002a. Reregistration Eligibility Science Chapter for Atrazine Environmental Fate and Effects Chapter. U.S. Environmental Protection Agency. [Online]. Available: www.epa.gov/oppsrrd1/reregistration/atrazine/efed_redchap_22apr02.pdf [accessed Oct. 10, 2003].

EPA (U.S. Environmental Protection Agency). 2002b. EPA Issues Aquaculture Permit for Maine Fish Farm. New England Press Releases: February 21, 2002. U.S. Environmental Protection Agency, Region 1: New England. [Online]. Available: http://www.epa.gov/region1/pr/2002/feb/020213.html [accessed May 16, 2003].

EPA (U.S. Environmental Protection Agency). 2003a. Clean Water Act. Office of Laws and Regulations, U.S. Environmental Protection Agency, Region 5. [Online]. Available: http://epa.gov/region5/water/cwa.htm [accessed May 16, 2003].

EPA (U.S. Environmental Protection Agency). 2003b. Endangered Species Act. Major Environmental Laws. U.S. Environmental Protection Agency, Region 5. [Online]. Available: http://www.epa.gov/Region5/defs/html/esa.htm [accessed May 16, 2003].

EPA (U.S. Environmental Protection Agency). 2003c. About EPA's Pesticides Program. Office of Pesticides Program, U.S. Environmental Protection Agency. [Online]. Available: http://www.epa.gov/pesticides/about/aboutus.htm [accessed May 16, 2003].

EPA (U.S. Environmental Protection Agency). 2003d. What "Integrated Pest Management" Means. Pesticides and Food. Human Health Issues. Office of Pesticides, U.S. Environmental Protection Agency. [Online]. Available: http://www.epa.gov/pesticides/food/ipm.htm [accessed May 16, 2003].

EPA (U.S. Environmental Protection Agency). 2003e. How the Government Regulates Pesticides. Pesticides and Food. Human Health Issues. Office of Pesticides, U.S. Environmental Protection Agency. [Online]. Available: http://www.epa.gov/pesticides/food/govt.htm [accessed May 16, 2003].

EPA (U.S. Environmental Protection Agency). 2003f. Section 404 of the Clean Water Act: An Overview. Wetlands Facts Sheets. Office of Wetlands, Oceans and Watersheds, Office of Water, U.S. Environmental Protection Agency. [Online]. Available: http://www.epa.gov/owow/wetlands/facts/fact10.html [accessed May 16, 2003].

Fairchild, W.L., E.O. Swansburg, J.T. Arsenault, and S.B. Brown. 1999. Does an association between pesticide use and subsequent declines in catch of Atlantic salmon (*Salmo salar*) represent a case of endocrine disruption? Environ Health Perspect. 107(5):349–358.

Farmer, G.J., R.L. Saunders, and T.R. Goff. 1989. Some physiological responses of Atlantic salmon (*Salmo salar*) exposed to soft, acidic water during smolting. Aquaculture 82(1/4):229–244.

FDA (U.S. Food and Drug Administration). 1990. Color Additives. U.S. Food and Drug Administration, U.S. Department of Health and Human Services. February 1990. [Online]. Available: http://www.fda.gov/opacom/backgrounders/coloradd.html [accessed May 21, 2003].

FDA (U.S. Food and Drug Administration). 2001. Laws, Regulations, Policies and Procedures for Drug Applications. Center for Drug Evaluation and Research, U.S. Food and Drug Administration. [Online]. Available: http://www.fda.gov/cder/regulatory/applications/laws.htm [accessed May 21, 2003].

FDA (U.S. Food and Drug Administration). 2003. Food Ingredients and Packaging: Approval and Notification Programs. Center for Food Safety and Applied Nutrition, U.S. Food and Drug Administration, College Park, MD. [Online]. Available: http://www.cfsan.fda.gov/~lrd/foodadd.html [accessed May 21, 2003].

Felt, L., B. Neis, and B.J. McCay. 1997. Comanagement. Pp. 185–194 in Northwest Atlantic Groundfish: Perspectives on a Fishery Collapse, J. Boreman, B.S. Nakashima, J.A. Wilson, and R.L. Kendall, eds. Bethesda, MD: American Fisheries Society.

FERC (Federal Energy Regulatory Commission). 2003. Hydropower. Federal Energy Regulatory Commission, Washington, DC. [Online]. Available: http://www.ferc.gov/hydro/docs/waterpwr.htm [accessed May 21, 2003].

Ferguson, M.M., R.G. Danzman, and F.W. Allendorf. 1988. Developmental success of hybrids between two taxa of salmonid fishes with moderate structural gene divergence. Can. J. Zool. 66:1389–1395.

FHWA (Federal Highway Administration). 2003. Federal Highway Administration Maine Division. U.S. Department of Transportation. [Online]. Available: http://www.fhwa.dot.gov/mediv/maine.htm [accessed May 21, 2003].

Finstad, B., P.A. Bjørn, A. Grimnes, and N.A. Hvidsten. 2000. Laboratory and field investigations of salmon lice [*Lepeophtheirus salmonis* (Krøyer)] infestation on Atlantic salmon (*Salmo salar* L.) post-smolts. Aquacult. Res. 31(11):795–804.

Fisher, J.P., and W.G. Pearcy. 1988. Growth of juvenile coho salmon (*Onchorynchus kisutch*) off Oregon and Washington, USA, in years of differing coastal upwelling. Can. J. Fish. Aquat. Sci. 45:1036–1044.

Fisk, A. 2002. Review of Aquaculture Leasing in Maine. Department of Marine Resources. January 2002. [Online]. Available: http://www.state.me.us/dmr/aquaculture/lease_report.htm [accessed Dec. 18, 2002].

Flagg, T.A., and C.E. Nash. 1999. A Conceptual Framework for Conservation Hatchery Strategies for Pacific Salmonids. Seattle, WA: U.S. Dept. of Commerce, National Oceanic and Atmospheric Administration, National Marine Fisheries Service, Northwest Fisheries Science Center.

Flagg, T.A, B.A. Berejikian, J.E. Colt, W.W. Dickhoff, L.W. Harrell, D.J. Maynard, C.E. Nash, M.S. Strom, R.N. Iwamoto, and C.V.W. Mahnken. 2000. Ecological and Behavioral Impacts of Artificial Production Strategies on the Abundance of Wild Salmon Populations: A Review of Practices in the Pacific Northwest. NOAA Technical Memorandum NMFS-NWFSC 41. Seattle, WA: U.S. Dept. of Commerce, National Oceanic and Atmospheric Administration, National Marine Fisheries Service, Northwest Fisheries Science Center.

Fleming, I.A. 1996. Reproductive strategies of Atlantic salmon: Ecology and evolution. Rev. Fish Biol. Fish. 6(4):379–416.

Fleming, I.A., and S. Einum. 1997. Experimental tests of genetic divergence of farmed from wild Atlantic salmon due to domestication. ICES J. Mar. Sci. 54(6):1051–1063.

Fleming, I.A., and M.R. Gross. 1989. Evolution of adult female life history and morphology in Pacific salmon (Coho: *Onchorhynchus kisutch*). Evolution 43(1):141–157.

Fleming, I.A., and M.R. Gross. 1993. Breeding success of hatchery and wild coho salmon (*Oncorhynchus kisutch*) in competition. Ecol. Appl. 3(2):101–121.

Fleming, I.A., and E. Petersson. 2001. The ability of released, hatchery salmonids to breed and contribute to the natural productivity of wild populations. Nord. J. Freshwater. Res. 75:71–98.

Fleming, I.A., B. Jonsson, and A. Lamberg. 1996. An experimental study of the reproductive behavior and success of farmed and wild Atlantic salmon. J. Appl. Ecol. 33(4):893–905.

Fleming, I.A., K. Hindar, I.B. Mjølnerød, B. Jonsson, T. Balstad, and A. Lamberg. 2000. Lifetime success and interactions of farm salmon invading a native population. Proc. R. Soc. Lond Biol. Sci. 267(1452):1517–1523.

Fleming, I.A., T. Agustsson, B. Finstad, J.I. Johnsson, and B.T. Björnsson. 2002. Effects of domestication on growth physiology and endocrinology of Atlantic salmon (*Salmo salar*). Can. J. Fish. Aquat. Sci. 59(8):1323–1330.

Fletcher, J.S., R.M. Jordan, and K.F. Beland. 1982. The Machias river. Pp. 1–38 in Machias & East Machias, K.F. Beland, N.R. Duber, J.S. Fletcher, and R.M. Jordan, eds. Atlantic Salmon River Management Report No. 6. Bangor, ME: State of Maine Atlantic Sea Run Salmon Commission.

Foster, D.R. 1999. Thoreau's Country: Journey through a Transformed Landscape. Cambridge, MA: Harvard University Press.

Foster, D.R., and J.F. O'Keefe. 2000. New England Forests through Time: Insights from the Harvard Forest Dioramas. Cambridge, MA: Harvard University Press.

Foye, R.E., C.F. Ritzi, and R.P. AuClair. 1969. Fish Management in the Kennebec River. Augusta, ME: Maine Dept. of Inland Fisheries and Game.

Frank, K.T., J.E. Carscadden, and J.E. Simon. 1996. Recent excursions of capelin (*Mallottus villosus*) to the Scotian Shelf and Flemish Cap during anomalous hydrographic conditions. Can. J. Fish. Aquat. Sci. 53(7):1473–1486.

Friedland, K.D. 1994. History of salmon fisheries and management in the North Atlantic. Pp. 6–22 in Ninth ICES Dialogue Meeting, "Atlantic Salmon: A Dialogue" ICES Cooperative Research Report 197. International Council for the Exploration of the Sea, Copenhagen, Denmark. March 1994.

Friedland, K.D. 1994a. Marine survival of restoration stocks. Pp. 223–239 in a Hard Look at Some Tough Issues, S. Calabi, and A. Stout, eds. Camden, ME: Silver Quill Books.

Friedland, K.D. 1998. Ocean climate influences on critical Atlantic salmon (*Salmo salar*) life history events. Can. J. Fish. Aquat. Sci. 55(Suppl. 1):119–130.

Friedland, K.D., and L.W. Haas. 1988. Emigration of juvenile Atlantic menhaden, *Brevoortia tyrannus* (Pisces: Clupeidae), from the York River estuary. Estuaries 11(1):45–50.

Friedland, K.D., and D.G. Reddin. 2000. Growth patterns of Labrador Sea Atlantic salmon postsmolts and the temporal scale of recruitment synchrony for North American salmon stocks. Can. J. Fish. Aquat. Sci. 57(6):1181–1189.

Friedland, K.D., D.G. Reddin, and J.F. Kocik. 1993. Marine survival of North American and European Atlantic salmon: Effects of growth and environment. ICES J. Mar. Sci. 50(4): 481–492.

Friedland, K.D., L.P. Hansen, and D.A. Dunkley. 1998. Marine temperatures experienced by post-smolts and the survival of Atlantic salmon, *Salmo salar* L., in the North Sea Area. Fish. Oceanogr. 7:22–34.

Friedland, K.D., J.D. Dutil, and T. Sadusky. 1999. Growth patterns in postmolts and the nature of the marine juvenile nursery for Atlantic salmon, *Salmo salar*. Fish. Bull. 97(3):472–481.

Friedland, K.D., L.P. Hansen, D.A. Dunkley, and J.C. MacLean. 2000. Linkage between ocean climate, post-smolt growth, and survival of Atlantic salmon (*Salmo salar* L.) in the North Sea area. ICES J. Mar. Sci. 57(2):419–429.

Friedland, K.D., D.G. Reddin, and M. Castonguay. 2003. Ocean thermal conditions in the post-smolt nursery of North American Atlantic salmon. ICES J. Mar. Sci. 60(2):343–355.

FSIS (Food Safety and Inspection Service). 2001. Protecting the Public from Foodborne Illness: The Food Safety and Inspection Service. Backgrounders, Food Safety and Inspection Service, U.S. Dept. of Agriculture, Washington, DC. [Online]. Available: http://www.fsis.usda.gov/OA/background/fsisgeneral.htm [accessed May 19, 2003].

FWS (U.S. Fish and Wildlife Service). 1998. Branch of Federal Activities Homepage. Division of Habitat Conservation, U.S. Fish and Wildlife Service, Arlington, VA. [Online]. Available: http://www.fws.gov/r9dhcbfa/wwwcover.htm [accessed May 14, 2003].

FWS (U.S. Fish and Wildlife Service). 2003a. Mission Statement. Message from the Director and Deputy Director. Strategic Plan-1999–2003. Government Performance and Results Act. U.S. Fish and Wildlife Service. [Online]. Available: http://planning.fws.gov/Mission.html#miss [accessed May 14, 2003].

FWS (U.S. Fish and Wildlife Service). 2003b. Endangered Species Habitat Conservation Planning. U.S. Fish and Wildlife Service. [Online]. Available: http://endangered.fws.gov/hcp/index.html [accessed May 14, 2003].

FWS (U.S. Fish and Wildlife Service). 2003c. Permits for Native Species. The Endangered Species Program, U.S. Fish and Wildlife Service. [Online]. Available: http://endangered.fws.gov/permits/index.html [accessed May 14, 2003].

Garant, D., I.A. Fleming, S. Einum, and L. Bernatchez. 2003. Alternative male life-history tactics as potential vehicles for speeding introgression of farm salmon traits into wild populations. Ecol. Lett. 6(6):541–549.

Garcia-Vazquez, E., P. Moran, J.L. Martinez, J. Perez, B. de Gaudemar, and E. Beall. 2001. Alternative mating strategies in Atlantic salmon and brown trout. J. Hered. 92(2):146–149.

Gaston, P.B. 1988. Atlantic Salmon Culture for Restoration. U.S. Fish and Wildlife Service, Newton Corner, MA.

Getchell, R. 1997. ISA investigation continues: New disease challenges fish. Fish Farming News. 5(Oct):12.

Gharrett, A.J., and W.W. Smoker. 1991. Two generations of hybrids between even- and odd-year pink salmon a test for outbreeding depression? Can. J. Fish. Aquat. Sci. 48(9):1744–1749.

Gharrett, A.J., and W.W. Smoker. 1993. A perspective on the adaptive importance of genetic infrastructure in salmon populations to ocean ranching in Alaska. Fish. Res. 18(1/2):45–58.

Ghent, A.W., and B.P. Hanna. 1999. Statistical assessment of Huntsman's 3-y salmon-rainfall correlation, and other potential correlations, in the Miramichi fishery, New Brunswick. Am. Midland Nat. 142(1):110–128.

Gilpin, M.E., and M.E. Soulé. 1986. Minimum viable populations: Processes of species extinction. Pp. 13–34 in Conservation Biology: The Science of Scarcity and Diversity, M.E. Soulé, ed. Sunderland, MA: Sinauer Associates.

Gjedrem, T., H.M. Gjoen, and B. Gjerde. 1991. Genetic origin of Norwegian farmed Atlantic salmon. Aquaculture 98(1/3):41–50.

Gjerde, B. 1984. Response to individual selection for age at sexual maturity in Atlantic salmon. Aquaculture 38(3):229–240.

Glebe, B.D. 1998. East Coast Salmon Aquaculture Breeding Programs: History and Future. DFO Canadian Stock Assessment Secretariat Research Document 98/157. Dept. of Fisheries and Oceans, Canada.

Grahn, M., A. Langesfors, and T. von Schantz. 1998. The importance of mate choice in improving viability in captive populations. Pp. 341–363 in Behavioral Ecology and Conservation Biology, T.M. Caro, ed. New York: Oxford University Press.

Griffith, D.M., and C.L. Alerich. 1996. Forest Statistics for Maine, 1995. Resource Bulletin NE-135. Radnor, PA: U.S. Dept. of Agriculture, Forest Service, Northeastern Forest Experiment Station.

Griffiths, S.G. 1994. Characterization of bacteria associated with cold water vibriosis of Atlantic salmon in eastern Canada. Pp. 64 in 3rd International Marine Biotechnology Conference Program, Abstracts and List of Participants, Tromsø. Tromsø University, Tromsø, Norway.

Grimnes, A., and P.J. Jakobsen. 1996. The physiological effects of salmon lice infection on post-smolt of Atlantic salmon. J. Fish Biol. 48(6):1179–1194.

Gross, M.R. 1985. Disruptive selection for alternative life histories in salmon. Nature 313:47–48.

Gross, M.R. 1996. Alternative reproductive strategies and tactics: Diversity within sexes. Trends Ecol. Evolut. 11(2):92–98.

Gross, M.R. 1998. One species with two biologies: Atlantic salmon (*Salmo salar*) in the wild and in aquaculture. Can. J. Fish Aquat. Sci. 55(Supp.1):131–144.

Gross, M.R., and J. Repka. 1998a. Game theory and inheritance in the conditional strategy. Pp. 168–187 in Game Theory and Animal Behavior, L.A. Dugatkin, and H.K. Reeve, eds. New York: Oxford University Press.

Gross, M.R., and J. Repka. 1998b. Stability with inheritance in the conditional strategy. J. Theor. Biol. 192(4):445–453.

Gudjónsson, S., S.M. Einarsson, T. Antonsson, and G. Gudbergsson. 1995. Relation of grilse/salmon ratio to environmental changes in several wild stocks of Atlantic salmon (*Salmo salar*) in Iceland. Can. J. Fish. Aquat. Sci. 52(7):1385–1398.

Gulf of Maine Council on the Marine Environment. 2003. Gulf of Maine Council on the Marine Environment. [Online]. Available: http://www.gulfofmaine.org/index.html [accessed May 8, 2003].

Haines, T.A., S.A. Norton, J.S. Kahl, C.W. Fay, S.J. Pauwels, and C.H. Jagoe. 1990. Intensive Studies of Stream Fish Populations in Maine. EPA/600/3-90/043. Washington, DC: U.S. Environment Protection Agency, Office of Acid Deposition, Environmental Monitoring and Quality Assurance.

Hallerman, E. 2002. Coadaptation and outbreeding depression. Pp. 239–259 in Population Genetics: Principles and Practices for Fisheries Scientists, E.M. Hallerman, ed. Bethesda, MD: American Fisheries Society.

Hamilton, W.D., and M. Zuk. 1982. Heritable true fitness and bright birds: A role for parasites? Science 218(4570):384–387.

Hansen, L.P., and J.A. Jacobsen. 2000. Distribution and migration of Atlantic salmon, *Salmo salar* L., in the sea. Pp. 75–87 in the Ocean Life of Atlantic Salmon: Environmental and Biological Factors Influencing Survival, D.H. Mills, ed. Malden, MA: Fishing News Books.

Hansen, L. P., and B. Jonsson. 1985. Downstream migration of hatchery-reared smolts of Atlantic salmon (*Salmo salar* L.) in the river Imsa, Norway. Aquaculture 45: 237–248.

Hansen, L.P., and B. Jonsson. 1991. The effect of timing of Atlantic salmon smolt and post-smolt release on the distribution of adult return. Aquaculture 98(1/3):61–67.

Hansen, L.P., and T.P. Quinn. 1998. The marine phase of the Atlantic salmon (*Salmo salar*) life cycle, with comparisons to Pacific salmon. Can. J. Fish. Aquat. Sci. 55(Suppl. 1): 104–118.

Hansen, M.M., D.E. Ruzzante, E.E. Nielsen, and K-L.D. Mensberg. 2000. Microsatellite and mitochondrial DNA polymorphism reveals life-history dependent interbreeding between hatchery trout and wild brown trout (*Salmo trutta* L.). Mol. Ecol. 9(5):583–594.

Hara, M. 2003. Co-management of natural resources: Theory and the attendant assumptions. Pp. 13–36 in Waves of Change: Coastal and Fisheries Co-management in Southern Africa, M. Hauck, and M. Sowman, eds. Cape Town: University of Cape Town Press.

Hare, S.R., N.J. Mantua, and R.C. Francis. 1999. Inverse production regimes: Alaskan and West Coast Salmon. Fisheries 24(1):6–14.

Håstein, T., and T. Lindstad. 1991. Diseases in wild and cultured salmon: Possible interaction. Aquaculture 98(1-3):277–288.

Havey, K.A. 1968. A Study of Standing Crops of Juvenile Salmon and Their Competitors in a Salmon Spawning and Nursery Stream. Job Completion Report No. 1, Proj. F-14-R-9. Maine Dept. of Inland Fisheries & Game, Augusta, ME.

Haya, K., B.A. Waiwood, and L. Van Eeckhaute. 1985. Disruption of energy metabolism and smoltification during exposure of juvenile Atlantic salmon (*Salmo salar*) to low pH. Comp. Biochem. Physiol. C. 82(2):323–329.

Heath, D.D., C. Busch, J. Kelly, and D.Y. Atagi. 2002. Temporal change in genetic structure and effective population size in steelhead trout (*Oncorhynchus mykiss*). Mol. Ecol. 11(2):197–214.

Hedrick, P.H. 2001. Conservation genetics: Where are we now? Trends Ecol. Evol. 16(11): 629–636.

Hedrick, P.W., V.K. Rashbrook, and D. Hedgecock. 2000. Effective population size of winter-run chinook salmon based on microsatellite analysis of returning spawners. Can. J. Fish. Aquat. Sci. 57(12):2368–2373.

Heggberget, T.G., R.A. Lund, N. Ryman, and G. Ståhl. 1986. Growth and genetic variation of Atlantic salmon (*Salmo salar*) from different sections of the river Alta, north Norway. Can. J. Fish. Aquat. Sci. 43(10):1828–1835.

Heinz Center. 2002. Dam Removal: Science and Decision Making. Washington, DC: The John Heinz III Center for Science, Economics, and the Environment.

Helfrich, L., J. Parkhurst, and R. Neves. 2000. Liming Acidified Lakes and Ponds. Virginia Cooperative Extension of Virginia Polytechnic Institute and State University. [Online]. Available: http://www.ext.vt.edu/pubs/fisheries/420-254/420-254.html [accessed Oct. 10, 2003].

Hennessey, T.M. 1994. Governance and adaptive management for estuarine ecosystems: The Case of Chesapeake Bay. Coast. Manage. 22(2):119–145.

Heuch, P.A, and T.A. Mo. 2001. A model of salmon louse production in Norway: Effects of increasing salmon production and public management measures. Dis. Aquat. Organ. 45(2):145–152.

Hindar, K., N. Ryman, and F. Utter. 1991. Genetic effects of cultured fish on natural fish populations. Can. J. Fish. Aquat. Sci. 48(5):945–957.

Hoar, W.S. 1965. The endocrine system as a chemical link between the organism and its environment. Trans. Roy. Soc. Canada 3 (Ser. 4):175–200.

Hoar, W.S. 1988. The physiology of smolting salmonids. Pp. 275–343 in Fish Physiology, Vol. 11. The Physiology of Developing Fish—Part B. Viviparity and Posthatching Juveniles, W.S. Hoar, and D.J. Randall, eds. New York: Academic Press.

Hockett, K. S. 1994. The Effects of Cover and a Model Kingfisher on the Anti-avian Predator Response of Juvenile Atlantic Salmon (*Salmo salar*). M.S. Thesis, University of Maine, Orono. 135pp.

Holbrook, S.H. 1938. Holy Old Mackinaw: A Natural History of the American Lumberjack. New York: The Macmillan Company.

Holling, C.S. 1978. Adaptive Environmental Assessment and Management. Chichester: Wiley. 377 pp.

Holm, M., J.C. Holst, and L.P. Hansen. 2000. Spatial and temporal distribution of post-smolts of Atlantic salmon (*Salmo salar* L) in the Norwegian Sea and adjacent area. ICES J. Mar. Sci. 57(4):955–964.

Holmes, M.J. 2001. ISA virus found at two more aquaculture sites in Maine. The Quoddy Tides April 27, 2001.

Holst, J.C., R. Shelton, M. Holm, and L. Hansen. 2000. Distribution and possible migration routes of post-smolt Atlantic salmon in the north-east Atlantic. Pp. 65–74 in the Ocean Life of Atlantic Salmon: Environmental and Biological Factors Influencing Survival, D.H. Mills, ed. Malden, MA: Fishing News Books.

Holtby, L.B., and D.P. Swain. 1992. Through a glass, darkly: A response to Ruzzante's reappraisal of mirror image stimulation studies. Can. J. Fish. Aquat. Sci. 49:1968–1969.

Holtby, L.B., B.C. Andersen, and R.K. Kadowaki. 1990. Importance of smolt size and early ocean growth to interannual variability in marine survival of coho salmon (*Oncho-rynchus kisutch*). Can. J. Fish. Aquat. Sci. 47(11):2181–2194.

Hornbeck, J.W., M.B. Adams, E.S. Corbett, E.S. Verry, and J.A. Lynch. 1993. Long-term impacts of forest treatments on water yield: A summary for northeastern USA. J. Hydrol. 150(2/4):323–344.

Hornbeck, J.W., C.W. Martin, and C. Eager. 1997. Summary of water yield experiments at Hubbard Brook Experimental Forest, New Hampshire. Can. J. For. Res. 27(12):2043–2052.

Hoskin, G.P. 1993. A Speech Presented by G.P. Hoskin, Associate Director, Office of Seafood on March 4, 1993 Before the Joint Subcommittee on Agriculture. U.S. Food and Drug Administration, Center for Food Safety and Applied Nutrition. [Online]. Available: http://vm.cfsan.fda.gov/~ear/FDA-AQUA.html [accessed Dec.18, 2002].

Huntsman, A.G. 1938. North American Atlantic Salmon. Rapports et Proces-Verbaux, Conseil Permanent International pour l'Exploration de la Mer. 101:11–15.

Hurrell, J.W., and H. van Loon. 1997. Decadal variations in climate associated with the North Atlantic oscillation. Climate Change 36(3/4):301–326.

Hutchings, J.A., and M.E.B. Jones. 1998. Life history variation and growth rate thresholds for maturity and Atlantic salmon, *Salmo salar*. Can. J. Fish. Aquat. Sci. 55(suppl. 1):22–47.

Hutchinson, P., and D. Mills. 2000. Executive summary. Pp. 7–18 in the Ocean Life of Atlantic Salmon: Environmental and Biological Factors Influencing Survival, D.H. Mills, ed. Malden, MA: Fishing News Books.

Hvidsten, N.A., A.J. Jensen, H. Viväs, Ø. Bakke, and T.G. Heggbergot. 1995. Downstream migration of Atlantic salmon smolts in relation to water flow, water temperature, moon phase and social interaction. Nordic J. Freshwater Res. 70:38–48.

ICES (International Council for the Exploration of the Sea). 2002. Report of the Working Group on North Atlantic Salmon. CM 2002/ACFM:14. Copenhagen, Denmark: International Council for Exploration of the Sea. [Online]. Available: http://www.ices.dk/reports/ACFM/2002/WGNAS/WGNAS2002.pdf [accessed Jan. 6, 2003].

Inada, Y., and N. Taniguchi. 1991. Spawning behavior and after-spawning survilli in induced triploid ayu Plecoglossus altivelis. Nihon Suisan Gakkai Shi 57(12): 2265–2269.

Irland, L.C. 1998. Maine's Forest Area, 1600–1995: Review of Available Estimates. Misc. Pub. No.736. Orono, ME: College of Natural Sciences, Forestry and Agriculture.

Irland, L.C. 1999. The Northeast's Changing Forest. Cambridge, MA: Harvard University Press.

Iwata, M. 1995. Downstream migration behavior of salmonids and the relationship with cortisol and thyroid hormones: A review. Aquaculture 135:131–139.

Jackson, H. 2002. Pleasant River Time Series Study. Maine Board of Pesticides Control, Augusta.

Jackson, H. 2003. Drift Study of Two Aerially Applied Blueberry Pesticides. Maine Board of Pesticides Control, Augusta.

Jacobsen, J.A., and L.P. Hansen. 2000. Feeding habits of Atlantic salmon at different life stages at sea. Pp. 170–192 in the Ocean Life of Atlantic Salmon: Environmental and Biological Factors Influencing Survival, D.H. Mills, ed. Malden, MA: Fishing News Books.

Jagoe, C., and T. Haines. 1997. Changes in gill morphology of Atlantic salmon (*Salmo salar*) smolts due to addition of acid and aluminum to stream water. Environ. Pollut. 97(1): 137–146.

Jansson, H., and T. Öst. 1997. Hybridization between Atlantic salmon (*Salmo salar*) and brown trout (*S. trutta*) in a restored section of the River Dälalven, Sweden. Can. J. Fish. Aquat. Sci. 54(9):2033–2039.

Jentoft, S. 2000. Co-managing the coastal zone: Is the task too complex? Ocean Coast. Manage. 43(6):527–535.

Jentoft, S., and B.J. McCay. 1995. User participation in fisheries management: Lessons drawn from international experience. Mar. Pol. 19(3):227–246.

JGIWG (Joint Government/Industry Working Group). 2000. Final Report of the Joint Government/Industry Working Group on Infectious Salmon Anaemia (ISA) in Scotland. [Online]. Available: http://www.marlab.ac.uk/PDFs/Specialist/ JGIWGReport.pdf [accessed Dec. 18, 2002].

Johnsen, B.O., and A.J. Jensen. 1986. Infestations of Atlantic salmon, *Salmo salar,* by *Gyrodactylus salaris* in Norwegian rivers. J. Fish Biol. 29:233–241.

Johnsen, B.O., and A.J. Jensen. 1991. The *Gyrodactylus* story in Norway. Aquaculture 98(1/3):289–302.

Johnsen, B.O., and A.J. Jensen. 1994. The spread of furunculosis in salmonids in Norwegian rivers. J. Fish Biol. 45(1):47–55.

Johnsson, J.I., and M.V. Abrahams. 1991. Interbreeding with domestic strain increases foraging under threat of predation in juvenile steelhead trout (*Oncorhynchus mykiss*): An experimental study. Can. J. Fish. Aquat. Sci. 48(2):243–247.

Johnsson, J.I., E. Petersson, E. Jönsson, B.T. Björnsson, and T. Järvi. 1996. Domestication and growth hormone alter antipredator behavior and growth patterns in juvenile brown trout, *Salmo trutta.* Can. J. Fish. Aquat. Sci. 53(7):1546–1554.

Johnsson, J.I., J. Höjesjö, and I.A. Fleming. 2001. Behavioral and heart rate responses to predation risk in wild and domesticated Atlantic salmon. Can. J. Fish. Aquat. Sci. 58(4):788–794.

Jonsson, B. 1997. A review of ecological and behavioral interactions between cultured and wild Atlantic salmon. ICES J. Mar. Sci. 54(6):1031–1039.

Jonsson, B., and J. Ruud-Hansen. 1985. Water temperature as the primary influence on timing seaward migrations of Atlantic salmon (*Salmo salar*) smolts. Can. J. Fish Aquat. Sci. 42:593–595.

Jonsson, N. 1991. Influence of water flow, water temperature and light on fish migration in rivers. Nordic Journal of Freshwater Research 66:20–65.

Johnston, C.E., and R.L. Saunders. 1981. Parr-smolt transformation of yearling Atlantic salmon (*Salmo salar*) at several rearing temperatures. Can. J. Fish. Aquat. Sci. 38:1189–1198.

Juanes, F., B.H. Letcher, and G. Gries. 2000. Ecology of stream fish: Insights gained from an individual-based approach to juvenile Atlantic salmon. Ecol. Freshwater Fish 9(1-2):65–73.

Juda, L. 1999. Considerations in developing a functional approach to the governance of large marine ecosystems. Ocean Dev. Int. Law 30(2):89–125.

Judd, R.W. 1997. Common Lands, Common People: The Origins of Conservation in Northern New England. Cambridge, MA: Harvard University Press.

Kaill, W.M., K. Rawson, and T. Joyce. 1990. Retention rates of half-length coded wire tags implanted in emergent pink salmon. Pp. 253–258 in Fish-Marking Techniques, N.C. Parker, A.E. Giorgi, R.C. Heidinger, D.B. Jester Jr., E.D. Prince, and G.A. Winans, eds. American Fisheries Society Symposium 7. Bethesda, MD: American Fisheries Society.

Kapuscinski, A.R. 2001. Controversies in designing useful ecological assessments of genetically engineered organisms. Pp. 385–415 in Genetically Engineered Organisms: Assessing Environmental and Human Health Effects, D.K. Letourneau, and E. Burrows, eds. Boca Raton: CRC Press.

Kapuscinski, A.R., and D.J. Brister. 2001. Genetic impacts of aquaculture. Pp. 128–153 in Environmental Impacts of Aquaculture, K.D. Black, ed. Sheffield: Sheffield Academic Press.

Keeney, R.L., and H. Raiffa. 1976. Decisions with Multiple Objectives: Preferences and Value Tradeoffs. New York: Wiley.

Kendall, W.C. 1935. The Fishes of New England: The Salmon Family, Part 2. The Salmons. Memoirs of the Boston Society of Natural History: Monographs on the Natural History of New England, Vol. 9(1). Boston, MA: Boston Society of Natural History.

Kincaid, H., L. Mengel, and J. Johnson. 1994. Meristic and Morphometric Evaluation of Parr from Atlantic Salmon Stocks Collected from the Downeast Rivers of Maine in 1992–1993. Preliminary Report submitted to the Federal Atlantic Salmon Technical Working Group. National Biological Survey, Research and Development Laboratory, Wellsboro, PA.

King, T.L., W.B. Schill, B.A. Lubinski, M.C. Smith, M.S. Eackles, and R. Coleman. 1999. Microsatellite and Mitochondrial DNA Diversity in Atlantic Salmon with Emphasis on Small Coastal Drainages of the Downeast and Midcoast Regions of Maine. A Report to Region 5, USFWS, Hadley, MA, by USGS-BRD-Leetown Science Center, Kearneysville, WV. March 1999.

Kitamura, S., H. Ogata, and H. Onozato. 1991. Triploid male masu salmon *Oncorhynchus masou* shows normal courtship behavior. Nippon Suisan Gakk. 57:2157.

Kitchell, J.F., and S.W. Hewett. 1987. Forecasting forage demand and yields of sterile Chinook salmon (*Oncorhynchus tshawytscha*) in the Lake Michigan. Can. J. Fish. Aquat. Sci. 44(Suppl. 2):384–389.

Klyashtorin, L.B. 1998. Long-term climate change and main commercial fish production in the Atlantic and Pacific. Fish. Res. 37(1-3):115–125.

Kolpin, D.W., E.T. Furlong, M.T. Meyer, E.M. Thurman, S.D. Zaugg, L.B. Barber, and H.T. Buxton. 2002. Pharmaceuticals, hormones, and other organic wastewater contaminants in U.S. streams, 1999–2000: A national reconnaissance. Environ Sci. Technol. 36(6):1202–1211.

Krohn, W.B., R.B. Allen, J.R. Moring, and A.E. Hutchinson. 1995. Double-crested cormorants in New England: Population and management histories. Pp. 99–109 in the Double-Crested Cormorant: Biology, Conservation, and Management, D.N. Nettleship, and D.C. Duffy, eds. Washington, DC: Colonial Waterbird Society.

Lacroix, G.L., and P. McCurdy. 1996. Migratory behavior of post-smolt Atlantic salmon during initial stages of seaward migration. J. Fish. Biol. 49(6):1086–1101.

Lamberg, A., P. Fiske, and N.A. Hvidsten. 2001. Monitoring anadromous salmonids by means of video systems in rivers. [in Norwegian]. NINA Oppdragsmelding 715: 1–26.

Leary, R.F., F.W. Allendorf, and G.K. Sage. 1995. Hybridization and introgression between introduced and native fish. Pp. 91–101 in Uses and Effects of Cultured Fishes in Aquatic Ecosystems, H.L. Schramm, and R.G. Piper, eds. American Fisheries Society Symposium 15. Bethesda, MD: American Fisheries Society.

Lee, K.N. 1995. Deliberately seeking sustainability in the Columbia River Basin. Pp. 214–238 in Barriers and Bridges to the Renewal of Ecosystems and Institutions, L.H. Gunderson, C.S. Holling, and S.S. Light, eds. New York: Columbia University Park.

Leivestad, H., and J.P. Muniz. 1976. Fish kill at low pH in a Norwegian river. Nature 259:391–392.

Likens, G.E., F.H. Bormann, R.S. Pierce, J.S. Eaton, and N.M. Johnson. 1977. Biogeochemistry of a Forested Ecosystem. New York: Springer.

Lura, H., and H. Sægrov. 1991. Documentation of successful spawning of escaped farmed female Atlantic salmon, *Salmo salar*, in Norwegian rivers. Aquaculture 98(1/3):151–159.

Lura, H., and H. Sægrov. 1993. Timing of spawning in cultured and wild Atlantic salmon (*Salmo salar*) and brown trout (*Salmo trutta*) in the river Vosso, Norway. Ecol. Freshwater Fish 2(4):167–172.

LURC (Maine Land Use Regulation Commission). 2000a. Rules and Standards. Maine Land Use Regulation Commission, Department of Conservation, Alberta, ME. October 17, 2000. [Online]. Available: http://www.state.me.us/doc/lurc/resources/rules/rulestoc.htm [accessed June 5, 2002].

LURC (Maine Land Use Regulation Commission). 2000b. Chapter 15. Guidelines for Private Roads or Ways in the Land Use Regulation Commission's Management Districts, Supplement to Section 6 of the Land Use Handbook. Maine Land Use Regulation Commission, Department of Conservation, Alberta, ME. October 17, 2000. [Online]. Available: http://www.state.me.us/doc/lurc/resources/rules/R15.htm [accessed May 12, 2003].

LURC (Maine Land Use Regulation Commission). 2001. Statute Applicable To Maine Land Use Regulation Commission Department of Conservation, As Amended through the First Session of the 120th Legislature. [Online]. Available: http://www.state.me.us/doc/lurc/resources/statutes/statutes01.htm [accessed Oct. 22, 2003].

LURC (Maine Land Use Regulation Commission). 2002a. Land Use Regulation Commission, Maine Department of Conservation, Alberta, ME. [Online]. Available: http://www.state.me.us/doc/lurc/lurchome.htm [accessed June 6, 2002].

LURC (Maine Land Use Regulation Commission). 2002b. Chapter 10—Land Use Districts and Standards. Maine Land Use Regulation Commission, Department of Conservation, Alberta, ME. August 29, 2002 [Online]. Available: http://www.state.me.us/doc/lurc/resources/rules/10toc.htm [accessed May 14, 2003].

Lynch, M., and M. O'Hely. 2001. Captive breeding and the genetic fitness of natural populations. Conserv. Gen. 2(4):363–378.

MacCrimmon, H.R., and B.L. Gots. 1979. World distribution of Atlantic salmon, *Salmo salar*. J. Fish. Res. Board Can. 36(4):422–457.

MacGregor, B.W., R.M. Peterman, B.J. Pyper, and M.J. Bradford. 2002. A decision analysis framework for comparing experimental designs of projects to enhance Pacific salmon. N. Am. J. Fish. Manage. 22(2):509–527.

MacLean, N., and R.J. Laight. 2000. Transgenic fish: An evaluation of benefits and risks. Fish and Fisheries 1(2):146–172.

Magee, J.A., T.A. Haines, J.F. Kocik, K.F. Beland, and S.D. McCormick. 2001. Effects of acidity and aluminum on the physiology and migratory behavior of Atlantic salmon smolts in Maine, USA. Water Air Soil Pollut. 130(1-4):881–886.

Magee, J.A., M. Obedzinski, S.D. McCormick, and J.F. Kocik. 2003. Effects of episodic acidification on Atlantic salmon (*Salmo salar*) smolts. Can. J. Fish. Aquat. Sci. 60(2):214–221.

Maguire, L.A. 1986. Using decision analysis to manage endangered species populations. J. Environ. Manage. 22:345–360.

Maine Aquaculture Innovation Center. 2003. Information [Online]. Available: http://www.maineaquaculture.org/ [accessed May 12, 2003].

Maine Atlantic Salmon Task Force. 1997. Atlantic Salmon Conservation Plan for Seven Maine Rivers. March 1997 [Online]. Available: http://www.state.me.us/asa/ascp all.htm [accessed Dec. 18, 2002].

Maine Department of Agriculture. 2002. Maine Department of Agriculture. [Online]. Available: http://www.state.me.us/agriculture/ [accessed May 14, 2003].

Maine Revenue Services. 2003a. List of Unorganized Townships. State of Maine Unorganized Territory, Maine Revenue Services Property Tax Division, Augusta, ME. [Online]. Available: http://www.state.me.us/revenue/propertytax/unorganized%20 page/unorganized%20Territory%20towns.htm [accessed Oct. 22, 2003].

Maine Revenue Services. 2003b. Map of Unorganized Territory. State of Maine Unorganized Territory, Maine Revenue Services Property Tax Division, Augusta, ME. [Online]. Available: http://www.state.me.us/revenue/propertytax/Unorganized%20 Page/Unorganized%20Territory%20Page.htm [accessed Oct. 22, 2003].

Maine State Planning Office. 2003. Maine State Planning Office. [Online]. Available: http://www.state.me.us/spo/ [accessed May 14, 2003].

Mantua, N.J., S.R. Hare, Y. Zhang, J.M. Wallace, and R.C. Francis. 1997. A Pacific interdecadal climate oscillation with impacts on salmon production. Bull. Am. Meteor. Soc. 78:1069–1079.

Martin, J.H.A., and K.A. Mitchell. 1985. Influence of sea temperature upon the numbers of grilse and multi-sea-winter Atlantic salmon (*Salmo salar*) caught in the vicinity of the River Dee (Aberdeenshire). Can. J. Fish. Aquat. Sci. 42(9):1513–1521.

Martinez, J.L., P. Moran, J. Perez, B. De Gaudemar, E. Beall, and E. Garcia-Vazquez. 2000. Multiple paternity increases effective size of southern Atlantic salmon populations. Mol. Ecol. 9(3):293–298.

Marvinney, R.G., and W.B. Thompson. 2000. A Geologic History of Maine. Maine Geological Survey, Department of Conservation, Augusta, ME. [Online]. Available: http://www.state.me.us/doc/nrimc/pubedinf/factsht/bedrock/megeol.htm [accessed Nov. 8, 2002].

MASC (Maine Atlantic Salmon Commission). 2000. Atlantic Salmon Conservation Plan for Seven Maine Rivers. 2000 Annual Progress Report. Maine Atlantic Salmon Commission, Augusta, ME. [Online]. Available: http://www.state.me.us/asa/2000AnnRpt.pdf [accessed Nov. 13, 2002].

MASC (Maine Atlantic Salmon Commission). 2001. Report of the Maine Atlantic Salmon Commission to the Maine Legislature, Fisheries and Wildlife Joint Standing Committee for the Period June 1999 through December 2000. Maine Atlantic Salmon Commission, Augusta, ME.

MASC (Maine Atlantic Salmon Commission). 2002. Report of the Maine Atlantic Salmon Commission to the Maine Legislature Fisheries and Wildlife Joint Standing Committee for the Period January through December 2001. Maine Atlantic Salmon Commission, Augusta, ME. [Online]. Available: http://www.state.me.us/asa/2001%20Leg.%20 Report.pdf [accessed Dec. 18, 2002].

MASC (Maine Atlantic Salmon Commission). 2003. Salmon Efforts in Maine. Maine Atlantic Salmon Commission, Augusta, ME. [Online]. Available: http://www.state.me.us/asa/effortsinme.htm [accessed May 9, 2003].

Maxwell, J., C. Edwards, M. Jensen, S. Paustian, H. Parrott, and D. Hill. 1995. A Hierarchical Framework of Aquatic Ecological Units in North America (Nearctic Zone). Gen. Tech. Rep. NC-176. St. Paul, MN: U.S. Dept. of Agriculture, Forest Service, North Central Forest Experiment Station.

Maynard, D.J., T.A. Flagg, and C.V.W. Mahnken. 1995. A review of seminatural culture strategies for enhancing the postrelease survival of anadromous salmonids. Pp. 307–316 in Uses and Effects of Cultured Fishes in Aquatic Ecosystems, H.L. Schramm, and R.G. Piper, eds. American Fisheries Society Symposium 15. Bethesda, MD: American Fisheries Society.

Maynard, D.J., T.A. Flagg, and C.V.W. Mahnken. 1996. Development of a Natural Rearing System to Improve Supplemental Fish Quality, 1991–1995: Process Report. DOE/BP-20651-1. Portland, OR: Bonneville Power Administration.

McCormick, S.D., L.P. Hansen, T.P. Quinn, and R.L. Saunders. 1998. Movement, migration, and smolting of Atlantic salmon (*Salmo salar*). Can. J. Fish. Aquat. Sci. 55(suppl.):77–92.

McCormick, S.D., R.A. Cunjak, M.F. O'Dea, and J.B. Carey. 1999. Temperature-related loss of smolt characteristics in Atlantic salmon (*Salmo salar*) in the wild. Can. J. Fish. Aquat. Sci. 56(9):1649–1658.

McCormick, S.D., J.M. Shrimpton, S. Moriyama, and B.T. Björnsson. 2002. Effects of an advanced temperature cycle on smolt development and endocrinology indicate that temperature is not a zeitgeber for smolting in Atlantic salmon. J. Exp. Biol. 205(Pt22): 3553–3560.

McDaniel, T.R., K.M. Pratt, T.R. Meyers, T.D. Ellison, J.E. Follet, and J.A. Burke. 1994. Alaska Sockeye Salmon Culture Manuel. Special Fisheries Report No. 6. Alaska Department of Fish and Game, Juneau, AK. 39pp.

McGinnity, P., C. Stone, J.B. Taggart, D. Cooke, D. Cotter, R. Hynes, C. McCamley, T. Cross, and A. Ferguson. 1997. Genetic impact of escaped farmed Atlantic salmon (*Salmo salar* L.) on native populations: Use of DNA profiling to assess freshwater performance of wild, farmed, and hybrid progeny in a natural river environment. ICES J. Mar. Sci. 54(6):998–1008.

McGinnity, P., P. Prodöhl, A. Ferguson, R. Hynes, N. Ó Maoiléidigh, N. Baker, D. Cotter, B. O'Hea, D. Cooke, G. Rogan, J. Taggart, and T. Cross. 2003. Fitness reduction and potential extinction of wild populations of Atlantic salmon, *Salmo salar*, as a result of interactions with escaped farm salmon. Proc. Roy. Soc. Lond. B [Online DOI 10.1098/rspb.2003.2520].

MDOT (Maine Department of Transportation). 2002a. Fish Passage Policy and Design Guide. Maine Department of Transportation, Augusta, ME. March 2002. [Online]. Available: http://www.state.me.us/mdot/finalfishpassage5.pdf [accessed May 14, 2003].

MDOT (Maine Department of Transportation). 2002b. Maine Department of Transportation Website. [Online]. Available: http://www.state.me.us/mdot/ [accessed May 14, 2003].

MDOT (Maine Department of Transportation). 2003. Community Service Division. Maine Local Road Center. [Online]. Available: http://www.state.me.us/mdot/planning/csd/mlrc.htm [May 14, 2003].

Meister, A.L. 1982. The Sheepscot River: An Atlantic Salmon River Management Report. Atlantic Salmon River Management Report No. 8. Bangor, ME: State of Maine Atlantic Sea Run Salmon Commission.

Mesa, M.G. 1991. Variation in feeding, aggression, and position choice between hatchery and wild cutthroat trout in an artificial stream. Trans. Am. Fish. Soc. 120(6):723–727.

Meyers, T. 1994. The program to restore Atlantic salmon in the Connecticut River. Pp. 11–21 in Hard Look at Some Tough Issues: Proceedings of the New England Atlantic Salmon Management Conference, S. Calibi, and A. Stout, eds. Newburyport, MA: New England Salmon Association.

MFS (Maine Forest Service). 1999. The State of the Forest and Recommendations for Forest Sustainability Standards. Final Report to the Joint Standing Committee of the 119th Legislature on Agriculture, Conservation, and Forestry. Department of Conservation, Maine Forest Service, Augusta, ME. [Online]. Available: http://www.state.me.us/doc/mfs/pubs.htm [accessed Nov. 11, 2002].

MFS (Maine Forest Service). 2000. 1999 Silvicultural Activities Report. Maine Forest Service, Maine Department of Conservation, Augusta, ME. [Online]. Available: http://www.state.me.us/doc/mfs/pubs/htm/silvi/99silvi.htm [accessed Nov. 11, 2002].

MFS (Maine Forest Service). 2002a. Maine Forest Service. Maine Department of Conservation, Augusta, ME. [Online]. Available: http://www.state.me.us/doc/mfs/mfshome.htm [accessed July 13, 2002].

MFS (Maine Forest Service). 2002b. Division of Forest Health and Monitoring, Maine Forest Service, Maine Department of Conservation, Augusta, ME. [Online]. Available: http://www.state.me.us/doc/mfs/idmhome.htm [accessed July 13, 2002].

MFS (Maine Forest Service). 2002c. Forest Policy and Management Division, Maine Forest Service, Maine Department of Conservation, Augusta, ME. [Online]. Available: http://www.state.me.us/doc/mfs/fpmhome.htm [accessed July 13, 2002].

MFS (Maine Forest Service). 2002d. Forest Protection Division, Maine Forest Service, Maine Department of Conservation, Augusta, ME. [Online]. Available: http://www.state.me.us/doc/mfs/ffchome.htm [accessed July 13, 2002].

Miller, L.M., and A.R. Kapuscinski. 2002. Genetic guidelines for hatchery supplementation programs. Pp. 329–355 in Population Genetics: Principles and Practices for Fisheries Scientists, E.M. Hallerman, ed. Bethesda, MD: American Fisheries Society.

Mills, D.H. 1989. Ecology and Management of Atlantic Salmon. London: Chapman and Hall. 351pp.

Milner, J.W. 1874. The progress of fish-culture in the United States. Pp. 523–558 in Report of the Commissioner for 1872 and 1873, Part 2, S.F. Baird, ed. Washington, DC: U.S. Govt. Print. Off.

Minns, C.K., R.G. Randall, M.E. Chadwick, J.E. Moore, and R. Green. 1995. Potential impact of climate changes on the habitat and population dynamics of juvenile Atlantic salmon (*Salmo salar*) in eastern Canada. Pp. 699–708 in Climate Change and Northern Fish Populations, R.J. Beamish, ed. Can. Spec. Publ. Fish Aquat. Sci. No. 121. Canada: NRC-Research Press.

Mjølnerød, I.B., L.H. Refseth, F. Karlsen, T. Balstad, K.S. Jakobsen, and K. Hinda. 1997. Genetic differences between two wild and one farmed population of Atlantic salmon (*Salmo salar*) revealed by three classes of genetic markers. Hereditas 127(3):239–248.

MMA (Maine Municipal Association). 2002. Home Rule. Maine Municipal Association, Augusta, ME [Online]. Available: http://www.memun.org/resources/public/HTML/Local_Govt/home_rule.htm [accessed May 7, 2003].

Møller, A.P., and R.V. Alatalo. 1999. Good-genes effects in sexual selection. Proc. R. Soc. London B. Biol. Sci. 266(1414):85–91.

Montevecchi, W.A., G.K. Davoren, I.J. Stenhouse, C. Burke, and A. Hedd. 2002. Learning from Avian Predators—Natural Assays of Fish Stock and Ecosystem Conditions. Presentation at the Ocean Management Research Network National Conference, October 25–27, 2002, Ottawa, Ontario.

Moore, A., and C.P. Waring. 1998. Mechanistic effects of a triazine pesticide on reproductive endocrine function in mature male Atlantic salmon (*Salmo salar* L.) parr. Pest. Biochem. Physiol. 62(1):41–50.

Moring, J. 2000a. Atlantic Salmon, Restoration, Rehabilitation, and Federal Protection. College of Natural Sciences, Forestry, and Agriculture White Papers, Series 3-1: Maine and the Atlantic Salmon. University of Maine, Orono, ME. 2pp.

Moring, J.R. 2000b. Fisheries history—the creation of the first public salmon hatchery in the United States. Fisheries 25(7):6–12.

Moring, J.R., J. Marancik, and F. Griffiths. 1995. Changes in stocking strategies for Atlantic salmon restoration and rehabilitation in Maine, 1871–1993. Pp. 38–46 in Uses and Effects of Cultured Fishes in Aquatic Ecosystems, H.L. Schramm, and R.G. Piper, eds. American Fisheries Society Symposium 15. Bethesda, MD: American Fisheries Society.

Morris, R.T. 1900. Sea and river fishing. Notes from Dennysville. Forest and Stream 55:69.

MPUC (Maine Public Utilities Commission). 2001. Vision. Maine Public Utilities Commission, Alberta, ME. [Online]. Available: http://www.state.me.us/mpuc/homepage/qmvision.htm [accessed June 5, 2002].

MRDC (Maine Rural Development Council). 2002. At Loggerheads—the State of Maine and the Wabanaki. Maine Rural Development Council, University of Maine Cooperative Extension. [Online]. Available: http://mrdc.umext.maine.edu/archive/wabanaki/loggerheads.htm [accessed May 8, 2003].

Muoneke, M.I., and W.M. Childress. 1994. Hooking mortality: A review for recreational fisheries. Rev. Fish. Sci. 2(22):123–156.

NAFO (Northwest Atlantic Fisheries Organization). 2003. About NAFO. Northwest Atlantic Fisheries Organization. [Online]. Available: http://www.nafo.ca/about/FRAMES/AbFrMand.html [accessed May 21, 2003].

Narayanan, S., J. Carscadden, J.B. Dempson, M.F. O'Connell, S. Prinsenberg, D.G. Reddin, and N. Shackell. 1995. Marine climate off Newfoundland and its influence on Atlantic salmon (*Salmo salar*) and capelin (*Mallotus villosus*). Pp. 461–474 in Climate Change and Northern Fish Populations, R.J. Beamish, ed. Can. Spec. Publ. Fish Aquat. Sci. No. 121. Canada: NRC-Research Press.

NASCO (North Atlantic Salmon Conservation Organisation). 2003a. Report of the Twentieth Annual Meetings of the Commissions. Edinburgh, Scotland. NASCO, Edinburgh, Scotland. Also available at NASCO's web site (http://www.nasco.int/; last accessed 4/2004).

NASCO (North Atlantic Salmon Conservation Organisation). 2003b. Summary of Regulatory Measures Agreed to by NASCO for the West Greenland Salmon Fishery. Available at http://www.nasco.int/pdf/nasco_res_reggreenland.pdf. Last accessed 4/2004.

Newman, D., and D. Pilson. 1997. Increased probability of extinction due to decreased genetic effective population size: Experimental populations of Clarkia pulchella. Evolution 51(2):354–362.

NEFMC (New England Fisheries Management Council). 2003. New England Fisheries Management Council. [Online]. Available: http://www.nefmc.org/ [accessed May 21, 2003].

Nickelson, T.E., M.F. Solazzi, and S.L. Johnson. 1986. Use of hatchery coho salmon (*Oncorhynchus kisutch*) presmolts to rebuild wild populations in Oregon coastal streams. Can. J. Fish. Aquat. Sci. 43:2443–2449.

Nielsen, J.L. 1998. Population genetics and the conservation and management of Atlantic salmon (*Salmo salar*). Can. J. Fish. Aquat. Sci. 55(Suppl. 1):145–152.

Nightingale, E.R. 2000. Atlantic Salmon Chronicles: The Fish, the Rivers, the Fly-Fishing Techniques and Equipment. Boulder, CO: Sycamore Island Books.

NMFS (National Marine Fisheries Service). 1998. NOAA's Aquaculture Policy. Southwest Regional Office, National Marine Fisheries Service, Long Beach, CA. [Online]. Available: http://swr.ucsd.edu/fmd/bill/aquapol.htm [accessed May 16, 2003].

NMFS (National Marine Fisheries Service). 2000a. Wild Atlantic Salmon in Maine Protected as Endangered Species. NMFS Northeast Region News, November 11, 2000. [Online]. Available: http://www.nefsc.noaa.gov/press_release/news00.30.html [accessed May 16, 2003].

NMFS (National Marine Fisheries Service). 2000b. Salmonid Travel Time and Survival Related to Flow in the Columbia River Basin. White Paper, March 2000. Northwest Fisheries Science Center, National Marine Fisheries Service, Seattle, WA. [Online]. Available: http://www.nwfsc.noaa.gov/pubs/white/whiteflow.pdf [accessed May 22, 2003]

NMFS (National Marine Fisheries Service). 2002. Recreational Fisheries Program. Office of Constituent Services, National Marine Fisheries Service, Silver Spring, MD. [Online]. Available: http://www.nmfs.noaa.gov/irf/irf.html [accessed May 16, 2003].

NMFS and FWS (National Marine Fisheries Service and U.S. Fish and Wildlife Service). 1999. Status Review of Anadromous Atlantic Salmon in the United States. National Marine Fisheries Service, and U.S. Fish and Wildlife Service, Washington, DC.

NOAACSC (National Oceanic and Atmospheric Administration Coastal Service Center). 2003. Atlantic States Marine Fisheries Commission. [Online]. Available: http://www.csc.noaa.gov/cms/asmfc.htm [accessed May 8, 2003].

Norris, A.T., D.G. Bradley, and E.P. Cunningham. 1999. Microsatellite genetic variation between and within farmed and wild Atlantic salmon (*Salmo salar*) populations. Aquaculture 180(3/4):247–264.

NRC (National Research Council). 1983. Risk Assessment in the Federal Government: Managing the Process. Washington, DC: National Academy Press.

NRC (National Research Council). 1995. Science and the Endangered Species Act. Washington, DC: National Academy Press.

NRC (National Research Council). 1996a. Upstream: Salmon and Society in the Pacific Northwest. Washington, DC: National Academy Press.

NRC (National Research Council). 1996b. Understanding Risk: Informing Decisions in Democratic Society, P.C. Stern, and H.V. Fineberg, eds. Washington, DC: National Academy Press.

NRC (National Research Council). 1999. Hormonally Active Agents in the Environment. Washington, DC: National Academy Press.

NRC (National Research Council). 2000. Scientific Frontiers in Developmental Toxicology and Risk Assessment. Washington, DC: National Academy Press.

NRC (National Research Council). 2002a. Genetic Status of Atlantic Salmon in Maine. Interim Report. Washington, DC: National Academy Press.

NRC (National Research Council). 2002b. Letter from Michael Clegg, Chair, NRC Committee on Atlantic Salmon in Maine, to Mamie Parker, Acting Regional Director, U.S. Fish and Wildlife Service, Northwest Regional Office, Handley, MA; Pat Kurkul, Regional Administrator, NOAA, National Marine Fisheries Service, Gloucester, MA, and Fred Kircheis, Executive Director, Maine Atlantic Salmon Commission, Augusta Central Office, Augusta, ME. March 29, 2002.

NRC (National Research Council). 2002c. Animal Biotechnology: Science Based Concerns. Washington, DC: National Academy Press.

NRC (National Research Council). 2002d. Riparian Areas: Functions and Strategies for Management. Washington, DC: National Academy Press.

NRC (National Research Council). 2002e. The Drama of the Commons. Washington, DC: National Academy Press.

NRCS (Natural Resources Conservation Service). 2003. Natural Resources Conservation Service, U.S. Department of Agriculture. [Online]. Available: http://www.nrcs.usda.gov [accessed May 19, 2003].

NWPPC (Northwest Power Planning Council). 2000. Return to the River. Northwest Power Planning Council, Portland, OR. [Online]. Available: http://www.nwppc.org/library/return/2000-12.htm [accessed Nov. 13, 2002].

Olafsen, J.A., and R.J. Roberts. 1993. Salmon disease: The microbial ecology of fish aquaculture. Pp. 166–186 in Salmon Aquaculture, K. Heen, R.L. Monahan, and F. Utter, eds. Oxford: Fishing News Books.

Oliver, C.D., and B.C. Larson. 1990. Forest Stand Dynamics. New York: McGraw-Hill Pub.

O'Neil, S., J. Ritter, and K. Robichaud-LeBlanc, eds. 2000. Proceedings of a Workshop on Research Strategies into the Causes of Declining Atlantic Salmon Returns to North American Rivers, June 12–14, 2000, Halifax, Nova Scotia. Proceedings Series 2000/18. Fisheries and Oceans, Diadromous Fish Division, Bedford Institute of Oceanography, Dartmouth, Canada. August 2000. [Online]. Available: http://www.mar.dfo-mpo.gc.ca/science/rap/internet/meetings2000.htm [accessed May 9, 2003].

Orciari, R.D., G.H. Leonard, and E.C. Schluntz. 1994. Survival, growth, and smolt production of Atlantic salmon stocked as fry in a southern New England stream. N. Am. J. Fish. Manage. 14(3):588–606.

OSTP/CEQ (Office of Science and Technology Policy/ Council on Environmental Quality). 2001. CEQ/OSTP Assessment: Case Studies of Environmental Regulation for Biotechnology, Case Study No. 1: Salmon Plus Sidebar. [Online]. Available: http://www.ostp.gov./html/012201.html [accessed May 27, 2003]

Pakkasmaa, S. 2000. Morphological and Early Life History Variation in Salmonid Fishes. Helsinki: Oy Edita Ab.

Patric, J.H. 1976. Soil erosion in eastern forests. J. For. 74(10):671–677.

Pearcy, W.G. 1992. Ocean Ecology of North Pacific Salmonids. Seattle, WA: University of Washington Press.

Peltz, L., and J. Miller. 1990. Performance of half-length coded wire tags in a pink salmon hatchery marking program. Pp. 244–252 in Fish-Marking Techniques, N.C. Parker, A.E. Giorgi, R.C. Heidinger, D.B. Jester Jr., E.D. Prince, and G.A. Winans, eds. American Fisheries Society Symposium 7. Bethesda, MD: American Fisheries Society.

Penn, D.J., and W.K. Potts. 1999. The evolution of mating preferences and major histocompatibility complex genes. Am. Nat. 153(2):145–164.

Peterman, R. M., and C.N. Peters. 1998. Decision analysis: Taking uncertainties into account in forest resource management. Pp. 105–127 in Statistical Methods for Adaptive Management Studies, V. Sit and B. Taylor, eds. Land Management Handbook No. 42. Victoria: British Columbia, Ministry of Forest Research Program.

PFRCC (Pacific Fisheries Resources Conservation Council). 2003. Presentation of the Pacific Fisheries Resources Conservation Council to the House of Commons, Standing Committee on Fisheries, Ottawa, Feb. 20, 2003. [Online]. Available: http://www.fish.bc.ca/reports/pfrcc_house_brief_2003-02-20.pdf [accessed May 23, 2003].

Pickart, R.S., T.K. McKee, D.J. Torres, and S.A. Harrington. 1999. Mean structure and interannual variability of the slopewater system of Newfoundland. J. Phys. Oceanogr. 29(10):2541–2558.

Pinkerton, E. 1994. Local fisheries co-management: A review of international experiences and their implications for salmon management in British Columbia. Can. J. Fish. Aquat. Sci. 51(10):2363–2378.

Pinkerton, E., and M. Weinstein. 1995. Fisheries That Work: Sustainability through Community-Based Management. Vancouver: The David Suzuki Foundation. 199pp.

Policansky, D. 2002. Catch-and-release recreational fishing: A historical perspective. Pp. 74–94 in Recreational Fisheries: Ecological, Economic, and Social Evaluation, T. Pitcher, and C. Hollingworth, eds. Oxford, UK: Blackwell Science.

Power, G. 1981. Stock characteristics and catches of Atlantic salmon (*Salmo salar*) in Quebec, and Newfoundland Labrador in relation to environmental variables. Can. J. Fish. Aquat. Sci. 38(12):1601–1611.

Pusey, A., and M. Wolf. 1996. Inbreeding avoidance in animals. Trends Ecol. Evolut. 11(5): 201–206.

Reddin, D.G. 1988. Ocean life of Atlantic salmon (*Salmo salar* L.) in the Northwest Atlantic. Pp. 483–511 in Atlantic Salmon: Planning for the Future: The Proceedings of the Third International Atlantic Salmon Symposium, October 21–23, 1986, Biarritz, France. London: Croom Helm.

Reddin, D.G., and P.B. Short. 1991. Postmolt Atlantic salmon (*Salmo salar*) in the Labrador Sea. Can. J. Fish Aquat. Sci. 48(1):2–6.

Reddin, D.G., W.M. Shearer, and R.F. Burfitt. 1984. Inter-continental migrations of Atlantic salmon. ICES C.M. 1984/M:11.

Reddin, G., J. Helbig, A. Thomas, B.G. Whitehouse, and K.D. Friedland. 2000. Survival of Atlantic salmon (*Salmo salar* L.) related to marine climate. Pp. 88–91 in the Ocean Life of Atlantic Salmon: Environmental and Biological Factors Influencing Survival, D.H. Mills, ed. Malden, MA: Fishing News Books.

Rees, C.B., S.D. McCormick, J.P. Vanden Heuvel, and W. Li. 2003. Quantitative PCR analysis of CYP1A induction in Atlantic salmon (*Salmo salar*). Aquat. Toxicol. 62(1):67–78.

Reid, P.C., and B. Planque. 2000. Long-term planktonic variations and the climate of the north Atlantic. Pp. 153–169 in the Ocean Life of Atlantic Salmon: Environmental and Biological Factors Influencing Survival, D.H. Mills, ed. Malden, MA: Fishing News Books.

Reinhart, K.G., A.R. Eschner, and G.R. Trimble, Jr. 1963. Effect on Streamflow of Four Forest Practices in the Mountains of West Virginia. U.S. Forest Service Research Paper No. 1. Upper Darby, PA: Northeastern Forest Experiment Station, Forest Service, U.S. Dept. of Agriculture. 79 pp.

Reisenbichler, R.R. 1997. Genetic factors contributing to declines of anadromous salmonids in Pacific Northwest. Pp. 223–244 in Pacific Salmon and Their Ecosystems: Status and Future Options, D.J. Stouder, P.A. Bisson, and R.J. Naiman, eds. New York: Chapman and Hall.

Reynolds, J.D., and M.R. Gross. 1992. Female mate preference enhances offspring growth and reproduction in a fish, *Poecilia reticulata*. Proc. R. Soc. London B. Biol. Sci. 250(1327):57–62.

Richardson, J. 2003. Penobscot to lose dams, gain salmon. Portland Press Herald October 7, 2003. [Online]. Available: http://www.pressherald.com/news/state/031007dam.shtml [accessed Nov. 3, 2003].

Ritter, J.A. 1989. Marine Migration and Natural Mortality of North American Atlantic salmon (*Salmo salar* L.). Canadian Manuscript Report of Fisheries and Aquatic Sciences No. 2041. Halifax, NS: Fisheries and Oceans, Canada.

RMA (Risk Management Agency). 2003. About RMA, Organizational Chart. Risk Management Agency, U.S. Department of Agriculture. [Online]. Available: http://www.rma.usda.gov/aboutrma/who/RMA_Org_Chart.pdf [accessed May 19, 2003].

Rodriguez-Clark, K.M. 1999. Genetic theory and evidence supporting current practices in captive breeding for conservation. Pp. 47–74 in Genetics and the Extinction of Species. DNA and the Conservation of Biodiversity, L.F. Landweber, and A.P. Dobson, eds. Princeton, NJ: Princeton University Press.

Rosseland, B.O., and O.K. Skogheim. 1986. Neutralization of acidic brook-water using a shell-sand filters or sea-water: Effects on egg, alevins and smolts of salmonids. Aquaculture 58(1-2):99–110.

Rosseland, B.O., O.K. Skogheim, H. Abrahamsen, and D. Matzow. 1984. Survival and Reduction of Physiological Stress of Atlantic Salmon (*Salmo salar* L.) Smolts in an Acid River Through Slurry Liming. Report from the Norwegian Liming Project 17/84. 24pp.

Rounsefell, G.A., and L.H. Bond. 1949. Salmon Restoration in Maine. Research Report No. 1. Augusta, ME: Atlantic Sea Run Salmon Commission. 52 pp.

Russell, H.S. 1980. Indian New England before the Mayflower. Hanover, NH: University Press of New England.

Ruzzante, D.E. 1991. Variation in agonistic behavior between hatchery and wild populations of fish: A comment on Swain and Riddell (1990). Can. J. Fish. Aquat. Sci. 48:519–520.

Ruzzante, D.E. 1992. Mirror image stimulation, social hierarchies, and population differences in agonistic behavior: A reappraisal. Can. J. Fish. Aquat. Sci. 49:1966–1968.

Ruzzante, D.E. 1994. Domestication effects on aggressive and schooling behavior in fish. Aquaculture 120(1-2):1–24.

Ryman, N. 1994. Supportive breeding and effective population size: Differences between inbreeding and variance effective numbers. Conserv. Biol. 8(3):888–890.

Ryman, N., and L. Laikre. 1991. Effects of supportive breeding on the genetically effective population size. Conserv. Biol. 5(3):325-329.

Ryman, N., and G. Ståhl. 1980. Genetic changes in hatchery stocks of brown trout (*Salmo trutta*). Can. J. Fish. Aquat. Sci. 37:82-87.

Ryman, N., P.E. Jorde, and L. Laikre. 1995b. Supportive breeding and variance effective population size. Conserv. Biol. 9(6):1619–1628.

Saccheri, I., M. Kuussaari, M. Kankare, P. Vikman, W. Fortelius, and I. Hanski. 1998. Inbreeding and extinction in a butterfly metapopulation. Nature 392(6675):491–494.

Salminen, M., S. Kuikka, and E. Erkamo. 1995. Annual variability in survival of sea ranched Baltic salmon, *Salmo salar* L.: Significance of smolt size and marine conditions. Fish. Manage. Ecol. 2:171–184.

Saunders, R.L. 1981. Atlantic salmon (*Salmo salar*) stocks and management implications in the Canadian Atlantic provinces and New England, USA. Can. J. Fish. Aquatic Sci. 38(12):1612–1625.

Saunders, R.L., E.B. Henderson, B.D. Glebe, and E.J. Loudenslager. 1983a. Evidence of a major environmental component in determination of the grilse: Lager salmon ratio in Atlantic salmon (*Salmo salar*). Aquaculture 33(1-4):107–118.

Saunders, R.L., E.B. Henderson, P.R. Harmon, C.E. Johnston, and J.G. Eales. 1983b. Effects of low environmental pH on smolting of Atlantic salmon (*Salmo salar*). Can. J. Fish. Aquat. Sci. 49:1203–1211.

Scarnecchia, D.L. 1984. Climatic and oceanic variations affecting yield of Icelandic stocks of Atlantic salmon (*Salmo salar*). Can. J. Fish. Aquat. Sci. 41:917–935.

Scarnecchia, D.L., A. Isaksson, and S.E. White. 1991. Effects of the Faroese long-line fishery, other oceanic fisheries and oceanic variations on age at maturity of Icelandic north-coast stocks of Atlantic salmon (*Salmo salar*). Fish. Res. 10:207–228.

SCIWC (St. Croix International Waterway Commission). 2003. St. Croix International Waterway Management Plan. St. Croix International Waterway: A Heritage—A Future. [Online]. Available: http://www.asf.ca/OrgsNB/SCIWC/watplan.html [accessed Nov. 10, 2003].

Sessions, J., and J.B. Sessions. 1997. Scheduling and Network Analysis Program: SNAP II+ and III. Forest Engineering Department, Oregon State University, Corvallis, OR.

Shaklee, J.B., and K.P. Currens. 2002. Genetic stock identification and risk assessment. Pp. 291–328 in Population Genetics: Principles and Practices for Fisheries Scientists, E.M. Hallerman, ed. Bethesda, MD: American Fisheries Society.

Sigholt, T., T. Asgard, and M. Staurnes. 1998. Timing of parr-smolt transformation in Atlantic salmon (*Salmo salar*): Effects of changes in temperature and photoperiod. Aquaculture 160(1):129–144.

Smiley, C.W. 1884. Pp. 825–915 in Report of the Commissioner for 1881. U.S. Commission of Fish and Fisheries, Washington, D.C.

Smith, D.M., B.C. Larson, M.J. Kelty, and P.M.S. Ashton. 1997. The Practice of Silviculture: Applied Forest Ecology, 9th Ed. New York: John Wiley. 537pp.

Smith, T.R., and T.A. Haines. 1995. Mortality, growth, swimming activity and gill morphology of brook trout (*Salvelinus fontinalis*) and Atlantic salmon (*Salmo salar*) exposed to low pH with and without aluminum. Environ. Pollut. 90(1):33–40.

Smoker, W.W., A.J. Gharrett, and M.S. Stekoll. 1998. Genetic variation of return date in a population of pink salmon: Consequence of fluctuating environment and dispersive selection? Alaska Fish. Res. Bull. 5(1):46–54.

Solbakken, V.A., T. Hansen, and S.O. Stefansson. 1994. Effects of photoperiod and temperature on growth and parr-smolt transformation in Atlantic salmon (*Salmo salar* L) and subsequent performance in seawater. Aquaculture 121(1/3):13–27.

Solomon, D. 1978. Some observations on salmon smolt migration in a chalkstream. J. Fish Biol. 12(6):571–574.

Specker, J.L., J.G. Eales, M. Tagawa, and W.A. Tyler. 2000. Parr-smolt transformation in Atlantic salmon: Thyroid hormone deiodination in liver and brain and endocrine correlates of change in rheotactic behavior. Can. J. Zool.-Rev. Can. Zool. 78:696–705.

Spidle, A.P., S.T. Kalinowski, B.A. Lubinski, D.L. Perkins, K.F. Beland, J.F. Kocik, and T.L. King. 2003. Population structure of Atlantic salmon in Maine with reference to populations from Atlantic Canada. Trans. Am. Fish. Soc. 132(2):196–209.

Ståhl, G. 1983. Differences in the amount and distribution of genetic variation between natural populations and hatchery stocks of Atlantic salmon. Aquaculture 33:23-32.

Ståhl, G. 1987. Genetic population structure of Atlantic salmon. Pp. 121–140 in Population Genetics & Fishery Management, N. Ryman and F. Utter, eds. Seattle: University of Washington Press.

Staurnes, M., G. Lysjord, L.P. Hansen, and T.G. Heggberget. 1993. Recapture rates of hatchery-reared Atlantic salmon (*Salmo salar*) related to smolt development and time of release. Aquaculture 118(3/4):327–337.

Staurnes, M., L.P. Hansen, K. Fugelli, and O. Haraldstad. 1996. Short-term exposure to acid water impairs osmoregulation, seawater tolerance, and subsequent marine survival of smolts of Atlantic salmon (*Salmo salar* L.). Can. J. Fish Aquat. Sci. 53(8):1695–1704.

Stegeman, J.J., J.J. Schlezinger, J.E. Craddock, and D.E. Tillitt. 2001. Cytochrome P450 1A expression in midwater fishes: Potential effects of chemical contaminants in remote oceanic zones. Environ. Sci. Technol. 35(1):54–62.

Stewart, T.R. 2000. Uncertainty, judgment, and error in prediction. Pp. 41–57 in Prediction: Science, Decision Making, and the Future of Nature, D. Sarewitz, R.A. Pielke, Jr., and R. Byerly, Jr., eds. Washington, DC: Island Press.

Stickney, R.R. 1996a. Aquaculture in the United States: A Historical Survey. New York: John Wiley & Sons.

Stickney, R.R. 1996b. Fish production and distribution in the United States: 1 – Government initiatives. World Aquaculture 27(1):31–41.

Stickney, R.R. 2002. Impacts of cage and net-pen culture on water quality and benthic communities. Pp. 105–118 in Aquaculture and the Environment in the United States, J.R. Tomasso, ed. Baton Rouge, LA: U.S. Aquaculture Society, a Chapter of the World Aquaculture Society.

Stoddard, J.L., D.S. Jeffries, A. Lükewille, T.A. Clair, P.J. Dillon, C.T. Driscoll, M. Forsius, M. Johannessen, J.S. Kahl, J.H. Kellogg, A. Kemp, J. Mannio, D.T. Montieth, P.S. Murdoch, S. Patrick, A. Rebsdorf, B.L. Skjelkvale, M.P. Stainton, T. Traaen, H. van Dam, K.E. Webster, J. Wieting, and A. Wilander. 1999. Regional trends in aquatic recovery from acidification in North America and Europe. Nature 401(6753):575–578.

Stokesbury, M.J., and G.L. Lacroix. 1997. High incidence of hatchery origin Atlantic salmon in the smolt output of a Canadian river. ICES J. Mar. Sci. 54(6):1074–1081.

Stolte, L. 1981. The Forgotten Salmon of the Merrimack. Washington DC: U.S. Dept. of the Interior, Northeast Region.

Stolte, L.W. 1994. Atlantic salmon restoration in the Merrimack River Basin. Pp. 22–35 in A Hard Look at Some Tough Issues: Proceedings of the New England Atlantic Salmon Conference, Danvers, MA, April 22–23, 1994, S. Calabi, and A. Stout, eds. Newburyport, MA: New England Salmon Association.

Swain, D.P., and B.E. Riddell. 1990. Variation in agonistic behavior between newly emerged juveniles from hatchery and wild populations of coho salmon, *Oncorhynchus kisutch.* Can. J. Fish. Aquat. Sci. 47(3):566–577.

Swain, D.P., and B.E. Riddell. 1991. Domestication and agonistic behavior in coho salmon: Reply to Ruzzante. Can. J. Fish. Aquat. Sci. 48:520–522.

Task Force of Tribal-State Relations. 1997. At Loggerheads—the State of Maine and the Wabanaki. Final Report, January 15, 1997. Maine Indian Tribal-State Commission, Hallowell, ME.

Taylor, E.B. 1991. A review of local adaptation in Salmonidae, with particular reference to Pacific and Atlantic salmon. Aquaculture 98(1–3):185–207.

Tessier, N., and L. Bernatchez. 2000. A genetic assessment of single versus double origin of landlocked Atlantic salmon (*Salmo salar*) from Lake Saint-Jean, Quebec, Canada. Can. J. Fish. Aquat. Sci. 57(4):797–804.

Thomas, G.A. 2001. Special series: Habitat conservation planning, II: Incorporating adaptive management and the precautionary principle into HCP design. Endangered Species Update 18(2):32–40.

Tucker, S., I. Pazzia, D. Rowan, and J.B. Rasmussen. 1999. Detecting pan-Atlantic migration in salmon (*Salmo salar*) using 137Cs. Can. J. Fish. Aquat. Sci 56(12):2235–2239.

Tully, O., P. Gargan, W.R. Poole, and K.F. Whelan. 1999. Spatial and temporal variation in the infestation of sea trout (*Salmo trutta* L.) by the caligid copepod *Lepeophtheirus salmonis* (Kroyer) in relation to sources of infection in Ireland. Parasitology 119:41–51.

USACE (U.S. Army Corps of Engineers). 1993. The Highway Methodology Workbook. Integrating Corps Section 404 Permit Requirements with Highway Planning and Engineering and the NEPA EIS Process. NEDEP-360-1-30. U.S. Army Corps of Engineers, New England District. October 1993. [Online]. Available: http://www.nae.usace.army.mil/reg/hwy.pdf [accessed Dec. 18, 2002].

USACE (U.S. Army Corps of Engineers). 2002. Hydroelectric Power. Services for the Public, U.S. Army Corps of Engineers. [Online]. Available: http://www.usace.army.mil/public.html#Hydroelectric [accessed May 19, 2003].

USASAC (U.S. Atlantic Salmon Assessment Committee). 1996. Annual Report of the U.S. Atlantic Salmon Assessment Committee: Report No. 8-1995 Activities. 1996/8. U.S. Atlantic Salmon Assessment Committee, Nashua, NH.

USASAC (U.S. Atlantic Salmon Assessment Committee). 1997. Annual Report of the U.S. Atlantic Salmon Assessment Committee: Report No. 9-1996 Activities. 1997/9. U.S. Atlantic Salmon Assessment Committee, Nashua, NH.

USASAC (U.S. Atlantic Salmon Assessment Committee). 1999. Annual Report of the U.S. Atlantic Salmon Assessment Committee. Report No. 11-1998 Activities. Prepared for U.S. Section to NASCO, by USASAC, Gloucester, MA. March 1–4, 1999. [Online]. Available: http://www.fws.gov/r5crc/stuff/ascom99.html [October 27, 2001].

USASAC (U.S. Atlantic Salmon Assessment Committee). 2000. Annual Report of the U.S. Atlantic Salmon Assessment Committee. Report No. 12-1999 Activities. Prepared for U.S. Section to NASCO, by USASAC, Gloucester, MA. March 6–9, 2000. [Online]. Available: http://www.fws.gov/r5crc/documents/2000/ascrpt00.pdf [accessed Oct. 22, 2003].

USASAC (U.S. Atlantic Salmon Assessment Committee). 2001. Annual Report of the U.S. Atlantic Salmon Assessment Committee Report No. 13-2000. Prepared for U.S. Section to NASCO, by USASAC, Nashua, New Hampshire, March 26, 2001.

USASAC (U.S. Atlantic Salmon Assessment Committee). 2002. Annual Report of the U.S. Atlantic Salmon Assessment Committee. Report 14-2001 Activities. Prepared for U. S. Section to NASCO, by USASAC, Concord, NH, March 5–9, 2002 [Online]. Available: http://www.fws.gov/r5cneafp/atsasscom.htm [accessed Nov. 14, 2002].

USASAC (U.S. Atlantic Salmon Assessment Committee). 2003. Annual Report of the U.S. Atlantic Salmon Assessment Committee. Report No. 15-2002 Activities. Prepared for U. S. Section to NASCO, by USASAC, East Orland, ME, February 25–27, 2003. [Online]. Available: http://www.nefsc.noaa.gov/USASAC/2003%20USASAC%20Report/ASSCOM.rpt.03.PDF [accessed Oct 10, 2003].

USASAC (U.S. Atlantic Salmon Assessment Committee). 2004. Annual Report of the U.S. Atlantic Salmon Assessment Committee. Report No. 16-2003 Activities. Prepared for U. S. Section to NASCO, by USASAC, Woods Hole, MA, February 23–26, 2003. [Online]. Available: http://www.nefsc.noaa.gov/USASAC/2004%20USASAC%20Report/ASSCOM.rpt16.pdf [accessed June 22, 2004].

U.S. Coast Guard. 2003a. Recreational Boating Safety, Maritime Safety. Coast Guard Missions. U.S. Coast Guard. [Online]. Available: http://www.uscg.mil/overview/ [accessed May 21, 2003].

U.S. Coast Guard. 2003b. Maritime Mobility. Coast Guard Missions. U.S. Coast Guard. [Online]. Available: http://www.uscg.mil/overview/ [accessed May 21, 2003].

U.S. Coast Guard. 2003c. Passenger Vessel Safety. Fact File. U.S. Coast Guard. [Online]. Available: http://www.uscg.mil/hq/g-cp/comrel/factfile/index.htm [accessed May 21, 2003].

USDA (U.S. Department of Agriculture). 2002. Infectious Salmon Anemia. Animal and Plant Inspection Service, U.S. Department of Agriculture. January 2002. [Online]. Available: http://www.aphis.usda.gov/oa/pubs/tnisa.html [accessed Dec. 18, 2002].

USDA (U.S. Department of Agriculture). 2003a. Northeastern Area State & Private Forestry, Forest Service, U.S. Department of Agriculture. [Online]. Available: http://www.na.fs.fed.us/ [accessed May 8, 2003].

USDA (U.S. Department of Agriculture). 2003b. Durham Field Office. Forest Service, U.S. Department of Agriculture. [Online]. Available: http://www.fs.fed.us/na/durham/ [accessed May 8, 2003].

USDA (U.S. Department of Agriculture). 2003c. Strategic Plan for 2000–2004 Connecting Forestry and People. Northeastern Area, State and Private Forestry, Forest Service, U.S. Department of Agriculture. [Online]. Available: http://www.na.fs.fed.us/plan/plan.htm [accessed May 8, 2003].

USDA (U.S. Department of Agriculture). 2003d. State and Private Forestry. Forest Service, U.S. Department of Agriculture. [Online]. Available: http://www.fs.fed.us/spf/ [accessed May 19, 2003].

USDA (U.S. Department of Agriculture). 2003e. Natural Resources and Environment. U.S. Department of Agriculture. [Online]. Available: http://www.usda.gov/yourusda/nretext.htm [accessed May 19, 2003].

Utter, F. 2000. Patterns of subspecific anthropogenic introgression in two salmonid genera. Rev. Fish Biol. Fish. 10(3):265–279.

Vail, W.J. et al. 1995. Four Rivers Working Group. Governor's Maine Atlantic Salmon Task Force, Augusta, ME.

Veneman, A.M. 2001. Declaration of Emergency Because of Infectious Salmon Anemia. Department of Agriculture, Office of the Secretary, Washington, DC. Docket No. 01-082-1 [Online]. Available: http://aquanic.org/news/2001/isa.htm [accessed April 8, 2004].

Verry, E.S. 1986. Forests Harvesting and Water: The Lake States Experience. Water Resour. Bull. 22(6).

Verry, E.S., J.W. Hornbeck, and C.A. Dolloff, eds. 2000. Riparian Management in Forests of the Continental Eastern United States. Boca Raton, FL: Lewis. 402 pp.

Verspoor, E. 1998. Genetic Impacts of Wild Atlantic Salmon (*Salmo salar* L.) Stocks of Escaped Farm Conspecifics: An Assessment of Risk. DFO Canadian Stock Assessment Secretariat Research Document 98/156. Department of Fisheries and Oceans Canada.

Verspoor, E., N.H.C. Fraser, and A.F. Youngston. 1991. Protein polymorphism in Atlantic salmon within a Scottish river: Evidence for selection and estimates of gene flow between tributaries. Aquaculture 98(1-3):217–230.

Virtanen, E., L. Söderholm-Tana, A. Soivio, L. Forsman, and M. Muona. 1991. Effect of physiological condition and smoltification status at smolt release on subsequent catches of adult Atlantic salmon. Aquaculture 97(2/3):231–257.

Volpe, J.P., B.W. Glickman, and B.R. Anholt. 2001. Reproduction of aquaculture Atlantic salmon in a controlled stream channel on Vancouver Island, British Columbia. Trans. Am. Fish. Soc. 130(3):489–494.

Waknitz, F.W., T.J. Tynan, C.E. Nash, R.N. Iwamoto, and L.G. Rutter. 2002. Review of Potential Impacts of Atlantic Salmon Culture on Puget Sound Chinook Salmon and Hood Canal Summer-Run Chum Salmon Evolutionarily Significant Units. NOAA Tech. Memorandum NMFS-NWFSC-53. U.S. Dept. of Commerce. [Online]. Available: http://www.nwfsc.noaa.gov/pubs/tm/tm53/tm53.pdf [accessed Jan. 6, 2003].

Walters, C.J. 1986. Adaptive Management of Renewable Resources. New York: Macmillan.

Walters, C., R.D. Goruk, and D. Radford. 1993. Rivers Inlet Sockeye salmon: An experiment in adaptive management. N. Am. J. Fish. Manage. 13(2):253–262.

Wan, M.T., R.G. Watts, and D.J. Moul. 1988. Evaluation of the acute toxicity to juvenile pacific salmonids of hexazinone and its formulated products: Pronone 10G, Velpar L, and their carriers. Bull. Environ. Contam. Toxicol. 41(4):609–616.

Wang, J., and N. Ryman. 2001. Genetic effects of multiple generations of supportive breeding. Conserv. Biol. 15(6):1619–1631.

Waples, R.S. 1991a. Genetic interactions between hatchery and wild salmonids: Lessons from the Pacific Northwest. Can. J. Fish. Aquat. Sci. 48(Suppl. 1): 124-133.

Waples, R.S. 1991b. Definition of "Species" Under the Endangered Species Act: Application to Pacific Salmon. NOAA Technical Memorandum NMFS F/NWC-194. U.S. Department of Commerce, National Oceanic and Atmospheric Administration, National Marine Fisheries Service, Seattle. WA.

Waples, R.S. 1995. Genetic effects of stock transfers of fish. Pp. 51–69 in Protection of Aquatic Biodiversity: Proceedings of the World Fisheries Congress, Theme 3, D.P. Philipp, J.M. Epifanio, J.E. Marsden, J.E. Claussen, and R.J. Wolotira, eds. Lebanon, NH: Science Pub.

Waples, R.S. 1999. Fish culture perspective: Dispelling some myths about hatcheries. Fisheries 24(2):12–21.

Waples, R.S., and C. Do. 1994. Genetic risks associated with supplementation of Pacific salmonids. Can. J. Fish. Aquat. Sci. 51(Suppl.1):310–329.

Waples, R.S., M.J. Ford, and D. Schmitt. In press. Empirical results of salmon supplementation: A preliminary assessment. In Ecological and Genetical Implications of Aquaculture Activities, T.M. Bert, ed. New York: Kluwer Academic.

Watt, W., and L. Hinks. 1999. Acid rain devastation. Atlantic Salmon Journal 48(4) [Online]. Available: http://www.asf.ca/Journal/1999/Wint99/acidrain.html [accessed June 6, 2003].

WBANA (Wild Blueberry Association of North America). 2001. The Wild Advantage, Bar Harbor, ME. 7pp.

Webb, J.H., and H.A. McLay. 1996. Variation in the time of spawning of Atlantic salmon (*Salmo salar*) and its relationship to temperature in the Aberdeenshire Dee, Scotland. Can. J. Fish. Aquat. Sci. 53(12):2739–2744.

Webb, J.H., D.W. Hay, P.D. Cunningham, and A.F. Youngson. 1991. The spawning behavior of escaped farmed and wild adult Atlantic salmon in a northern Scottish river. Aquaculture 98(1/3):97–110.

Webb, J.H., I.S. McLaren, M.J. Donaghy, and A.F. Youngson. 1993. Spawining of farmed Atlantic salmon, *Salmo salar* L., in the second year after their escape. Aquacul. Fish. Manage. 24(4):557–561.

Wedekind, C. 2002. Sexual selection and life-history decisions: Implications for supportive breeding and the management of captive populations. Conserv. Biol. 16(5)1204–1211.

Wedekind, C., R. Müller, and H. Spicher. 2001. Potential genetic benefits of mate selection in whitefish. J. Evol. Biol. 14(6):980–986.

Weigmann, D.L., L.A. Helfrich, and D. C. Josephson, and R.M. Speenburgh. 1993. Guidelines for Liming Acidified Streams and Rivers. Virginia Water Resources Research Center, Virginia Polytechnic Institute and State University, Blacksburg, VA. [Online]. Available: http://www.vwrrc.vt.edu/publications/educatio.htm [accessed Oct. 10, 2003].

Whalen, K.G., D.L. Parrish, and M.E. Mather. 1999a. Effect of ice formation on selection of habitats and winter distribution of post-young-of-the-year Atlantic salmon parr. Can. J. Fish. Aquat. Sci. 56(1):87–96.

Whalen, K.G., D.L. Parrish, and S.D. McCormick. 1999b. Migration timing of Atlantic salmon smolts relative to environmental and physiological factors. Trans. Am. Fish. Soc. 128(2): 289–301.

Whitney, G.G. 1994. From Coastal Wilderness to Fruited Plain: A History of Environmental Change in Temperate North America from 1500 to the Present. Cambridge, UK: Cambridge University Press.

Whoriskey, F.G., and J.W. Carr. 2001. Returns of transplanted adult, escaped, cultured Atlantic salmon to the Magaguadavic River, New Brunswick. ICES J. Mar. Sci. 58(2):504–509.

Wiese, R.J., and K. Willis. 1999. Population Management Plan Handbook. Forth Worth, TX: Forth Worth Zoo.

Wilkie, M.P., K. Davidson, M.A. Brobbel, K.D. Kieffer, R.K. Booth,, A.T. Bielak, and B.L. Tufts. 1996. Physiology and survival of wild Atlantic salmon following angling in warm summer waters. Trans. Am. Fish. Soc. 125(4):572–580.

Wilkie, M.P., M.A. Brobbel, K.G. Davidson, L. Forsyth, and B.L. Tufts. 1997. Influences of temperature on the post exercise physiology of Atlantic salmon (*Salmo salar*). Can. J. Fish. Aquat. Sci. 54(3):503–511.

Williams, R.L. 1976. The Loggers. New York: Time-Life Books.

Wilson, J. 2000. Impact of the Atlantic Salmon Industry on Maine Ecology. College of Natural Sciences, Forestry, and Agriculture White Papers, Series 3: Maine and the Atlantic Salmon No. 5. University of Maine, Orono. [Online]. Available: http://www.umaine.edu/mainesci/Archives/ResEconomics/salmon-economy.htm [accessed Nov. 13, 2002].

Wisconsin River Alliance. 2001. Shopiere Dam. [Online]. Available: http://www.wisconsinrivers.org/SmallDams/shopiere_dam.html] [accessed Nov. 10, 2003].

Wootton, R.J. 1995. Ecology of Teleost Fishes. London: Chapman & Hall. 404pp.

Wright, S. 1938. Size of population and breeding structure in relation to evolution. Science 87(2263):430–431.

WWF (World Wildlife Fund). 2001. The Status of Wild Atlantic Salmon: A River by River Assessment. [Online]. Available: http://www.worldwildlife.org/oceans/pdfs/atlantic_salmon.pdf [accessed Aug. 10, 2004].

Youngson, A.F., J.H. Webb, J.C. MacLean, and B.M. Whyte. 1997. Short communication: Frequency of occurrence of reared Atlantic salmon in Scottish salmon fisheries. ICES J. Mar. Sci. 54(6):1216–1220.

Appendix A

Endangered Species Act Basics

As part of the set of environmental laws enacted in the 1970s, the United States Congress passed and President Nixon signed the Endangered Species Act (ESA) in 1973. This statute is widely considered to embody the most stringent provisions of any wildlife protection law in the United States (Bean 1983). Its strength derives from the affirmative duty it imposes on federal agencies to protect listed species, its prohibitions against actions that kill, injure, or harm them, and the substantial penalties it imposes on violators.

The ESA is implemented by the Departments of Interior and Commerce through regulatory programs administered by the Fish and Wildlife Service (FWS) and the National Marine Fisheries Service (NMFS; collectively, the Services). The broad sweep of its protections comes to bear as a result of the initial regulatory action under Section 4 that determines if a species should be listed as threatened or endangered (NRC 1995). The term "species" has a broader meaning under the ESA than in its accepted scientific definition. In addition to taxonomic species, the ESA allows for the listing of any taxonomically described subspecies as well as any distinct population segment (DPS) of vertebrate animals. The policy of FWS for listing distinct population segments allows the agency broad discretion in its interpretation of distinctness criteria. Biological distinctness is not a requisite factor, as the agency is free to use geographical and political boundaries to describe a "species" eligible for listing.

A species is considered endangered if it is "in danger of extinction throughout all or a significant portion" of its range, while a threatened species, although not endangered, is likely to become so in the foreseeable future. As a practical matter, threatened species are often listed so that the ESA's protective measures may reduce their chance of declining further into endangered status. As of April 2000, 960 species were listed as endangered and 270 were classed as threatened in the United States.

The listing action is a regulatory process that is usually initiated by the Services, although the agencies must also take into account petitions for listing that may be submitted by private parties. The Services maintain a priority list of species that have enough supporting information to warrant a listing decision, and they are required to review the list of protected species every 5 years to determine if any changes in status should be made. Status changes (delisting or reclassification) also are regulatory actions. The listing process includes provisions for public review and comment to solicit input about the species' status, the nature and degree of threat, and protective actions that may be needed or are in place. A species may be listed if it is considered threatened or endangered for one or more of the following reasons: habitat reduction, overharvesting, disease or predation, absence of adequate protective measures, and other unspecified factors that may contribute to its imperiled condition.

Listing decisions are to be made solely on the basis of the best scientific and commercial information available at the time, without regard to political or economic interests. The Services must also consider protective measures that are in place by government agencies. For some time, there has been an extensive backlog of species whose listing action is pending because of funding constraints (Doremus 2000). Recent legal decisions that require the FWS to take action on many species that are eligible for listing have also added to the backlog. The result of this hourglass effect on the pace of listing decisions is that many species do not get listed until their populations are substantially reduced. Listing of Atlantic salmon that were designated as a DPS in eight Maine rivers is a recent example.

In response to congressional direction, the Services have adopted priority-setting guidelines to help manage the backlog. FWS guidelines consider the magnitude of threat, the imminence of the threats, and the taxonomic uniqueness of the species. Taxonomic uniqueness is applied hierarchically; first priority is given to monospecific genera, followed by full species, and then subspecies. NMFS guidelines do not include taxonomic factors, but their listing regulations invoke the concept of "evolutionary significant unit" (NRC 1995, Waples 1991b) to define eligible entities for listing below the species level.

At the time of listing, the Services are required to designate critical habitat to the extent that necessary information is available and it is prudent to do so. Critical habitat is the area within the species' range that contains those physical or biological features essential to the conservation of the species and that may require special management considerations or protection. Designation of critical habitat is the only decision under the ESA that must consider economic factors. Although it is a statutory requirement for the Services to designate critical habitat, in 1999 the FWS reported that it had designated critical habitat for only 9% of the species under its jurisdiction (Doremus 2000). To justify their lapse, the Services have asserted that critical habitat confers little additional protective advantage beyond the listing action itself. But their resistance is being overturned as a result of losing several recent legal decisions, and they recently have accelerated the rate of critical habitat determinations.

Section 4 of the ESA also provides for the Services to develop and implement recovery plans for listed species. These plans are ultimately designed to improve the species' status to the level where protections are no longer needed and upgrading or delisting may be warranted. Recovery plans may be prepared within either the FWS or the NMFS or they may be collaborative efforts involving outside experts. They are expected to contain measurable recovery criteria, such as population size and survival rate, replacement or recruitment rate, amount of available and occupied habitat, habitat in protected status, number and distribution of discreet populations, measures to alleviate threats, and other protective and conservation measures to be undertaken by other state and federal agencies. They also need to specify an intended schedule for needed recovery actions and to forecast projected costs. A recovery team has been formed to develop the recovery plan for Atlantic salmon but is still in the early stages of its deliberations. The earlier National Research Council (NRC 1995) report *Science and the Endangered Species Act* should provide guidance on crafting effective recovery plans.

Section 6 provides for revenue sharing with the states to assist them with implementation of conservation and recovery actions. The Services depend on local activities to complement theirs, and revenue sharing enhances the cooperation needed to carry out coordinated efforts.

The major protective features of the ESA are in Sections 7 and 9. Section 7 mandates all federal agencies to utilize their authorities to promote the conservation of listed species. It also prohibits actions by federal agencies to fund, authorize, or carry out activities that are likely to jeopardize the continued existence of listed species or result in adverse modification of their critical habitat. To determine whether jeopardy may result, the responsible federal agency engages in a three-stage consultation process with the Services. Consultation starts with a request for information about any species that may be affected by its proposed action. The second step is for the action agency to prepare a biological assessment that details the likely effect of the proposed action on listed species. The biological assessment initiates a formal consultation with one or both of the Services, the result of which is a biological opinion by the Services that must specify reasonable and prudent alternatives to the proposed action if jeopardy or adverse modification of critical habitat is likely. This Section 7 consultation process must be conducted according to well-defined regulatory procedures. It must also make use of the best scientific and commercial information available at the time. To expedite the process the regulations also allow for informal consultation and early consultation.

Section 7 will come into play for federal actions that may affect Atlantic salmon on DPS rivers. Issuance of Section 404 permits by the U.S. Army Corps of Engineers for dredge and fill activities in wetlands is the most likely trigger. Renewal of federal permits for dam operators, logging activities in national forests, research and habitat improvement measures funded or carried out by the National Oceanographic and Atmospheric Administration (NOAA), and authorization for activities that might impair navigation, such as aquaculture facilities, are also likely to undergo Section 7 consultation.

Section 9 is the part of the ESA that prohibits the taking of or commerce in listed species. "Take" is very broadly defined to include harassment or harm to the species in addition to shooting, trapping, collecting, and other actions. The somewhat vague concept of harm has been legally interpreted to include habitat modification to the extent that it kills or injures listed wildlife or interferes with their essential behavioral functions such as feeding, nesting, and breeding. Taking of endangered species is expressly prohibited in Section 9; that section allows for more flexibility in regulations covering take of threatened species. As the Services' regulatory programs are administered, the distinction between threatened and endangered species is almost moot except for a few species. Listed plants are not protected on private lands unless their take is banned under state law. In the case of Atlantic salmon, the take prohibition reinforces the earlier state-imposed ban on take—including catch-and-release angling—in DPS rivers.

Because the ban on taking is almost absolute, the Services also administer programs that allow for waivers of the taking prohibitions. Before "incidental take" provisions were included in the 1982 amendments to the ESA, actions that might cause take at a level below the jeopardy threshold could not proceed. Incidental take is defined as take that is incidental to but not the purpose of an otherwise lawful action. For federal actions covered by Section 7, the consulting service includes an incidental take statement in the biological opinion. The statement specifies reasonable and prudent measures that must be complied with in order for a certain level of take to be allowed. For nonfederal actions, a separate permitting process is available under Section 10. An applicant must prepare a Habitat Conservation Plan (HCP) in accordance with guidelines issued by the

Services that spells out the nature and extent of the proposed activity, the status of listed species potentially subject to take, and conservation measures to mitigate or offset the effects of the take. Issuance of the incidental take permit depends on Service findings that the taking will be minimized and mitigated to the maximum extent practicable, the taking will not result in jeopardy, and that adequate funding is available to carry out the measures called for in the HCP.

HCPs are becoming more widely used as the Services have developed guidelines to assist with their development. They are viewed skeptically by some environmentalists who question whether the conservation measures adequately protect against and mitigate for taking of protected species (Thomas 2001). Landowners are also often frustrated because of the bureaucratic process involved and inconsistent implementation of the guidelines.

In an effort to reduce the tension and uncertainty of the administrative process surrounding HCPs, the FWS has recently made greater use of specific provisions in the ESA that provide greater flexibility in its implementation. Reintroduced species can be designated as "experimental populations" according to Section 10(j), which provides for treating an experimental population as threatened even if its donor population is endangered. The FWS has also adopted several innovative measures designed to provide greater assurance to landowners that the HCP process protects their interests as well as the species'. The flexibility allowed in Section 4(d) for species protected under state law has enabled adoption of regional natural community conservation plans in California that spell out permitted and prohibited activities that could result in take of threatened species listed under this statutory provision. The "no surprises policy" adopted by the FWS has provided greater assurance for holders of incidental take permits that the conditions agreed to at the time of issuance will remain in place over the life of the permit. HCPs may cover unlisted species so that they are incorporated in the permit if they are listed in the future, provided that the conservation measures specified are sufficient to protect them adequately. Finally, the "safe harbors policy" adopted by the FWS protects landowners who allow listed species to colonize currently unoccupied habitat on their land by authorizing them to carry out certain preexisting activities without an HCP.

HCPs could have a role to play in the conservation and recovery of Atlantic salmon in Maine rivers. The need for private landowners to obtain incidental taking permits depends on the enforcement policy and posture of the Services. Most land uses that may affect Atlantic salmon, such as logging and farming in the watershed, do not cause direct mortality. However, if the Services show that the uses appreciably reduce the likelihood of survival or reproduction as a result of habitat modification through water withdrawals or pollution, they could come under the regulatory sweep of the ESA. This interpretation could also apply to aquaculture activities that reduce the genetic integrity or threaten natural populations of Atlantic salmon in DPS rivers with disease or infestation by parasites.

Appendix B

Governance

UNORGANIZED AREAS

Several areas in Maine are not organized and governed by local townships. For a list and map of the unorganized and deorganized areas in Maine, see the sources listed below. Human activities in these unorganized areas, such as land use and forestry, are regulated by state agencies, including the State of Maine Land Use Regulation Commission and the Department of Environmental Protection, which are described below.

Sources:
List of names: Maine Revenue Services 2003a
Map: Maine Revenue Services 2003b

NATIVE AMERICAN TRIBAL GOVERNMENT

The term for Native Americans in Maine is the Wabanaki People. The Wabanaki include representatives from four tribes: the Passamaquoddy Tribe in Washington County, the Penobscot Indian Nation based at Indian Island on the Penobscot River, the Houlton Band of Maliseets, and the Aroostook Band. The relationship between the Penobscot Nation and the State of Maine is governed by a federal act, the Maine Indian Claims Settlement Act of 1980 and a state act, an Act to Implement the Maine Indian Claims Settlement (Chapter 732 of Maine public laws of 1979). The federal settlement act allowed both the Passamaquoddy Tribe and the

Penobscot Nation to reacquire land. It recognized the applicability of state laws to the tribes and to Indian people, lands, and resources except where otherwise provided in the act. The settlement act provided federal recognition for the Houlton Band of Maliseets but did not define a special relationship with the state of Maine. Also, the act did not include the Aroostook Band of Micmacs, nor did it include the Maliseet People who were not members of the Houlton Band. In late 1991, the Aroostook Band of Micmacs won federal recognition.

Pursuant to the federal and Maine settlement acts, the Penobscot Nation reservation encompasses the islands and related water and fishing rights within the Penobscot River from Indian Island, near Old Town, Maine, northward. The tribe has exclusive jurisdiction over "internal tribal matters," but those matters are not clearly defined. Recognition of reservations as entities with extraordinary municipal rights and responsibilities is also included in these acts. The Maine Indian Tribal-State Commission (MITSC), an independent commission made up of tribal and state representatives, has exclusive authority to promulgate regulations governing fishing within any section of a river both sides of which are within the reservation or trust lands (lands owned by the United States and held in trust for the tribe).

The Penobscot Nation has exclusive authority within its reservation to regulate sustenance fishing by tribal members, and sustenance fishing is a reserved right under the terms of the settlement acts. However, the capacity of the Penobscot Nation to fully exercise its sustenance fishing rights has been constrained in recent years by pollution of the Penobscot River by Lincoln Pulp and Paper Company (Bisulca 1996).

The findings of the Task Force on Tribal-State Relations (1997) examined the attitudes and concerns related to the settlement act and found that tribal members generally do not think the settlement act works. The complaints recorded in the report include a complaint by the director of natural resources of the Penobscot Nation about the settlement act and its use by the attorney general to claim that the Penobscot cannot take salmon from the river. The Task Force Report recommends that the state, the tribes and the MITSC treat the act as an "organic and living" document. It notes that over time, changes have taken place, including the development of cooperative law enforcement, fish and game, and environmental agreements.

Sources: Bisulca 1996; Department of the Secretary of State 2002; DIFW 2002a; Kaign Smith, Counsel for Penobscot Nation, personal communication, April 7, 2003; MRDC 2002

STATE GOVERNMENT AGENCIES

Maine Atlantic Salmon Commission (MASC)

In 1945, the Maine legislature created a single administrative unit with authority to manage Atlantic salmon in freshwater and saltwater. The Atlantic Sea Run Salmon Commission was superseded by the Atlantic Salmon Authority in 1995 and by the current Maine Atlantic Salmon Commission (MASC) in 1999. Its purpose is to protect, conserve, restore, manage and enhance Atlantic salmon habitat, populations, and sport fisheries within historical habitat in all (inland and tidal) waters of Maine. Its activities have included surveys of Atlantic salmon habitat; habitat improvement; fish-passage improvements; the elimination of commercial fishing; progressively restrictive sportfishing regulations, culminating in the current prohibition of any recreational fishing for anadromous Atlantic salmon in Maine; and various stocking programs. The commission also conducts research on salmon life histories, population status and trends, stocking methods and practices, effects of natural predation, and studies of migration. The commission's decades-long database of the results of Carlin tagging helped to document the high exploitation rates of Maine salmon in distant-water commercial fisheries. Current management strategies and progress are provided in the Conservation Plan for Seven Maine Rivers (Maine Atlantic Salmon Task Force 1997) as well as MASC's updates of the plan, reports to the Maine legislature, and other documents.

Sources: MASC 2002, Maine Atlantic Salmon Task Force 1997

Maine Department of Inland Fisheries and Wildlife (DIFW)

The DIFW regulates recreational fishing and boating and monitors and investigates salmon health problems in aquaculture facilities. More generally, DIFW is responsible for establishing and enforcing the rules and regulations that govern fishing, propagation and stocking of fish, registration of watercraft and all terrain vehicles, and issuing of licenses (hunting, fishing, trapping, guide, etc.) and permits. The DIFW enforce the rules adopted by the MASC. The Department's Bureaus of Resource Management and of Warden Service (the enforcement arm of the department), execute these responsibilities. In addition, the DIFW operates the Fish Health Laboratory, and monitors and investigates fish health problems such as infectious salmon anemia (ISA), a viral disease of farmed Atlantic salmon.

Sources: DIFW 2002b,c,d,e; 2003a,b

Maine Department of Marine Resources (DMR)

The DMR regulates marine aquaculture operations, marine fisheries, and recreational boating and operates programs for research and monitoring of living marine resources. For salmon aquaculture, DMR issues permits for aquaculture sites, enforces the Aquaculture Lease Law, administers the Finfish Aquaculture Monitoring Program (FAMP), and monitors for toxic contaminants under and in net-pens. For fisheries, DMR issues fishing licenses, enforces saltwater fishing laws and regulations, and operates research and habitat conservation programs.

DMR bears the statutory responsibility, among others, to conduct and sponsor scientific research on marine resources, to conserve and develop the utilization of marine and estuarine resources, and to restore diadromous fish resources to the rivers of Maine; to protect public health by ensuring sanitation of shellfish harvesting areas, harvesting, processing, and distribution; and to provide education and outreach. The DMR's Bureau of Marine Patrol enforces marine fisheries laws, boating registration laws, and safety laws, and conducts search and rescue operations on coastal waters. The DMR's Bureau of Resource Management conducts research and monitoring programs to support efforts to conserve, restore and manage the marine and estuarine resources of the state.[1]

Sources: DMR 2001, 2002; Fisk 2002

Maine Department of Environmental Protection (DEP) (Bureau of Land and Water Quality)

Several state statutes provide the DEP with the authority to govern a wide range of human activities, including hydropower and dams, natural resource protection, shoreline zoning, site development, erosion and sedimentation control, wastewater discharge, and others. The principal governance actions involve issuing permits and enforcing standards that apply to these activities. With respect to hydropower projects, the DEP, in cooperation with the Land Use Regulatory Commission (LURC), issues permits for the construction, reconstruction or the structural alteration of a hydropower project and enforces state laws concerning unapproved hydropower projects. With respect to salmon aquaculture, the DEP tests water for effluent quality from aquaculture sites and issues permits as

[1]DMR jurisdiction is within all waters of the state within the rise and fall of the tide and within the marine limits of the state but does not include areas above any fishway or dam when the fishway or dam is the dividing line between tide water and freshwater. The latter areas fall within the jurisdiction of the Maine DIFW.

part of the Maine Pollution Discharge Elimination System (MPDES). In addition, the DEP issues permits for activities on land adjacent to any freshwater wetland, great pond, river, stream, or brook that could wash harmful material into these resources. In addition, the DEP operates programs to monitor water quality (groundwater, lakes and streams, and coastal waters).

Sources: DEP 1996a,b,c; 2000; 2003

Land Use Regulatory Commission (LURC) (Department of Conservation)

In addition to the regulation of hydropower projects, the LURC regulates land use in the state's townships, plantations, and unorganized areas. Its objectives are to preserve public health, safety, and welfare; to encourage the well-planned, multiple uses of natural resources; to promote orderly development; and to protect natural and ecological values using land-use planning and zoning tools. One of LURC's important objectives is to protect groundwater in order to conserve important fish and wildlife habitats. The LURC issues permits for construction of roads and bridges, and sets standards for several uses of land (roads, agriculture, timber harvesting, filling and grading, applications of pesticides, etc.) and for the cutting of trees near water bodies.

Sources: LURC 2000a,b; 2001; 2002a,b

Maine Forest Service (MFS) (Department of Conservation)

The principal responsibilities of the MFS are to protect the state's forest resources from fire, disease, and pests. In addition, MFS aims to enhance forest resources through technical assistance; education; and outreach to the public, forest landowners, forest products processors and marketers, and municipalities. MFS encourages forest landowners to use the services of a consulting forester to help implement forest management projects on their woodlot.

Sources: MFS 2002a,b,c,d

Maine Public Utilities Commission (MPUC)

The MPUC has jurisdiction over water utilities, electric utilities, water carriers, gas utilities, telephone utilities, and resellers of telephone services. It is responsible for the enforcement of all state laws that apply to public utilities, such as hydropower dams. However, the MPUC shares these responsibilities with the DEP and the LURC, the two agencies that issue permits for the construction, reconstruction or the structural alteration of a hydropower project and enforces state laws concerning unapproved hydropower projects. Most hydropower dams are subject to regulation by the Federal Energy Regulatory Commission. Licenses are issued for waterpower projects for up to 50 years; at expiration, a dam may be relicensed or taken over by the federal government.

Source: MPUC 2001

Maine Department of Transportation (DOT)

The DOT is responsible for designing, building and maintaining many of the roads, highways and bridges in the state. It is also the main oversight agency for projects involving roads, railroads and associated facilities. The Maine DOT has developed a framework for integrating environmental and transportation decision making throughout the department. The framework interfaces planning, location, design, right-of-way, construction, maintenance, and environmental operations by fully integrating the decision-making processes of Maine's Sensible Transportation Policy Act (STPA); the National Environmental Policy Act (NEPA); and state and federal environmental permitting programs, especially the U.S. Army Corps of Engineers' (New England District) highway methodology (USACE 1993). Maine DOT has also developed the Fish Passage Policy and Design Guide, issued in March 2002.

The DOT restores habitat by addressing nonpoint source pollution associated with transportation facilities located in salmon watersheds. Maine DOT provides technical assistance to maintenance crews in salmon watersheds to implement erosion and sedimentation best management practices. It has also developed detailed geographic information systems (GIS)-based watershed maps identifying all DOT owned and operated facilities as a tool for workers to easily identify critical areas.

Sources: MDOT 2002a,b; 2003

Maine Department of Agriculture, Food and Rural Resources

The Maine Department of Agriculture regulates the use of pesticides and implements pest management and soil and water management programs.

Source: Maine Department of Agriculture 2002

Maine State Planning Office

In general, the Maine State Planning Office provides information, analysis, and guidance to policy makers about Maine's economy, resources, and governance. The duties of the State Planning Office include coordinating the development of the state's economy and energy resources with the conservation of its natural resources; providing technical assistance to the governor and legislature by undertaking special studies and plans and preparing policy alternatives; providing technical assistance to local and regional planning groups; and conducting continuing economic analysis.

Source: Maine State Planning Office 2003

FEDERAL GOVERNMENT AGENCIES

U.S. Fish and Wildlife Service (FWS, U. S. Department of the Interior)

The FWS and the National Marine Fisheries Service (NMFS) share responsibility for administration of the Endangered Species Act. The FWS implements ESA (ESA) programs and regulations for terrestrial and freshwater species, while NMFS implements programs and regulations for marine and anadromous species.

In general, the FWS operates programs to protect and restore fish and wildlife resources and their habitats. The FWS manages over 500 national wildlife refuges, and operates the National Fish Hatchery System, which consists of 70 fish hatcheries, 7 fish technology centers, and 9 fish health centers. The hatcheries are part of an effort to recover endangered species and restore native aquatic populations. In Maine, the FWS operates two national fish hatcheries, Craig Brook and Green Lake.

The FWS investigates, evaluates, and makes recommendations on permit and license activities of several federal agencies including the U.S. Army Corps of Engineers, Federal Energy Regulatory Commission, and the U.S. Forest Service. In addition, the FWS enforces federal wildlife laws.

Sources: FWS 1998; 2003a,b,c.

National Marine Fisheries Service (NMFS, National Oceanic and Atmospheric Administration, U.S. Department of Commerce)

The NMFS Office of Protected Resources is charged with the implementation of the ESA for marine and anadromous species. The NMFS develops, implements, and administers programs for the protection, conservation, and recovery of species protected under the ESA. The Office of Protected Resources also develops and implements policies, procedures, and regulations for permits to take listed species according to the ESA. In addition, the NMFS establishes cooperative agreements with states regarding listed species management and protection and identifies endangered species research needs to collect appropriate information for management decisions. The NMFS and FWS share responsibilities for listing endangered species and approving recovery plans for listed species under the ESA. The NMFS is primarily responsible for recovery actions in the marine environment, and the FWS is primarily responsible for recovery actions in the terrestrial environment.

In addition, the NMFS implements marine fishery management plans that have been approved by the Secretary of Commerce. This includes the fishery management plan for Atlantic salmon (*Salmo salar*), which was implemented by the NMFS on March 17, 1988. This fish management plan established explicit U.S. management authority over all Atlantic salmon of U.S. origin to complement state management programs in coastal and inland waters and federal management authority over salmon on the high seas conferred as a signatory nation to the North Atlantic Salmon Conservation Organization (NASCO). The fish management plan disallows any commercial fishery for Atlantic salmon, directed or incidental, in federal waters (3–200 miles) and prohibits the possession of Atlantic salmon taken from federal waters.

Through the Magnuson-Stevens Fishery Conservation and Management Act of 1996, the NMFS has regulatory responsibilities that affect aquaculture development in the Exclusive Economic Zone. The fishery management councils are involved in the decision-making process for offshore aquaculture permits. To date, this process has included granting a lease to an experimental scallop culture project off the coast of Massachusetts through an amendment to the New England Scallop Fishery Management Plan and consideration of an experimental permit for the culture of red snapper in the Gulf of Mexico.

The NMFS promotes aquaculture through scientific research and technology development, financial assistance, and its regulatory programs. The NMFS' basic research on finfish and shellfish biology and reproduction, habitat utilization and restoration, environmental impact assessment, and fish pathology supports private and government aquaculture and marine enhancement activities. The NMFS has also played an integral role in the rearing of threatened and protected species for stock recovery.

Sources: Magnuson-Stevens Fishery Conservation and Management Act (1996); NMFS 1998, 2000a, 2002

Environmental Protection Agency (EPA)

The EPA works in Maine with federal, state, regional, and local partners to protect and restore Maine's environment and protect human health. The primary state partners of The EPA are the Maine Department of Environmental Protection (MDEP) and the Maine Department of Health Services (MDHS). The EPA provides environmental and public heath protection assistance as well as over $9 million annually of financial support for air, water, and waste programs at the MDEP; drinking water protection by the MDHS; and monitoring, protection, and restoration efforts for the Casco Bay Estuary Project.

The EPA has funded a $1.9 million cooperative agreement with the Gulf of Maine Council. The project will coordinate, encourage, and support cooperative efforts to protect and sustain regionally significant Gulf of Maine coastal and marine habitats. The funds will support pilot projects to identify and conserve regionally significant habitats; a Gulf-wide monitoring program; a marine debris reduction program; a coastal citizen monitoring network; workshops on shellfish habitat restoration techniques; community surveys on the spawning and juvenile habitat areas of commercial fish stocks; and the production of various public education and outreach materials.

The EPA enforces the Federal Insecticide, Fungicide, and Rodenticide Act (FIFRA), which involves the review and approval of pesticide products and labeling through a pesticide registration process. In this role, the EPA indirectly and directly affects Atlantic salmon farming and agriculture operations. For example, the EPA is responsible for approving and regulating the use of pesticides around, and for monitoring the effluent quality from, aquaculture facilities.

Under the authority of the Clean Water Act, the EPA establishes wastewater standards for industry water-quality standards for all contaminants in surface waters, monitoring of water-quality discharged from site. The EPA has also established management procedures for the protection of surface water quality and in-stream and riparian habitat. The state of Maine applies these management procedures to dam operations and sites near water bodies that require development permits.

Sources: Brennan 1999; EPA 2002b; 2003a,b,c,d,e,f

U.S. Army Corps of Engineers (USACE)

Under the Rivers and Harbors Act and the Clean Water Act, the USACE has the authority to regulate activities in navigable waterways, including authority over dredging and filling of waterways and authority to issue permits for dams and dikes to be placed in interstate waterways. Although not mandatory, the USACE also has developed criteria for safe operation of hydropower projects and dams that have been widely adopted by privately operated projects throughout the United States. Based on its authority to regulate activities in navigable waterways, the USACE regulates the location of aquaculture pens.

The USACE also enforces regulations that require the installation of suitable culverts and bridges, designed to withstand and prevent restriction of high flows and maintain existing low flows, for roads that cross bodies of water. Roads and bridges should not obstruct the movement of aquatic life indigenous to the water body beyond the actual duration of construction.

Sources: USACE 2002; 33 CFR 321

Department of Agriculture (USDA)

The USDA Animal and Plant Health Inspection Service (APHIS) provides agricultural producers with services for protecting the health of animals and plants. The APHIS programs currently serve both plant and animal aquaculture, especially those aspects involving disease, pest prevention, and wildlife damage management. The APHIS also has become involved in facilitating the importation and exportation of aquaculture products. The APHIS provides diagnostic assistance to aquaculture producers on diseases afflicting aquaculture species, disseminates information on how to meet the aquaculture industry's animal health needs, endorses animal health certifications for the export of live aquatic species and their products; and develops aquatic animal health monitoring and surveillance programs. The APHIS investigates consumer complaints regarding biologics used in aquaculture, and tests fish biologics at APHIS's National Veterinary Services Laboratories. In addition, APHIS provides funds for "payment of indemnity"' to producers in Maine for the salmon destroyed in the effort to control outbreaks of infectious salmon anemia.[2]

The USDA Natural Resources Conservation Service (NRCS) operates the Wildlife Habitat Incentives Program (WHIP), a voluntary program for individuals who want to develop and improve wildlife habitat primarily on private land. Through WHIP, USDA's NRCS provides both technical

[2]67 Fed. Reg. 17605 (2002).

assistance and up to 75% cost-share assistance to establish and improve fish and wildlife habitat.

The NRCS provides technical and financial assistance to local authorities for projects to rehabilitate or remove aging dams. Rehabilitation projects may be cost shared between the federal government and local sponsors. The NRCS provides 65% of the total cost of a rehabilitation project. Local sponsors can provide the remaining 35% through "in kind" costs for the value of land rights, project administration, and other planning and implementation costs associated with the project.

The Small Watershed Program (SWP) assists federal, state, and local agencies; local government sponsors; tribal governments; and program participants to protect watersheds from damage caused by erosion, floodwater, and sediment; to conserve and develop water and land resources; and to solve natural resource and related economic problems on a watershed basis. The SWP addresses problems of watershed protection; erosion and sediment control; water supply; water quality; wetland and water storage capacity; water needs for fish, wildlife, and forest-based industries; fish and wildlife habitat enhancement; wetlands creation and restoration; and public recreation in watersheds of 250,000 or fewer acres. The program provides both technical and financial assistance.

In addition, the NRCS administers the Forestry Incentives Program (FIP) jointly with the Forest Service. The FIP supports good forest management practices on privately owned, nonindustrial forestlands nationwide. FIP is designed to benefit the environment while meeting future demands for wood products. Eligible practices are tree planting, timber stand improvement, site preparation for natural regeneration, and other related activities. The FIP was originally authorized in 1978 to share up to 65% of the costs of tree planting, timber stand improvements, and related practices on nonindustrial private forestlands. FIP's forest maintenance and reforestation practices provide numerous natural resource benefits, including reducing wind and soil erosion, enhancing water quality and wildlife habitat, and helping to assure a reliable future supply of timber.

The Stewardship Incentive Program (SIP) provides technical and financial assistance to encourage nonindustrial private forest landowners to keep their lands and natural resources productive and healthy. Qualifying land includes rural lands with existing tree cover or land suitable for growing trees that is owned by a private individual, group, association, corporation, Indian tribe, or other legal private entity. Eligible landowners must have an approved FIP and own 1,000 or fewer acres of qualifying land.

Sources: AMS 2003; APHIS 2002, 2003; FSIS 2001; NRCS 2003; RMA 2003; USDA 2003a,b,c,d,e

Food and Drug Administration (FDA)

FDA is the principal regulatory agency responsible for the safety of the nation's domestically produced and imported foods, cosmetics, drugs, biologics, medical devices, and radiological products. FDA's authority extends to all domestic and imported food with the following exceptions. Meat; poultry; and frozen, dried, and liquid eggs are under the authority of USDA's Food Safety and Inspection Service (FSIS), and the U.S. Environmental Protection Agency (EPA) establishes tolerances for pesticide residues in foods and ensures the safety of drinking water. FDA regulates the production and distribution of cultivated salmon and therefore indirectly affects the nature and extent of Atlantic salmon aquaculture in Maine.

FDA regulates seafood, including farm salmon. FDA operates an oversight compliance program for fishery products related to the product's safety, wholesomeness, identity, and economic integrity. FDA conducts both mandatory surveillance and enforcement inspections of domestic seafood harvesters, growers, wholesalers, warehouses, carriers, and processors under the authority of the Federal Food, Drug, and Cosmetic Act. FDA conducts in-plant inspections of product safety and plant/food hygiene. There are FDA laboratories to analyze samples taken by its investigators. Further, FDA has the authority to set tolerances in food for artificial contaminants, except for pesticides, which are set by EPA. FDA regulates the use of food and color additives in seafood and feed additives and drugs in aquaculture. Finally, FDA has stated that it intends to regulate transgenic fish (and other transgenic animals) under the new animal drug provisions of the Federal Food, Drug, and Cosmetic Act (NRC 2002c).

FDA operates two additional regulatory programs directed specifically at seafood—the Salmon Control Plan and the National Shellfish Sanitation Program (NSSP), recently augmented by the Interstate Shellfish Sanitation Conference (ISSC). These are voluntary programs involving the individual states and the industry. The Salmon Control Plan is a voluntary, cooperative program among the industry, FDA, and the National Food Processors Association (NFPA). The plan is designed to provide control over processing and plant sanitation and to address concerns about decomposition in the salmon canning industry

FDA conducts research in support of its seafood program. This research is directed to understanding the nature and degree of severity posed by various safety hazards, and other defects that may affect quality and economic integrity. Research also finds means to detect and to control these identified hazards. FDA's Center for Veterinary Medicine, through its Office of New Animal Drug Evaluation, works with government agencies and aquaculture associations to increase the number of safe and effective drugs that can be used by the aquaculture industry.

Sources: CVM 1999, 2002; FDA 1990, 2001, 2003; Hoskin 1993; NRC 2002c

Federal Energy Regulatory Commission (FERC)

Under the Federal Power Act, FERC is authorized to issue licenses authorizing construction of a hydropower facility and continuance of existing projects. FERC issues preliminary permits for up to 3 years and authorizes developers to perform feasibility studies while maintaining priority to apply for a future license. FERC issues licenses for a period of up to 50 years after the review of engineering, environmental, and economic aspects of the proposal. In issuing a license, FERC is supposed to equally consider developmental and environmental values, including, for example, hydroelectric development and fish and wildlife resources (including their spawning grounds and habitat). By statute, FERC must require provisions in licenses to "protect, mitigate damage to, and enhance fish and wildlife (and their habitats) . . ." (FPA, section 10(j)). Small hydro plants that are 5 megawatts or less that use an existing dam or that utilize a natural water feature for headwater and existing projects that propose to increase capacity are exempt from FERC licensing. FERC is also responsible for monitoring dam safety.

Source: FERC 2003

U.S. Coast Guard

The U.S. Coast Guard is the nation's maritime law enforcement agency and has broad, multifaceted jurisdictional authority. The Coast Guard enforces fisheries laws at sea, such as the Magnuson-Stevens Fishery Conservation and Management Act, in conjunction with the National Marine Fisheries Service. Charged with ensuring a safe, efficient, and effective marine transportation system, the Coast Guard regulates and inspects commercial and private vessels, licenses merchant mariners, manages waterways, and protects the security of America's ports.

The Coast Guard helps to recover and maintain marine protected species populations. The Coast Guard enforces a wide variety of fishery regulations designed to reduce the bycatch of threatened and endangered species. As part of its mission to manage waterways, the Coast Guard participates in aquaculture leasing permit processes and ensures that offshore structures are not hazards to navigation.

Sources: U.S. Coast Guard 2003a,b,c

Federal Highway Administration (Department of Transportation)

The Federal Highway Administration (FHWA) is an operating administration of the U.S. Department of Transportation. The Maine Division of FHWA works in partnership with the Maine DOT metropolitan planning organizations in Bangor, Kittery, Lewiston-Auburn, and Portland.

Source: FHWA 2003

REGIONAL ORGANIZATIONS

New England Fisheries Management Council (NEFMC)

The NEFMC is one of eight regional fisheries management councils established by the Magnuson Fishery Conservation and Management Act of 1976 (renamed the Magnuson-Stevens Fishery Conservation and Management Act when amended on October 11, 1996). The councils manage the living marine resources within the U.S. Exclusive Economic Zone, an area extending from 3 to 200 miles offshore. The NEFMC's jurisdiction extends from Maine to southern New England, although some NEFMC-managed species range to the mid-Atlantic.

The council develops management plans that are submitted to the National Marine Fisheries Service (NMFS) and the Secretary of Commerce for approval and implementation. The council is tasked with making fisheries management decisions to impose regulations on the fishing industry, which include setting the size of the allowable catch, the length of the fishing season, the allocation of any quotas to states and fishers, and permitting and licensing provisions.

The NEFMC developed the Fishery Management Plan for Atlantic salmon (*Salmo salar*) that was implemented by NMFS on March 17, 1988. The plan explicitly established U.S. management authority over all Atlantic salmon of U.S. origin. Specifically, the plan prohibits any commercial fishery for Atlantic salmon, directed or incidental, in federal waters (3–200 miles) and prohibits the possession of Atlantic salmon from federal waters.

Source: NEFMC 2003

Atlantic State Marine Fisheries Commission (ASMFC)

The 15 Atlantic coast states (Maine through Florida, including Pennsylvania) formed the ASMFC in 1942 to assist in managing and conserving the states' shared coastal fishery resources. Each of the 15 states is represented on the commission by three commissioners, including the director for the state's marine fisheries management agency, a state legislator, and an individual representing fishery interests, appointed by the state governor.

The commission initiated its Interstate Fisheries Management Program (ISFMP) in 1981, with a cooperative agreement with the NMFS. The ISFMP aims to promote the cooperative management of marine, estuarine, and anadromous fisheries in state waters of the East Coast through interstate fishery management plans. The major objectives of the ISFMP are to (1) determine the priorities for interjurisdictional fisheries management in coastal state waters; (2) develop, monitor, and review fishery management plans; (3) recommend to states, regional fishery management councils, and the federal government management measures to benefit these fisheries; (4) provide an efficient structure for the timely, cooperative administration of the ISFMP; and (5) monitor compliance with approved fishery management plans.

The species managed under this program are American lobster, American shad and river herring, Atlantic croaker, Atlantic herring, Atlantic menhaden, Atlantic sturgeon, bluefish, northern shrimp, red drum, scup, Spanish mackerel, spot, spotted seatrout, striped bass, summer flounder, tautog, weakfish, and winter flounder. Fishery management plans currently under development include American eel and black sea bass. The fishery management plans impose restrictions on the commercial and recreational catch of the species covered by individual plans. The plans set quotas on catch, minimum sizes of fish that can be landed and sold, and restrict other aspects of fishing. The ASMFC does not have a fishery management plan for Atlantic salmon.

The commission's Research and Statistics Program coordinates commercial and recreational fisheries data collection programs and is active in the development and implementation of the Atlantic Coastal Cooperative Statistics Program.

To achieve the conservation and improvement of marine fish habitat, the commission ensures that habitat information and needs are specified in fishery management plans and disseminated to the agencies with regulatory authority for habitat. The education portion of the commission's Habitat Program complements these efforts by also providing this information to fishermen and the general public, along with advice about what individuals can do to protect fish habitat.

The commission's Sport Fish Restoration Program is aimed at improving fishery conservation and wise utilization of critical sport fisheries resources of the Atlantic. Through this program, the commission acts as a liaison between state and federal agencies and nongovernmental organizations to promote interstate and state and federal cooperation on marine recreational fisheries programs. These activities are coordinated through the Commission's Sport Fish Restoration Committee to ensure compatibility with, and integration into other programs of the Commission.

The commission's Law Enforcement Program assists the states in coordinating their law enforcement efforts through data exchange and problem identification. The program provides information on law enforcement issues, brings resolutions addressing enforcement concerns before the commission, coordinates enforcement efforts among states, and monitors the enforcement of measures incorporated into the commission's interstate fisheries management plans.

Sources: ASMFC 2003, NOAACSC 2003

INTERNATIONAL ORGANIZATIONS

The principal international organization governing Atlantic salmon is the North Atlantic Salmon Conservation Organization (NASCO).[3] NASCO, established in 1984, aims to contribute to the conservation, restoration, enhancement, and rational management of salmon stocks. There are seven contracting parties to NASCO, including the European Union, and 26 nongovernmental organizations with observer status. NASCO consists of a council, three regional commissions, and a secretariat.

The three regional commissions are the North American Commission, North-East Atlantic Commission, and the West Greenland Commission. The two members of the North American Commission are Canada and the United States; the four members of the West Greenland Commission are Canada, the United States, Denmark, and the European Union. Denmark, the European Union, Iceland, Norway, and the Russian Federation are the members of the Northeast Atlantic Commission.

The North American Commission requires each of its members to implement measures to minimize the bycatch of Atlantic salmon that originate in the rivers of other members. In addition, the commission requires that before a member allows the increase in catches of salmon that originate in the rivers of another party, the member must obtain the consent of that party.

Source: NAFO 2003

[3]The North Atlantic Fisheries Organization governs fisheries in the North Atlantic that exploit species other than Atlantic salmon.

Gulf of Maine Council on the Marine Environment

Established in 1989 by the region's governors and premiers, the Gulf of Maine Council on the Marine Environment is an international body that promotes and facilitates cross-border cooperation among government, academic, and private groups. The council aims to develop and implement a sustainable management strategy for the Gulf of Maine, an area that extends from Nantucket through the Bay of Fundy to Cape Sable, Nova Scotia. The council's activities in marine monitoring, habitat protection, public education, and pollution prevention are overseen by public and private representatives from Massachusetts, New Hampshire, Maine, New Brunswick, and Nova Scotia.

The Gulf of Maine council has developed an action plan for the protection and conservation of coastal and marine habitats in the Gulf of Maine. The action plan will guide state, provincial, and federal policy and budgeting decisions affecting the Gulf's coastal and marine environments. The governors of Maine, Massachusetts, and New Hampshire; the premiers of New Brunswick and Nova Scotia; and six federal agencies with mandates in the marine environment (Environment Canada; the Department of Fisheries and Oceans, Canada; the U.S. Fish and Wildlife Service; the U.S. Army Corps of Engineers; the U.S. Environmental Protection Agency; and the National Oceanic and Atmospheric Administration) have agreed to the action plan. The plan focuses on coastal and marine habitat and has five major goals: (1) protect and restore regionally significant coastal habitats; (2) restore shellfish habitats; (3) protect human health and ecosystem integrity from toxic contaminants in marine habitats; (4) reduce marine debris; and (5) protect and restore fishery habitats and resources.

Source: Gulf of Maine Council on the Marine Environment 2003

NONGOVERNMENTAL ORGANIZATION SALMON EFFORTS IN MAINE:

Atlantic Salmon Federation—Maine Council
Atlantic Salmon for Northern Maine
Atlantic Salmon Unlimited
Dennys River Sportsman's Club
Downeast Salmon Federation
Eddington Salmon Club
F.I.S.H. (Facilitators Improving Salmonid Habitat
Fishing In Maine
Friend of the Penobscot

Friends of Craig Brook
Kennebec Coalition
Maine Environmental Policy Institute
Maine Rivers
Marine Environmental Research Institute
Narraguagus Salmon Association
Natural Resources Council of Maine
Northern Penobscot Salmon Club
Penobscot County Association
Penobscot River Coalition
Penobscot Riverkeepers 2000
Penobscot Salmon Club
Pleasant River Fish and Game Conservation Association
Pleasant River Hatchery
Project SHARE
Quoddy Regional Land Trust
Saco River Salmon Club
St. Croix International Atlantic Salmon Assoc.
St. Croix International Waterway Commission
Sheepscot River Club
Sheepscot Valley Conservation Association
Trout Unlimited, George's River Chapter
Trout Unlimited, Kennebec Valley Chapter
Trout Unlimited Maine Council
Trout Unlimited Merrymeeting Bay Chapter
Union Salmon Association
Veazie Salmon Club
Wild Salmon Resource Center

WATERSHED COUNCILS

Cove Brook
Dennys River
Ducktrap River (Coastal Mountain Land Trust)
East Machias River
Kennebec River (Friends of the Kennebec Salmon)
Machias Rivers
Narraguagus River
Pleasant River
Sheepscot River

Appendix C

Supportive Breeding, Effective Population Size, and Inbreeding

Release of hatchery fish is practiced extensively in fisheries management either to protect weak populations from extinction or to sustain sport or commercial fisheries. To avoid the risk of compromising the genetic integrity of the recipient population, the fish released are frequently produced through captive breeding of a part of the wild population. A fraction of the wild fish are brought into captivity for reproduction, and their offspring are released into the natural habitat where they mix with wild individuals. This breeding practice, referred to as supportive breeding (Ryman and Laikre 1991), is intended to increase the census population size without introducing exogenous genes into the managed population.

Supportive breeding may increase the total population size through a higher reproductive output from the captive breeders than from those reproducing in the wild. In many situations, however, there is a trade-off between this demographic gain and the genetic "health" of the population, because the procedure of supportive breeding may be coupled with a reduction of the genetically effective population size (N_e), resulting in excessive inbreeding and loss of genetic variation. The basic reason for this reduction of N_e is that supportive breeding implies manipulating the reproductive rate of the captive (hatchery) segment of the population that results in a change of the variance of family size in the population as a whole (wild + captive), and this parameter is of critical importance to the effective size of the population (Ryman 1994, Ryman and Laikre 1991, Ryman et al. 1995b; Wang and Ryman 2001).

Rather complex theoretical considerations are sometimes required to predict the genetic dynamics of a population under supportive breeding, and for the purpose of this presentation, we confine the discussion to illustrating the basic problems by means of some worked examples. We focus on inbreeding (F) and on the corresponding parameter inbreeding effective size that is related to the rate of inbreeding per generation (ΔF) through $\Delta F = 1/(2N_e)$ (see Ryman et al. [1995b] or Wang and Ryman [2001 for details). The results presented below have been generated using the equations for inbreeding effective size of Wang and Ryman (2001).

Model: We consider a wild population with an even sex ratio that in a particular generation (t) consists of N individuals. Before mating, these N individuals (breeders) are distributed at random into a captive (c) and a wild (w) group of size N_c and N_w, which reproduce in captivity and in the wild, respectively ($N_c + N_w = N$; "enumeration" takes place at sexual maturity, and the unit of measurement refers to adults that are potential breeders). The mean and variance of the number of (adult) progeny per wild individual is μ_w and σ^2_w, respectively, and the corresponding quantities for the captive segment are μ_c and σ^2_c. The captive offspring are released into the natural habitat where they mix and breed with wild individuals. Mating is random within each of the wild and captive groups, and the entire process of selecting breeders for captive propagation and releasing their progeny may be repeated in generation $t + 1$ and subsequent generations. The wild population is of constant size when $\mu_w = 2$, it grows when $\mu_w > 2$, and it is declining if $\mu_w < 2$. When considering the effect of supportive breeding on the total population size (N), the operation is successful when $\mu_c > \mu_w$, it has no effect when $\mu_c = \mu_w$, and it is unfavorable when $\mu_c < \mu_w$.

Binomial distribution of family size: As an example, we consider a natural population of $N = 50$ individuals that is constant in size ($\mu_w = 2$). The organism can be bred in captivity, and under captive conditions, the average number of progeny is typically around 10 ($\mu_c = 10$). Ecological studies indicate that the present population size is far below carrying capacity, and the manager wants to raise the number through captive propagation of some of the individuals. It is assumed that the removal of some individuals will not affect the reproductive rate of those that are left to reproduce in the wild. A supportive breeding program is initiated, and during each of 10 generations, five randomly selected breeders are caught in the wild and brought into captivity for reproduction and subsequent release of all their offspring (initial $N = 50$, $N_c = 5$, and $N_w = N - 5 = 45$).

The number of progeny per individual (family size) follows a binomial distribution within each of the wild and captive groups. Under such conditions, the mean and variance of family size are approximately the same ($\sigma^2_c \approx \mu_c$ $\sigma^2_w \approx \mu_w$), and the census and effective population sizes are identical (which follows from the definition of effective size).

Without interference, the effective size of the wild population would remain constant at $N_e = N = 50$ for each generation, because family size is binomially distributed. Under supportive breeding, the total size (N) of the population (wild + captive) grows linearly over time, but effective size behaves quite differently (Figure C-1a). Although the total population size (N) grows quickly, there is a sudden reduction of N_e in generation 2 below what it would have been without supportive breeding ($N_{e,2} \approx$ 29). Further, although N_e grows larger after generation 2, the rate of increase is slower than that of the census size (N). The reason for this behavior of N_e is that the captive breeding program results in a change of demographic parameters that affect N_e. On one hand, the growing N tends to increase N_e. On the other, the differential contribution from wild and captive breeders (μ_w *vs.* μ_c) results in a variance for the population as a whole (σ^2) that is larger than that without supportive breeding ($\sigma^2 = \sigma^2_w \approx 2$), which tends to reduce effective size. As N grows, σ^2 declines asymptotically to $\sigma^2 \approx 2$, but in the first few generations, the increase of σ^2 outweighs that of N, and effective size is reduced (Wang and Ryman 2001).

The reduction of N_e during the early stage of supportive breeding creates a "genetic bottleneck" that results in accumulation of inbreeding at a rate higher than that of a population left on its own. Thus, the boost of census size is obtained at the cost of increased inbreeding, and the effect of this increase may persist for many subsequent generations. The reason for this extended effect is that the overall genetic success of a supportive breeding operation cannot be judged exclusively from the number of generations required for N_e to exceed the value it had before the program was launched. Rather, evaluation of the total genetic impact of a breeding program must be based on the amount of inbreeding that has accumulated during the program as a whole.

The cumulative effects of supportive breeding on inbreeding can be assessed from the harmonic mean of N_e as indicated in Figure C-1a. As noted above, the effective number would stay constant at $N_e = 50$ in the absence of supportive breeding. As an effect of the breeding program, however, N_e first drops to a minimum in generation 2 and then starts to increase as the total population grows, and N_e exceeds 50 already in generation 4 when $N_{e,4} \approx 58$ (Figure C-1a). It is important to note, however, that this by no means implies that the population has now "recovered" from the excessive rate of inbreeding caused by the reduced effective size during the initial phase of the program. The reason is that the amount of inbreeding accumulating over multiple generations of variable N_e is de-

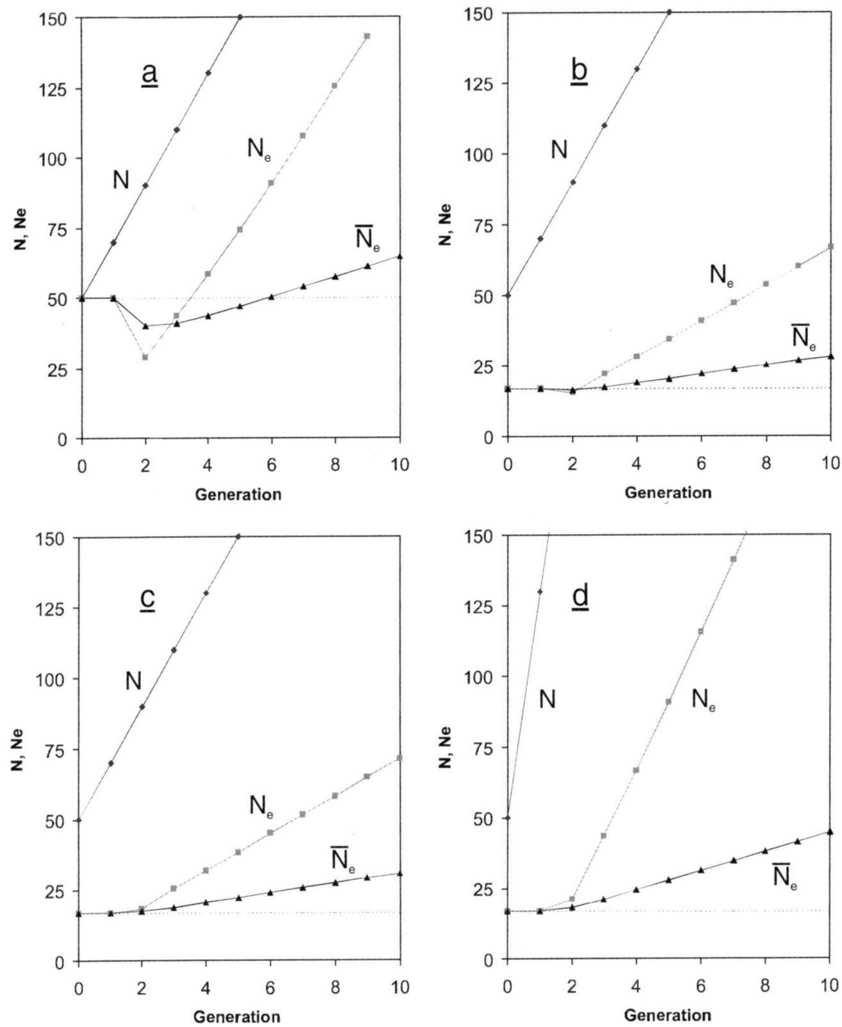

FIGURE C-1 Census size (N), effective size (N_e), and cumulative harmonic mean of effective size ($\overline{N}_e$) during 10 generations of supportive breeding in a population of 50 individuals that would be of constant size if left on its own (initial N = 50). In each generation a fixed number of N_c individuals are caught at random and brought into captivity for reproduction. The mean number of progeny per individual is $\mu_c = 10$ in captivity and $\mu_w = 2$ in the wild. Dashed lines indicate the effective size in the absence of supportive breeding. a, $N_c = 5$; the variance (σ^2) of the number of the progeny per individual in the wild and in captivity are approximately equal to their respective means, i.e., $\sigma^2_w \approx \mu_w = 2$, and $\sigma^2_c \approx \mu_c = 10$. b, $N_c = 5$, $\sigma^2_w = 10$, and $\sigma^2_c = 50$. c, $N_c = 5$, $\sigma^2_w = 10$ and $\sigma^2_c = 10$. d, $N_c = 20$, $\sigma^2_w = 10$, and $\sigma^2_c = 50$.

termined by the harmonic mean of N_e over generations, and the harmonic mean is most heavily affected by the smaller of the values being averaged. In generation 4, this average is only about 43, implying that the population has accumulated more inbreeding during the first four generations than it would in the absence of supportive breeding, i.e., if the effective number had stayed constant at $N_e = 50$. It is not until generation 7 (at $N_7 = 190$ and $N_{e,7} \approx 108$) that the harmonic mean of N_e exceeds 50, and the overall accumulation of inbreeding (total F) is less than it would have been had the population been left alone. Thus, although already the first six generations of this supportive breeding operation may be considered very successful in boosting both the census (N) and effective (N_e) numbers, the population still "suffers" from the genetic bottleneck that occurred during the first few generations of the program (Wang and Ryman 2001).

Under the present model the outcome of supportive breeding on N_e is qualitatively the same as that depicted in Figure C-1a in all situations where the number of progeny of wild and captive breeders is binomially distributed around their respective means of μ_w and μ_c (Ryman and Laikre 1991, Ryman et al. 1995b). The quantitative effects, however, i.e., the magnitude of the sudden drop in generation 2, the rate at which N_e increases in subsequent generations, and the time required for "recovery" depend on N_c, μ_w, μ_c, initial N, and the duration of the program. The severity of genetic bottleneck tends to be most pronounced when a small number of captive breeders are allowed to produce a large number of offspring over many generations. Expressed differently, the amount of "genetic damage" through accumulation of inbreeding is larger the more successful the support program is from a purely census perspective.

Nonbinomial distribution of family size: In natural populations of most organisms, including fishes, the variance of family size is frequently likely to be larger than binomial. The same is true for many captive populations unless active management measures are taken to reduce this variance. In such situations the effect of supportive breeding on N_e is more difficult to predict qualitatively, because the outcome now also depends on σ^2_w and σ^2_c, as exemplified in Figure C-1b. Here, the basic conditions are the same as in the previous example ($N_c = 5$; $N = 50$, and $N_w = 45$ at the start of the program; $\mu_w = 2$ and $\mu_c = 10$), except that the variances of family size are now five times larger than their binomial values with $\sigma^2_w = 10$ and $\sigma^2_c = 50$.

Under this latter scenario, the total (census) size of the population changes as in the previous example, as it should because N_c, μ_w, μ_c, and initial N are the same as before. The immediate reduction of N_e is quite small, however, and already in generation 3 the harmonic mean of N_e is larger (and the overall accumulation of inbreeding correspondingly smaller) than it would have been without supportive breeding. The reason is that effective size is always smaller than census size in a population where the variance of family size exceeds its binomial value. Because of the larger than binomial variance in the present case, the wild population of 50 individuals ($N = 50$) only maintains an effective size of 17 ($N_e \approx 17$) if left on its own, and the breeding program does not imply a change of the demographic parameters of the population as a whole (wild+captive) that brings N_e below its original value. Thus, for this particular population, the support program is successful both demographically and genetically, as it results in a rapid increase of both the census and the effective size.

The variance of family size can frequently be manipulated under captive conditions and efforts can be made to reduce this variance to increase the effective number of captive breeders (as is presently done in, for example, the Craig Brook hatchery). Reducing σ^2_c is expected to provide an increase of N_e in a population under supportive breeding, but the effect of such a reduction is minor in the present case. As indicated in Figure C-1c, bringing this variance down to $\sigma^2_c = 10$ (from its original value of $\sigma^2_c = 50$) only results in an increase of N_e from 15 to 18 in generation 2, and from 66 to 71 at the end of the program. For other combinations of wild and captive parameter values, the "genetic gain" of reducing σ^2_c may be substantial, however.

In contrast, for the combinations of mean and variances of family size in the present example (Figure C-1b) the breeding program could be made more successful by using more captive breeders, and Figure C-1d shows the result of increasing N_c from 5 to 20. Bringing a larger number of breeders into captivity not only boosts the census population more efficiently (as expected when $\mu_c > \mu_w$), but for the present parameters, it also results in a more rapid increase of the effective size and a correspondingly reduced accumulation of inbreeding (reflected in the larger harmonic mean N_e over the entire program).

Supportive breeding may drastically reduce the effective size of a population and thereby accelerate both the inbreeding and the loss of genetic diversity, also when the breeding program is successful in boosting the census population size. Predicting the genetic effects of a particular program may be quite complicated, however, requiring access to at least crude estimates of several parameters. The manager may have a quite good knowledge of the number of breeders brought into captivity and the mean and variance of the number of progeny they produce (N_c, μ_c, and σ^2_c), but the information on the corresponding quantities for the wild segment is frequently rather poor (but see Spidle et al (2003) for some recent estimates).

Clearly, the overall biological success of a supportive breeding program cannot be evaluated without assessing the impact on the effective size, and research efforts should be directed toward collection of data that are necessary for realistic predictions of that impact. In this context, preliminary appraisals that are based on a range of seemingly realistic parameter values may be helpful in identifying critical information that may be missing.

The above examples are aimed at illustrating some of the basic genetic problems associated with supportive breeding, rather than providing an inventory of all the factors that should be considered when designing a breeding program. Situations not discussed include unsuccessful supportive breeding (i.e., $\mu_c < \mu_w$), declining wild populations ($\mu_w < 2$), variable number of captive breeders, preferentially selecting individuals of wild or captive origin for captive breeding, overlapping generations, and populations that crash when the support program is terminated. Some of these aspects have been discussed in the literature, and in addition to the citations given above, interested parties should consult the papers by Duchesne and Bernatchez (2002), Hedrick et al. (2000), Lynch and O'Hely (2001), Waples and Do (1994), and references therein.

Appendix D

Supportive Breeding and Risks to Genetic Quality

There are many genetic risks associated with the production of salmon through supportive breeding programs. One genetic risk of artificial propagation that has attracted widespread attention is the loss of genetic diversity (Hedrick 2001). Ryman and Laikre (1991), and the discussion above, show how supportive breeding may reduce effective population size ($N_e < N$) and therefore accelerate the loss of genetic diversity within wild populations. This loss may reduce the viability of individuals, for example through reduced heterozygosity, and it may also impact the potential evolution of new adaptations by populations over the long term. Concern over the short- and long-term impacts has led managers of supportive breeding programs to develop breeding protocols that retain maximum genetic diversity. Artificial pairing may include genotyping and calculation of "mean kinship" to determine breeding value (Ballou and Foose 1996). Or individuals are randomly bred, and emphasis is placed on equal contribution from each individual by equalization of family size (e.g., Rodriguez-Clark 1999, Wiese and Willis 1999). Although these protocols may achieve the objective of maintaining genetic diversity, they are not based on natural breeding systems.

The emphasis in current supportive breeding programs is the artificial pairing of genetically unrelated individuals. In nature, however, breeding is usually not random with respect to genetics (Andersson 1994). Supportive breeding, as currently practiced, limits or even works against both sexual selection and life history decisions that are necessary for the maintenance of genetic quality within populations (e.g., Fleming and

Gross 1993, Grahn et al 1998, Wedekind 2002). Sexual selection in natural breeding systems is known to expose heritable genetic quality, through male competition and condition-dependent characters, that is targeted by female choice and increases offspring viability (Møller and Alatalo 1999). Life history decisions that are made by certain individuals, such as precocious maturity by higher quality males, will also expose heritable genetic quality that is not captured in supportive breeding programs (Gross 1996). If potential mates differ in heritable genetic quality, maximizing genetic diversity through preventing reproductive skew is unlikely to be the best conservation strategy (Wedekind 2002).

In natural breeding systems, "genetic quality" may have three components that are targets of female mate choice: good genes, compatible genes, and diverse genes. Good genes refer to the superior fitness provided to a bearer by some genes relative to others in a population. These genes may be those most appropriate for particular pathogens or parasites (e.g., Hamilton and Zuk 1982) or for producing the enzymes that best process local prey items. Female mate choice for good genes is made possible by condition-dependent traits in males, such as body size or ornamentation that is preferred by females. For example, Reynolds and Gross (1992) showed that progeny fathered by preferred males had faster growth rates and earlier age of maturity in guppies. Moller and Alatalo (1999) reviewed a wide variety of organisms and found that males with larger condition dependent characters, favored by females, increased offspring viability by 1.5% (even at early offspring stages in relatively benign laboratory environments). Wedekind et al. (2001) showed that female mate choice reduced pathogen-related egg mortality in whitefish, increasing egg survival by 12% relative to random mating.

Compatible genes refer to superior fitness provided to a bearer by the complementation of genes at individual loci as well as across loci. For instance, the deterioration in viability from inbreeding is often due to the expression of two deleterious alleles; compatible alleles at a locus would therefore include at least one nondeleterious allele. Female avoidance of matched deleterious alleles, through the avoidance of breeding with kin, is well known (Pusey and Wolf 1996). Female mate choice for compatible genes at the MHC locus (major histocompatibility complex) is also well known (Penn and Potts 1999). Females target opposite MHC carriers, and their heterozygote progeny have superior fitness due to disease and pathogen resistance (Carrington et al 1999). Finally, coadapted gene complexes, such as coordination of diverse body parts, is compatibility across loci and may underlie the avoidance of outbreeding in females of some species (Andersson 1994).

Wedekind (2002) discusses some advantages of incorporating mate preference into conservation breeding programs with whitefish. Since individuals differ in their heritable viability, minimizing reproductive skew and thereby maximizing N_e might not be the best conservation strategy, since it disrupts the correlation between viability traits and reproductive success. Resistance to a virulent egg parasite is influence by both maternal and paternal effects. Random breeding and equalization would reduce reproductive skew, increasing genetic variation in freshly fertilized eggs, but both this genetic variation and egg number may later be reduced by directed selection from the egg pathogens. Alternatively, allowing preferential breeding by preferred males would decrease genetic variation in freshly fertilized eggs but increase mean survival of offspring. In some cases, preferential breeding would sufficiently reduce the effects of selection by pathogens and result in higher overall N_e. Random breeding and equalization could even increase the size of the pathogen population, further threatening population viability. This suggests that the supportive breeding program needs to find a breeding protocol that incorporates the heritable fitness benefits that come with natural mate choice.

Another example of the importance of incorporating natural breeding systems is seen in the life history decisions of precocious maturity in male salmon (Gross 1985). There is good theoretical reason to believe that precocious males ("jacks" or "precociously mature parr") are those that have the best quality genes in the population and thus derive the highest fitness (Gross and Repka 1998a,b). This increased fitness results in the spread and maintenance of the high quality genes in the population. In current conservation genetics breeding protocols, these males would receive no more breeding advantage than the less fit delayed-maturity males ("hooknose" or "adult" males). This stalls the movement of high-quality genes into the population by unfairly increasing the relative fitness of poor-quality genes.

In summary, supportive breeding programs that focus on maximizing genetic diversity are unlikely to maintain long-term genetic quality in wild populations. Studies of natural breeding systems reveal that genetic quality consists of good genes, compatible genes, and appropriate rather than random genetic diversity. The domestication of wildlife for agricultural consumption by human breeding protocols has only demonstrated that we can produce organisms with high fitness in artificial environments. We do not have equivalent evidence for the capacity of conservation breeding programs to produce organisms that are adapted for their natural environments. Until evolutionary and conservation genetics has matured in its understanding of genetic quality, the use of supplementation and the potential for genetic interactions between hatchery and wild fish should be viewed as further threats to population viability.

Appendix E

Summary of Committee's Interim Report (Excerpted from NRC 2002a)

Atlantic salmon in Maine, once abundant but now seriously depleted, were listed as endangered under the federal Endangered Species Act (ESA) in November 2000. The listing covers the wild fish in eight Maine rivers as a single "distinct population segment." The controversy in Maine that accompanied the listing led Congress to request the National Research Council's (NRC's) advice on the science relevant to understanding and reversing the declines in Maine's salmon populations. The charge to the NRC's Committee on Atlantic Salmon in Maine included an interim report focusing on the genetic makeup of Maine Atlantic salmon populations. This is the interim report. Understanding the genetic makeup of Maine's salmon is important for recovery efforts, because the degree to which populations in Maine differ from adjacent populations in Canada and the degree to which populations in different Maine rivers and tributaries differ from each other affect the choice of recovery options that are most likely to be effective. This report focuses only on questions of genetic distinctiveness. The committee's final report will address the broader issues, such as the factors that have caused Maine's salmon populations to decline and the options for helping them to recover.

SALMON BIOLOGY

Naturally reproducing populations of Atlantic salmon occur in rivers and streams from southwestern Maine to northwestern Europe. Historically, they were found in the Hudson River in New York and north and east to the Canadian border but today are found only in Maine, from the lower Kennebec River to the Canadian border. The populations have declined drastically, from perhaps half a million adults returning to U.S. rivers each year in the early 1800s to about 1,000 in 2000.

Salmon spawn in freshwater, where the young hatch and grow for a year or 2 before migrating to sea. At sea, they grow faster in the rich marine environment and then return to the rivers where they hatched (called natal streams) to spawn. Most fish die after spawning, but some return to the ocean, and some of those return to spawn again. Adults return to their natal streams; only about 2% stray to other (usually nearby) streams.

The occasional straying is probably important evolutionarily, because it allows recolonization of a stream if the local population dies out and provides for small infusions of new genetic material for evolutionary adaptation to changing conditions. Their homing provides an opportunity for the salmon to adapt to environmental conditions in their natal streams. This complex life history pattern makes salmon vulnerable to environmental disruptions both at sea and in fresh water. It also complicates the understanding of the genetic makeup of salmon populations because of the relationship between local adaptations and exchange of genetic material through occasional straying.

HATCHERIES AND AQUACULTURE

Augmentation of wild populations of Maine salmon with hatchery releases began in the early 1870s. At first, young fish were obtained from Lake Ontario, and then the Craig Brook Hatchery, using eggs from Penobscot River fish in Maine, was the stocking source. By the 1920s, Canadian eggs were being used, followed in the 1940s by eggs from the Machias, Penobscot, and Dennys rivers of Maine. In the 1950s and 1960s some eggs of Canadian origin were used again, but by the late 1960s, eggs from Maine's Machias, Narraguagus, and Penobscot rivers were used. Fish reared in hatcheries derived from Penobscot River fish were used until late 1991, when the practice of river-specific stocking was adopted. The protocol used since involves catching young, actively feeding fish (parr) in the river, rearing them to maturity in the hatchery, mating them, and releasing the resulting fry into the rivers before they start to feed.

In addition to stocking, which at least until 1992 added to rivers many fish (and eggs) whose genotypes did not reflect adaptation to the local environment, aquaculture (farming) of Atlantic salmon began in Maine in the 1980s, the first fish for market being produced in 1987. Derived in part from European Atlantic salmon, the genetic strains used for fish farming are even more different from native strains than hatchery strains. Farm fish escape at all life stages, despite the efforts of producers to prevent escapes. In some years and in some rivers, more escaped farm fish return to spawn than wild fish. The impact of escapees on the genetics of wild populations is not well documented in Maine, but both hatchery- and pen-reared fish compete poorly in rivers with wild fish in other areas that have been studied. However, because there are so many escaped farm fish compared with wild fish in some rivers, some impact is likely to have occurred, especially as farm production has increased in recent years.

The addition of so many nonwild genotypes from hatcheries and possibly from aquaculture escapees has led some to conclude that the fish returning to spawn in Maine's rivers could not possibly represent anything more than some nonnative mix of genotypes from Europe, Canada, and Maine. If that were true, then options for conservation might be considerably different from those that might be undertaken if the wild fish in Maine were distinct, and that is why it is important to understand the genetic makeup of the wild salmon populations in Maine.

THE DATA ON GENETICS OF MAINE SALMON

The committee's focus in this interim report is on assessing how Maine salmon populations differ from other Atlantic salmon populations and among themselves. The committee has addressed the question at three levels. First, are North American Atlantic salmon genetically different from European salmon? Second, are Maine salmon distinct from Canadian salmon? Third, to what degree are salmon populations in the eight Maine rivers mentioned in the ESA listing distinct from each other?

Much of the evidence on genetic distinctiveness is based on laboratory analyses of variations in the gene products (proteins) and in the genetic material (DNA) itself. The preliminary evidence indicated distinctiveness at all three levels, and that indication led to the ESA listing. However, the evidence has been questioned on statistical, methodological, biological, and other grounds, and so it bears close evaluation.

The committee evaluated the original evidence, including technical reports, as well as newly published information. It reviewed earlier studies and studies of similar situations involving other locations and some other species of fishes in the salmon family and considered the questions raised about the evidence on Maine salmon. In addition, the committee considered the effect that overlapping generations[1] of salmon might have on the evidence.

---

[1]Before progeny hatched in a particular year can reproduce, progeny hatched in earlier years will reproduce. Thus, the generations overlap.

The evidence on distinctiveness of Maine salmon includes statistical studies on a variety of protein and DNA markers. The statistical significance of the results is so strong and the departures from random expectations are so large that the committee judged the results to be persuasive. Many appropriate questions have been raised about the evidence, and the most recent studies have benefited from criticisms of earlier work. Those criticisms could still be used to improve future work, but the general conclusions are so strongly supported by the evidence that they are not invalidated by imperfections in the data collections or analyses.

The committee concludes that North American Atlantic salmon are clearly distinct genetically from European salmon. In addition, despite the extensive additions of nonnative hatchery and aquaculture genotypes to Maine's rivers, the evidence is surprisingly strong that the wild salmon in Maine are genetically distinct from Canadian salmon. Furthermore, there is considerable genetic divergence among populations in the eight Maine rivers where wild salmon are found.

The heavy stocking of salmon in Maine's rivers and streams has included periods of heavy Canadian stocking, interspersed with strictly Maine stocking. Exactly how much Canadian genetic material has infiltrated Maine salmon populations is impossible to judge at this date.

It is thus appropriate to ask whether wild salmon in Maine reflect only (or mainly) the result of decades of hatchery stocking. That seems unlikely, because if that were so, Maine salmon should be more similar to Canadian salmon than they are. In addition, if their genetic makeup were largely due to stocking of non local salmon broadly across Maine's rivers, salmon populations within Maine would be genetically much more similar than they are. A related question is whether the genetic differences among the fish in the various Maine streams reflect natural processes that occur in watersheds that are connected in networks. More specifically, the issue concerns the relative importance of natural selection over long periods, which influenced the differentiation of Maine's original salmon populations, versus recent genetic drift (sampling effects) caused by small populations. This question cannot be answered at present, but the pattern of genetic variation seen among Maine streams is similar to patterns seen elsewhere in salmon and their relatives where no stocking has occurred. Maine streams have salmon populations that are genetically as divergent from Canadian salmon populations and from each other as would be expected in natural salmon populations anywhere else in the Northern Hemisphere.

Appendix F

Stocking Numbers, 1871–1995

SUMMARY OF NUMBERS OF ATLANTIC EGGS BY SOURCE FOR RESTORATION AND ENHANCEMENT STOCKING FROM 1872 THROUGH 1995

Year	River of Origin	Eggs Collected[a]
1871	Penobscot, Maine	73,200
1872	Penobscot, Maine	1,566,045
1873	Penobscot, Maine	2,321,935
1874	Penobscot, Maine	3,056,500
1875	Penobscot, Maine	2,020,000
1876–78	No eggs were collected	0
1879	Penobscot, Maine	211,690
1880	Penobscot, Maine	1,930,560
1881	Penobscot, Maine	2,693,010
1882	Penobscot, Maine	2,090,000
1883	Penobscot, Maine	2,535,000
1884	Penobscot, Maine	1,935,185
1885	Penobscot, Maine	2,422,600
1886	Penobscot, Maine	1,158,775
1887	Penobscot, Maine	1,184,000

Year	River of Origin	Eggs Collected[a]
1888	Penobscot, Maine	2,253,205
1889	Penobscot, Maine	1,904,000
1890	Penobscot, Maine	533,400
1891	Penobscot, Maine	1,203,285
1892	Penobscot, Maine	1,108,500
1893	Penobscot, Maine	806,000
1894	Penobscot, Maine	415,350
1895	Penobscot, Maine	1,027,355
1896	Penobscot, Maine	3,192,125
1897	Penobscot, Maine	3,506,640
1898	Penobscot, Maine	2,147,675
1899	Penobscot, Maine	1,881,610
1900	Penobscot, Maine	655,500
1901	Penobscot, Maine	832,300
1902	Penobscot, Maine	2,506,575
1903	Penobscot, Maine	3,484,000
1904	Penobscot, Maine	954,500
1905	Penobscot, Maine	2,310,430
1906	Penobscot, Maine	2,804,400
1907	Penobscot, Maine	2,714,500
1908	Penobscot, Maine	1,114,300
1909	Penobscot, Maine	1,456,800
1910	Penobscot, Maine	3,800,200
1911	Penobscot, Maine	2,149,455
1912	Penobscot, Maine	3,966,430
1913	Penobscot, Maine	3,149,655
1914	Penobscot, Maine	2,014,400
1915	Penobscot, Maine	1,953,400
1916	Penobscot, Maine	3,739,180
1917	Penobscot, Maine	3,024,930
1918	Penobscot, Maine	2,613,400
1919	Penobscot, Maine	797,610

Year	River of Origin	Eggs Collected[a]
1920	Penobscot, Maine	911,720
	Miramichi, New Brunswick	1,000,000
1921	Penobscot, Maine	572,040
	Miramichi, New Brunswick	600,000
1922	Miramichi, New Brunswick	1,000,000
1923	Miramichi, New Brunswick	500,000
1924	Miramichi New Brunswick	550,600
1925	Miramichi, New Brunswick	1,000,000
	Saguenay, Quebec	500,000
1926	Miramichi, New Brunswick	533,000
	Saguenay, Quebec	546,000
1927	Miramichi, New Brunswick	1,023,200
	Saguenay, Quebec	500,000
1928	Miramichi, New Brunswick	1,026,100
	Saguenay, Quebec	500,000
1929	Miramichi, New Brunswick	1,000,000
1930	Penobscot, Maine	4,500
1931	Miramichi, New Brunswick	4,000,000
1932	Miramichi, New Brunswick	1,000,000
1933	Miramichi, New Brunswick	1,000,000
1934		0
1935	Miramichi, New Brunswick	1,000,000
1936	Miramichi, New Brunswick	1,500,000
1937	Miramichi, New Brunswick	100,000
1938		0
1939	Dennys, Maine	113,000
1940	Penobscot, Maine	250,450
	Miramichi, New Brunswick	51,150
1941	Machias, Maine	268,480
	Miramichi, New Brunswick	50,750
1942	Penobscot, Maine	708,945
1943	Penobscot, Maine	157,240
1944	Penobscot, Maine	150,000
	Miramichi, New Brunswick	50,000
1945	Penobscot, Maine	307,400
1946	Machias, Maine	266,525

Year	River of Origin	Eggs Collected[a]
1947	Penobscot, Maine	324,475
1948	Machias, Maine	140,215
1949	Machias, Maine	558,815
	Miramichi, New Brunswick	305,000
1950	Machias, Maine	203,400
1951	Miramichi, New Brunswick	200,000
1952	Miramichi, New Brunswick	415,000
1953	Miramichi, New Brunswick	300,000
1954	Miramichi, New Brunswick	302,980
1955	Miramichi, New Brunswick	503,840
1956	Miramichi, New Brunswick	496,550
1957	Machias, Maine	137,535
	Miramichi, New Brunswick	509,080
1958	Machias, Maine	138,670
	Miramichi, New Brunswick	464,510
1959	Machias, Maine	133,155
	Miramichi, New Brunswick	700,940
1960	Machias, Maine	81,910
	Miramichi, New Brunswick	455,420
1961	Machias, Maine	71,785
	Miramichi, New Brunswick	511,220
1962	Narraguagus, Maine	72,375
	Miramichi, New Brunswick	226,350
	Saguenay, Quebec	296,820
1963	Machias, Maine	150,575
	Narraguagus, Maine	131,095
	Miramichi, New Brunswick	504,000
1964	Machias, Maine	139,810
	Narraguagus, Maine	162,020
	Miramichi, New Brunswick	315,030
1965	Machias, Maine	127,120
	Narraguagus, Maine	139,685
	St. John, New Brunswick	303,800
1966	Machias, Maine	287,950
	Narraguagus, Maine	142,440
	St. John, New Brunswick	259,000
1967	Narraguagus, Maine	146,940
	Orland, Maine	41,110
	St. John, New Brunswick	506,490

Year	River of Origin	Eggs Collected[a]
1968	Machias, Maine	76,580
	Narraguagus, Maine	182,205
	Orland, Maine	207,940
1969	Penobscot, Maine	155,265
	Machias, Maine	190,705
	Narraguagus, Maine	160,735
	Orland, Maine	20,595
1970	Penobscot, Maine	269,480
	Machias, Maine	70,750
	Narraguagus, Maine	46,485
1971	Penobscot, Maine	224,130
	Machias, Maine	169,000
	Narraguagus, Maine	119,200
1972	Penobscot, Maine	682,745
1973	Penobscot, Maine	831,090
1974	Penobscot, Maine	1,447,785
	Union, Maine	54,000
1975	Penobscot, Maine	972,965
	Union, Maine	179,250
1976	Penobscot, Maine	1,313,995
	Union, Maine	441,830
1977	Penobscot, Maine	710,880
	Union, Maine	490,030
1978	Penobscot, Maine	1,407,930
	Union, Maine	431,770
1979	Penobscot, Maine	1,117,360
	Union, Maine	88,670
	Kennebec, Maine	50,000
1980	Penobscot, Maine	1,506,050
	Union, Maine	90,840
1981	Penobscot, Maine	1,028,000
	Union, Maine	846,790
1982	Penobscot, Maine	1,549,600
	Union, Maine	435,410
1983	Penobscot, Maine	1,557,490
	Union, Maine	450,550
1984	Penobscot, Maine	2,351,800
	Union, Maine	192,950
1985	Penobscot, Maine	1,838,900
	Union, Maine	285,740

Year	River of Origin	Eggs Collected[a]
1986	Penobscot, Maine	2,376,100
	Union, Maine	211,010
	Saguenay, Quebec	98,500
1987	Penobscot, Maine	2,150,165
	Union, Maine	161,110
	Saguenay, Quebec	100,000
1988	Penobscot, Maine	1,610,700
	Union, Maine	80,710
1989	Penobscot, Maine	2,427,200
	Union, Maine	67,175
1990	Penobscot, Maine	2,041,700
	Union, Maine	103,040
1991	Penobscot, Maine	2,427,000
1992	Penobscot, Maine	2,448,000
	Dennys, Maine	38,000
	Machias, Maine	15,850
1993	Penobscot, Maine	1,881,870
	Dennys, Maine	27,930
	Machias, Maine	50,080
	St. Croix, Maine	114,000
1994	Penobscot, Maine	1,669,905
	Dennys, Maine	155,550
	Machias, Maine	207,175
	Narraguagus, Maine'	145,710
	St. Croix, Maine	80,000
1995	Penobscot, Maine	12,735,645
	Dennys, Maine	338,025
	East Machias, Maine	143,735
	Machias, Maine	512,000
	Narraguagus, Maine	394,435
	Sheepscot, Maine	122,880
	St. Croix, Maine	87,000
Total	Maine	147,473,780
	Canada	28,825,330

[a]From 1871 through 1993, all eggs were obtained from returning sea-run fish. Eggs from kelts—adults after spawning—were first reportedly taken in 1993. Eggs were also collected from captive broodstock beginning in 1994. The figures in this table include eggs from all sources.

SOURCE: Adapted from Baum 1997.

Appendix G

Biographical Sketches of the Committee's Members

COMMITTEE ON ATLANTIC SALMON IN MAINE

M. T. CLEGG (Chair) is a professor of genetics at the University of California, Riverside. He earned his Ph.D. from the University of California, Davis. He is a leading student of the evolution of complex genetic systems and is recognized internationally for his contributions to understanding the genetic and ecological basis for adaptive evolutionary changes within populations and at higher taxonomic levels. His research interests include the population genetics of plants, plant molecular evolution, and genetic conservation in agriculture. He has served on many U.S. national committees, NRC committees, and oversight groups, including the Commission on Life Sciences. He is a member and foreign secretary of the National Academy of Sciences.

PAUL K. BARTEN is an associate professor at the University of Massachusetts Department of Natural Resources Conservation. He is currently a Bullard Fellow in Forest Research at Harvard University (2003–2004). He earned his Ph.D. from the University of Minnesota. His research interests include forest and wetland hydrology, woody debris dynamics in riparian and lotic ecosystems, pathways and mechanisms of non-point-source pollution, retrospective modeling of water and sediment yield, and conservation planning for watershed management. He was a member of the National Research Council Committee to Review New York

City's Watershed Management Strategy (1997–1999). He also serves as co-chair of the Quabbin Science and Technical Advisory Committee for the largest component of the metropolitan Boston system water supply.

IAN A. FLEMING is an associate professor of diadromous and marine fish ecology at the Hatfield Marine Science Center of Oregon State University. He is also adjunct professor at the Norwegian Institute for Nature Research in Trondheim, Norway, where he worked for a decade before moving to Oregon in 2001. He earned his Ph.D. from the University of Toronto. His research integrates perspectives from ecology and evolution with fishery and conservation biology and his areas of expertise include salmonid behavioral and evolutionary ecology, reproduction, life history, maternal effects, and population biology. He has written extensively on interactions of hatchery and farm salmon with wild salmon in the Atlantic and Pacific oceans. He currently serves on the Northwest Power Planning Council's Artificial Propagation Assessment Committee and is a member of several professional societies, including the Society for the Study of Evolution, the International Society for Behavioral Ecology, and the American and British Fisheries Societies.

MART R. GROSS is a professor of conservation biology at the University of Toronto. He earned his Ph.D. in biology from the University of Utah and was a postdoctoral fellow at the University of Washington and a professor from 1982 to 1987 at Simon Fraser University in British Columbia. He has been at the University of Toronto since 1987. His research focuses on the conservation biology of fishes through the study of their evolution, ecology, behavior, and genetics. His current research includes colonization of the Great Lakes by exotic Pacific salmon (Chinook and coho), colonization of Chile by introduced salmonids, evaluation of the Living Gene Bank Program for wild steelhead in British Columbia, and development of alternative breeding designs for maintaining genetic quality in captive fish populations. He has published extensively on Atlantic salmon and Pacific salmon conservation issues involving hatcheries and fish farms. In addition to his university position, Dr. Gross is appointed by the Canadian government to the Committee on the Status of Endangered Wildlife in Canada (COSEWIC) as co-chair of Marine Fishes.

LEWIS S. INCZE is a senior research scientist at the University of Southern Maine Bioscience Research Institute in Portland, Maine. He earned his Ph.D. from the University of Washington. His research interests include coupled biological-physical interactions in the oceans; their effects on the spatial and temporal patterns of upper trophic level production; recruitment interactions between organisms, such as feeding relationships; and climate forcing of system change. His current research is focused on the production dynamics and transport of larval lobsters in the Gulf of Maine; the quantitative relationships between larval supply, settlement, and fisheries recruitment; and the ecology of larval cod relative to prey populations, mixing, and transport on the Georges Bank. Dr. Incze was a research scientist at the Bigelow Laboratory for Ocean Sciences from 1987 to 2002 and was laboratory director from 1991 through 1995. He currently serves as chief scientist for the Gulf of Maine Area Program of the Census of Marine Life.

ANNE R. KAPUSCINSKI is a professor of fisheries and conservation biology at the University of Minnesota. She is also the director of the Institute for Social, Economic, and Ecological Sustainability as well as director for the MacArthur Interdisciplinary Program on Global Change, Sustainability, and Justice. Dr. Kapuscinski earned her Ph.D. from Oregon State University. Her research interests include understanding the influence of genetic makeup on long-term sustainability and the evolutionary potential of managed populations of fish and shellfish. Recent research projects include a comparison of 20-year trends in genetic diversity and productivity in steelhead trout populations based on analysis of DNA polymorphisms; performance evaluations of different walleye populations in the same lakes; examination of genetic effects of hatchery rainbow trout on naturalized steelhead populations by testing survival of pure and hybrid crosses in isolated stream reaches; and testing population effects of gene flow from growth-enhanced transgenic fish to wild relatives. She has served on a number of national and international committees, including the NRC's Committee on Protection and Management of Pacific Northwest Anadromous Salmonids and the NRC's Committee on Biological Confinement of Genetically Engineered Organisms. She received a USDA Secretary of Agriculture Honor Award (1997) and a Pew Fellowship in Marine Conservation (2001) for her linkage of science to public policy regarding aquatic biotechnology and fish genetic conservation.

BARBARA NEIS is a professor at the Memorial University of Newfoundland's Department of Sociology in St. John's and co-director of Safety-Net, a Community Research Alliance on Health and Safety in Marine and Coastal Work. Her research efforts have focused on the Newfoundland and Labrador fisheries and she has recently begun linking that research with international fisheries-related developments. Her current research interests include the health impacts of restructuring in the Newfoundland and Labrador fisheries and local ecological knowledge and science. She has conducted research on many different aspects of the Newfoundland Fishery including gender relations, occupational health, technological change, industrial restructuring, social movements, and fisheries ecology.

PATRICK O'BRIEN is a consulting environmental scientist with Chevron-Texaco Energy Technology Company in Richmond, California. He has a Ph.D. in ecology from the University of California, Irvine. Dr. O'Brien has expertise in the following areas: environmental impact assessment, endangered species, conservation planning, ecological risk assessment, wetlands permitting and management, natural resource damage assessment, habitat restoration, and the environmental elements of oil spill contingency planning and response. He served as a member of the NRC's Committee on Scientific Issues in the Endangered Species Act and is a member of the NRC's Board on Environmental Studies and Toxicology.

NILS RYMAN is a professor at Stockholm University. He is a population geneticist whose research has focused on salmon. More specifically, his research focuses on problems related to the intraspecific genetic structure of natural populations, their evolutionary and reproductive relationships, and the factors determining the amount and distribution of genetic variation among them. His particular attention is directed toward conservation genetics and the genetic effects of human impacts on natural populations. Dr. Ryman has published papers on the genetic effects of escaped hatchery fish into natural populations.

PETER E. SMOUSE is the chairman of the Department of Ecology, Evolution and Natural Resources at Rutgers University. He earned his Ph.D. from North Carolina State University. His research interests include biometrics and population theory, spanning the fields of genetics, ecology, demography, epidemiology, anthropology, and systematics. He has served on a number of committees and panels including the Committee to Study Quantitative Genetics and Common Diseases, NIH, 1978; Population Biology and Physiological Ecology Panel, NSF 1980–83; NRC Ad Hoc Committee to Study Endangered Amphibians, 1990; DNA Subcommittee, New York State Forensic Commission 1995–98; U.S. National Committee—IUBS 1995–98; Population Biology Panel—Dissertation Grants, NSF 1998; Advisory Committee—Columbia Earth Institute 1998; and the Scientific Advisory Board, Cooperative Institute Fisheries Molecular Biology (FISHTEC) 1998–2001.

JENNIFER L. SPECKER is a professor of oceanography at the University of Rhode Island. She earned her Ph.D. in Fisheries and Wildlife from Oregon State University. She is a fish endocrinologist specializing in the developmental endocrinology of Pacific and Atlantic salmon, the metamorphosis of flounder, the reproductive biology of commercially important marine fishes, the endocrine regulation of the gut and gills during adaptation, and vitellogenins and mouthbrooding in African cichlids. She has published scientific papers on the parr-smolt transformation of anadromous salmon, including "Parr-smolt transformation in Atlantic salmon: Thyroid hormone deiodination in liver and brain and endocrine correlates of change in rheotactic behavior." She participated in the series of Workshops on Salmonid Smoltification from 1977 to 1992, co-editing the proceedings of the last one, which appeared in the journal *Aquaculture*.

ROBERT R. STICKNEY is the Sea Grant College Program director and professor at Texas A&M University Department of Oceanography. He earned his Ph.D. from the Florida State University. His professional interests include aquaculture and fishery science. He has written extensively on the history of aquaculture in the United States and on sustainable aquaculture. He is past president of the World Aquaculture Society and the Fish Culture Section of the American Fisheries Society. He edits the journal *Reviews in Fisheries Science*.

JON G. SUTINEN is a professor at the University of Rhode Island Department of Environmental and Natural Resource Economics. He earned his Ph.D. in economics from the University of Washington. His primary research interests are fisheries management and regulation with an emphasis on compliance and enforcement. He has extensive experience in international fisheries, including those in Latin America, Africa, Southeast Asia, and Europe. Most of this experience involved conducting research and supplying advice on fisheries policy. His current research focuses on several bioeconomic aspects of New England marine fisheries and the Northeast Large Marine Ecosystem. Dr. Sutinen was the founding editor of the journal *Marine Resource Economics* and served in that role for over a decade. He is a member of the NRC's Ocean Studies Board.